The World Wide Military Command and Control System

Evolution and Effectiveness

DAVID E. PEARSON

Air University Press
Maxwell Air Force Base, Alabama

June 2000

Library of Congress Cataloging-in-Publication Data

Pearson, David E. (David Eric), 1953-
The world wide military command and control system : evolution and effectivenss / David E. Pearson.
p. cm.
Includes bibliographical references and index.
ISBN 1-58566-078-7
1. Worldwide Military Command and Control System—History. I. Title.

UB212.P43 2000
355.3'3041'0973—dc21 99-462377

Disclaimer

Opinions, conclusions, and recommendations expressed or implied within are solely those of the author and do not necessarily represent the views of Air University, the United States Air Force, the Department of Defense, or any other US government agency. Cleared for public release: distribution unlimited.

For
My Children

Contents

Chapter *Page*

PART III

Implementation

Illustrations

About the Author

David E. Pearson, an Army veteran, graduated *magna cum laude* from the University of Massachusetts at Amherst, received his PhD from Yale University, and was a Fellow in International Security Studies at Ohio State University. He currently teaches sociology at the University of Texas at Brownsville, which is Texas's southmost college.

Preface

Let me begin by saying what this book is not. It is not the same as the books previously written on command and control by Paul Bracken, Bruce Blair, C. Kenneth Allard, or others. The World Wide Military Command and Control System (WWMCCS) is constituted of four general types of elements: sensors, command posts, computers, and communications networks. Whereas previous books dealt mainly with the first two types, this book is concerned far more strongly with the second two. Nor has any previous writer dealt with WWMCCS comprehensively. Allard, for instance, devotes only about two pages to WWMCCS, Blair refers to it on but three occasions, and Bracken does not mention it at all. Here, addressing what the system is and how it got that way are central concerns, and attention is paid to a number of key system factors and elements that received almost no attention in these earlier works.

This book is also not another study of cold war deterrence, nor is it an examination of the (hypothesized) interactivity between the American and Soviet command and control systems under conditions of crisis or war.

Perhaps the best single way to summarize it is to view the book as a bureaucratic or organizational history. What I do is to take three distinct historical themes—organization, technology, and ideology—and examine how each contributed to the development of WWMCCS and its ability (and frequent inability) to satisfy the demands of national leadership. Whereas earlier works were primarily descriptive, cataloguing the command and control assets then in place or under development, I offer more analysis by focusing on the issue of how and why WWMCCS developed the way it did. While at first glance less provocative, this approach is potentially more useful for defense decision makers dealing with complex human and technological systems in the post-cold-war era. It also makes for a better story and, I trust, a more interesting read.

By necessity, this work is selective. The elements of WWMCCS are so numerous, and the parameters of the system potentially so expansive, that a full treatment is impossible within the compass of a single volume. Indeed, a full treatment of even a single WWMCCS asset or subsystem—the Defense Satellite Communications System, Extremely Low Frequency Communications, the National Military Command System, to name but a few—could itself constitute a substantial work. In its broadest conceptualization, WWMCCS is the world, and my approach has been to deal with the head of the octopus rather than its myriad tentacles.

My initial interest in WWMCCS goes back—what seems like a long, long time—to my graduate school days and a class in national security issues taught at Yale by Garry Brewer, who later became a member of my (non-WWMCCS related) dissertation committee. I'm not quite certain whether to thank him or denounce him for starting me on what has proved to be the lengthiest research project I have ever undertaken. Now that the project is completed, thanks, I suppose, is more appropriate. Thanks also go to Charles Perrow, my former dissertation chair, who, at an early stage of my career, introduced me to organizational theory and helped me to think analytically about complex organizations. As to more recent history, this project was formally launched during a postdoctoral fellowship year at Ohio State University's Mershon Center, and a subsequent junior faculty leave from Lafayette College helped advance it. Special thanks go to Charles Hermann, Thomas Norton, and Howard Schneiderman for their encouragement and insight.

Introduction

In the late 1970s, in response to a lengthy series of failures and snafus in various components of the Defense Department's World Wide Military Command and Control System (WWMCCS; pronounced "wimex"), a remarkable and rather unlikely team of consultants was assembled at the Pentagon. Their job was to try to figure out how to make the vast, multi-billion dollar metasystem of sensors, command centers, and communications links work better. The group included 30 anthropologists and sociologists, mathematicians, control theorists and systems theorists, and representatives from a variety of other scholarly disciplines—"academics with a philosophical bent," as one writer described them. In a series of meetings with a similarly sized group of defense experts, the academics considered ways to deal with WWMCCS's many problems. All sorts of recommendations were offered up, but, according to one of the consultants, conspicuously lacking was any "critical examination of the dominant paradigm which condones the expenditure of vast resources without even a semblance of a conceptual rationale for the effort."[1]

However correct that assessment might have been, and some certainly disputed it, the whole experience was, in a word, unprecedented. The fact that academics had been invited to the Pentagon in the first place could be read as an admission that the vast assemblage of technologies and human organizations that was WWMCCS was not up to snuff as the cold war moved into its final tense decade. The meetings were also a not-so-implicit admission that the Pentagon's traditional problem-solving method in this area, the so-called evolutionary approach to command and control system development, had come up short; in fact, this approach itself might have represented a major impediment to the formulation of a coherent conceptual basis for the system. After some 20 years of development, the World Wide Military Command and Control System, even in the eyes of some of its most enthusiastic advocates, was judged to be less than effective.

The Pentagon's concern with command and control effectiveness was soon complemented by the concerns of defense analysts outside the government. By the early 1980s, these analysts were pointing out how such traditional measures of effectiveness as number of warheads, throw weights, damage expectancy, and surviving equivalent megatonnage tended to selectively focus attention on only a few critical aspects of the strategic balance. They pointed out that while the defense literature contained an abundance of missile duels, it offered far fewer "serious inquiries into the organizational, human, and technical requirements for minimal, essential command and control."[2] They noted that while the centrality of command and control in the implementation of US strategic policy was everywhere implicit, plans seldom reflected key vulnerabilities and real-world system limitations.

Numerous efforts were made throughout the remainder of the 1980s to identify the key concepts in this area to evolve unambiguous measures of effectiveness for WWMCCS. The result was an increasingly elaborate lexicon for articulating what is meant by command and control effectiveness—elaborate, but still far from adequate. Many of the terms that came to be incorporated into this burgeoning conceptual list, such as *standardization* or *end-to-end security,* were superficial and far from self-explanatory, and the requirement for them by no means self-evident.[3] The diversity and complexity of the concepts was great, and thus revealing. Rather than serving as an indicator of greater understanding, the proliferation of tenebrous terminology could be interpreted as something quite different, perhaps as nothing so much as a signal indicator of incomprehensibility and unmanageable complexity.[4]

Amidst the confusion there were naturally some areas of consensus. Among the generally agreed-upon criteria of effectiveness that eventually emerged within the defense community, it was held that command and control systems should be *interoperable*—meeting the demands of users, with a variety of interests and emphases—at all system locations. As an obvious concomitant, equipment, computerized data formats, and other common-user elements must be *compatible*. The point was to do away with the situation, endemic to large-scale military operations throughout the cold war era, in which the different missions of the services, their different requirements, vernaculars, and assets, led to major problems when the services were called upon to work together in joint operations.[5] Next, the systems should be *responsive*, able to provide rapid, direct connections and real-time relays whenever necessary with adequate capacity. The systems must be *flexible*, able to meet changing requirements in a dynamic environment. This in turn suggests that they should be *survivable* in case of attack, to be accomplished by emphasizing reconstitution of assets, redundancy, and design of command nodes and communications links. That is, effective systems should permit an assessment of friendly and adversary residual capabilities in a post-attack environment, and allow for variable response options. Indeed, the Reagan administration's strategic modernization program of the early 1980s specifically highlighted the requirement that command and control systems be as survivable as the weapons they supported.[6] Given that conflicts might persist for some time, effective systems should also be *endurable*, degrading gradually rather than experiencing catastrophic failure under conditions of stress and damage. They must be *reliable*, able to perform acceptably with imperfect information and under severe time constraints. Finally, they must be able to provide *secure* linkages between users under a wide range of conditions.[7]

It all sounded fine, but as soon as these concepts were considered in the context of WWMCCS, their meanings became problematical and contextual. Take, for example, the apparently unambiguous criterion of survivability. Since resources are not infinite, it might well be appropriate for some WWMCCS elements to be designed to function only in peacetime. Others might need to deal with minor emergencies, while others might have to function through major conventional war. Still others might have to function during a tactical nuclear war or throughout and even subsequent to a strategic nuclear exchange. Which systems should be made more or less survivable? How should this best be accomplished? Precisely who should make these determinations? The answers that were offered often depended upon nothing so much as who was asked the question.

Consensus on these and a host of related issues was necessary for the promulgation of clear and specific, broadly applicable measures of effectiveness. But in the real world in which WWMCCS programs were conceived and developed—one of multiple users, competing organizational subunits, goal dissensus, budgetary constraints, and a context of rapid technological change—consensus was difficult to achieve. Absent agreement on specific definitions, the meaning of such concepts as survivability, reliability, or any of the others became possible only on a fairly general rhetorical level; that is, they became official goals, necessarily lacking specific human or technological referents to what the system was supposed to accomplish.[8] Who, after all, could possibly disagree with the general proposition of having more survivable, reliable, or flexible systems for the command and control of America's military forces? But in practice, such measures were frequently little more than sophisticated sloganeering of a politically expedient sort. In the end, these terms offered little guidance for determining whether the World Wide Military Command and Control System was an effective system, under what conditions, and from whose point of view.

Problems with properly conceptualizing effectiveness were hardly unique to WWMCCS. Among organizational analysts, interest in effectiveness had been persistent; it had also been persistently frustrated by a similar conceptual ambiguity.[9] Since the 1960s numerous highly divergent models of effectiveness had been advanced. Almost as quickly as they appeared, they were subjected to pointed criticism by writers who viewed their assumptions as either dubious or of limited applicability. It was pointed out in the mid-1960s that most of what had been written on the topic was highly judgmental, filled with advice that "seems sagacious but is tautological and contradictory."[10] But despite the doubts, the theoretical importance of the concept ensured an ongoing effort to promulgate an acceptable operational definition. And not without reason; after all, effectiveness represents the ultimate dependent variable in any organizational analysis.[11]

But as the literature in this area burgeoned, as alternative definitions of effectiveness continued to be propounded, it became increasingly apparent to many analysts that this goal was chimerical or simply misguided. One described it as a sort of trudging after an "ever-shifting rainbow's end."[12] By the end of the 1970s, efforts to come to grips with the concept had reached an apparent intellectual impasse. It was recognized that the many shortcomings of effectiveness research were attributable to the fact that the concept being addressed was ambiguous in the extreme. Two analysts lamented in summing up the state of the literature, "There are no definitive theories. There is no agreement on a definition for organizational effectiveness; the number of definitions varies with the number of authors who have been preoccupied with the concept."[13]

Things changed little with the passage of additional years, and well into the 1980s scholarly journals continued to report the confusion that characterized scholarly writing on the topic, noting how "problems of definition, circumscription, and criteria identification plague most authors' work."[14] Indeed, the intellectual hurdles presented by the issue appeared so insurmountable and the distances separating perspectives so vast and imponderable that some researchers fled the field entirely, concluding that effectiveness is a retractably subjective phenomenon defying objective definition and analysis, not unlike the notions of truth and beauty.[15] With efforts to define the concept mired in and beset by numerous and apparently hopeless contingencies, scholarly interest predictably declined.

The impasse represented by the increasingly widespread recognition that effectiveness is a complex and multidimensional concept can also be seen to represent a sort of watershed in the academic literature. Beginning in the late 1970s, research began shifting away from the earlier emphasis on conceptualization and operationalization. Its focus thereafter turned more to what have been described as the "contradictions" inherent in the concept, its emphasis on elaborating its conceptual complexity and cataloguing the normative, temporal, organizational, and environmental constraints presumed to render any definition of effectiveness of only limited utility.[16] What seems to have emerged as we move toward the present is not consensus concerning any single model's validity but rather a more-or-less widespread recognition that dissensus is the norm. For understanding WWMCCS's evolution and its many problems, it is a dissensus that will command our closest attention.

Three principal themes, or, perhaps better said, historical streams of action appear to have governed the development of WWMCCS. They are summarized by such terms as *technology*, *organization*, and *ideology*. Technological changes throughout the cold war dramatically altered the nature of warfare, and technological push would be a defining process in the development of defense systems, including WWMCCS. These changes in turn necessitated changes in organization, in particular the movement toward a more centralized defense management structure—something actively resisted by a number of powerful defense constituencies, most notably the military services. To allay doubts and overcome resistance, considerable authority for WWMCCS's development was ceded to the services, who proceeded to define system requirements in ways genial to their interests—a sort of technological "user pull."[17] Thus, from the outset WWMCCS has been a "subunit-dominated organization," emphasizing the services' needs and requirements over those of other elements, over the interests of the system as a whole, and, not infrequently, over the national interest.

A sense of WWMCCS's subunit-dominated character, of its fundamental ambiguity and fractiousness, was captured well by a former deputy director for defense research and engineering who, in the mid-1960s, pointed out: "We are talking about a picture which is constantly changing in different ways—in the functions performed, the people performing, and the equipment being used."[18] It was apparent at the end of that decade in the remarks of a House Military Operations Subcommittee staff administrator who exclaimed to the director of one WWMCCS subunit, "You have so many systems here, no wonder you need a systems engineering analysis setup."[19] It was clear when one defense journal described WWMCCS as "somewhat of a Rube Goldberg concoction consisting of Army, Navy and Air Force systems linked together with commercial carriers."[20] It was clear in the 1970s when the Defense Communications System, a key WWMCCS element, was described as "merely an association of facilities tied together and attempting to act in concert, but with no central authority to direct its actions."[21] In addition, it was apparent a decade later in a General Accounting Office evaluation of WWMCCS's automated data-processing program's management structure, when it was pointed out that things were so nebulous that no one could be found who had a thorough general understanding of the program. The difficulties with defining what is meant by WWMCCS are perhaps best summed up in the repeated references to it throughout the years as a "loosely knit federation," "more of a federation of systems than a single system," a "federation of subsystems," and various similar characterizations.

This condition of rampant organizational suboptimization was validated and ultimately institutionalized by the "evolutionary approach," an increasingly pervasive ideology within defense circles asserting that command and control system development is best conducted incrementally, by system subunits; and the reason it gained such wide currency probably lies more in its bureaucratic utility than in its ability to create an optimal system for command and control. Those interested in maintaining a decentralized defense status quo embraced the approach because it maintains that the decision-making process is situationally contingent and unknowable in advance. Centralized decision makers thus cannot adequately specify the sorts of information they require, with whom they might need to communicate, or precisely what type of system best suits their needs. In light of this ignorance at the center, the logical course of action is to devolve authority toward the periphery, thus providing greater flexibility for system development to lower-level system subunits. Thus the services, fully cognizant of the defensewide trend toward greater centralization and acutely sensitive to the loss of autonomy and authority it portended, perceived in the evolutionary approach a way to maintain some (though surely not all) of their earlier autonomy and authority. Unable to stop the juggernaut of defense centralization, they saw in the evolutionary approach a way to make the most out of a bad situation. Branch offices of the secretary of defense they would not be, and, by embracing it, they were able in substantial measure to co-opt the development of WWMCCS in ways they considered advantageous.

For those interested in advancing the cause of greater centralization, the evolutionary approach also had its appeal, mollifying as it did the opposition of the services, who otherwise might be expected to vigorously oppose any centralizing initiative. Whatever the other merits or liabilities of this approach, it appears to have held a certain Machiavellian appeal to the proponents of greater centralization (at least initially) because, as with the services, it was perceived as a way to advance their interests. But as things turned out, it also represented the classic deal with the devil, for the price paid by the centralizers turned out to be disproportionately high. Adopting the evolutionary approach certainly helped to diminish service resistance; but the price ultimately paid was nothing less than the very soul of the centralized WWMCCS concept.

The historical lack of any organizational center of gravity for WWMCCS and the serious lack of coordination between its constituent elements resulted in a multiplicity of problems and occasionally major failures when the system was called upon to function in coordinated, joint-service fashion. Focusing on process rather than on result, emphasizing what the sociologist Max Weber called formal rather than substantive rationality, WWMCCS's subunit-dominated structure and the evolutionary approach that validated it thus set the stage for an ongoing series of falls. Those who enjoy ironies may find this one especially delicious: the same conditions that cleared the way for the establishment of WWMCCS and that permitted its subsequent growth simultaneously guaranteed that it would not be able to function effectively. In structural terms, we might conclude that the World Wide Military Command and Control System was born to fail. The remainder of this work documents how this interplay of organization, technology, and ideology shaped the development of WWMCCS during the cold war's three final tense decades.

Notes

1. William J. Broad, "Philosophers at the Pentagon," *Science* 210 (24 October 1980): 409, 412.
2. Garry D. Brewer and Paul Bracken, "Some Missing Pieces of the C^3I Puzzle," unpublished manuscript, Yale University, 1983.
3. Van C. Doubleday, "WWMCCS in Transition: An Air Force View," *Signal* 30 (August 1976): 71.
4. Charles Perrow, *Normal Accidents: Living With High-Risk Technologies* (New York: Basic Books, 1984).

5. "Standardization: The Key to Increasing the Effectiveness of Tactical C&C," *Armed Forces Management,* July 1967, 44.

6. House, Committee on Armed Services, *Strategic Force Modernization Program,* 97th Cong., 1st sess., 1981 (Washington, D.C.: Government Printing Office [hereafter cited as GPO], October/November 1981), 211.

7. General Accounting Office, *The World Wide Military Command and Control System—Major Changes Needed in its Automated Data Processing Management and Direction, Report to the Congress,* LCD-80-22 (Washington, D.C.: GPO, 14 December 1979), 15–19.

8. Charles Perrow, "The Analysis of Goals in Complex Organizations," *American Sociological Review* 26, no. 4 (December 1961): 855.

9. Richard M. Steers, "Problems in Measurement of Organizational Effectiveness," *Administrative Science Quarterly,* 1975, 20.

10. Daniel Katz and Robert L. Kahn, *The Social Psychology of Organizations* (New York: Wiley, John, and Sons, Inc., 1966), 149.

11. Richard H. Hall, "Effectiveness Theory and Organizational Effectiveness," *Journal of Applied Behavioral Science* 16 (1980): 536.

12. Jeffrey Pfeffer and Gerald R. Salancik, *The External Control of Organizations: A Resource Dependence Perspective* (New York: Harper & Row, 1978), 8.

13. Paul S. Goodman and Johannes M. Pennnings, "Perspectives and Issues: An Introduction," in *New Perspectives on Organizational Effectiveness,* ed. Paul S. Goodman and Johannes M. Pennings (San Francisco: Jossey-Bass, 1977), 1–12.

14. Kim S. Cameron, "Effectiveness as Paradox: Consensus and Conflict in Conceptions of Organizational Effectiveness," *Management Science* 32, no. 5 (May 1986): 539–40.

15. Renee R. Anspach, "Everyday Methods for Assessing Organizational Effectiveness," *Social Problems* 38, no. 1 (February 1991): 2.

16. Hall, 245.

17. Blue Ribbon Commission on Defense Management, *A Quest for Excellence: Final Report to the President* (Washington, D.C.: GPO, 1986), 45.

18. "Dr. Fubini Stresses DDR&E's Desire for System Compatibility," *DATA* (February 1965): 9.

19. House, Committee on Government Operations, *Military Communications—1968,* 90th Cong. 2d sess. (Washington, D.C.: GPO, 1968), 6.

20. "Defense Communications Agency—The Future: Analog to Digital," *Armed Forces Management,* October 1968, 154.

21. House, Armed Services Investigating Subcommittee, Committee on Armed Services, *Review of Department of Defense Worldwide Communications, Phase I,* 92d Cong., 1st sess. (Washington, D.C.: GPO, May 1971), 24–25.

PART I

Conceptualization

Chapter 1

Centralizing the Defense Establishment

During the course of World War II, the development of military communications tended in a specific direction—toward systems that were *common user.* The phrase denotes general-purpose systems—those that serve the needs of a host of users at a number of geographic locations and can send message traffic of all types and precedences. The reasons for the wartime growth of this type of system are not difficult to appreciate. The global scale and rapid pace of the conflict necessitated large-scale coordination within the armed forces as well as between our forces and those of our Allies. Good communications naturally were vital to this coordination, and common-user systems, with their associated networks of tape relay centers and tributary stations, promised precisely the sort of flexibility that the exigencies of global war required.[1] Conversely, communications systems that were "dedicated" to a single use or user were frequently viewed as inherently limited and inflexible.

While common-user systems had their advantages, they were not universally lauded, and the reasons were equally easy to understand. Many users, notably the military services, were unhappy with them precisely because they were designed to serve the communications needs of others and were thus not fully under one's own control. In other words, common-user systems necessitated accommodation, and this was viewed as undesirable. Consider that within a common-user system the message's precedence level determines how rapidly it will be processed. Precedence level makes eminent sense in the abstract, but in a world characterized by bureaucratic parochialism, problems predictably arose. The messages of some users, especially those transmitting large volumes of lower precedence traffic, suffered substantial delays at times of heightened communications activity. Such delays being adjudged intolerable, there ensued an inflation of messages' precedence levels to speed up their transmission. This in turn produced the serious situation in which genuinely important, time-sensitive messages requiring immediate transmission were slowed down because large quantities of precedence-inflated traffic were choking the system.

Following the conclusion of World War II and the emergence of the United States as the preeminent global power, the military services assumed worldwide responsibilities commensurate with the nation's new role. Requiring worldwide communications capabilities, yet viewing the wartime regime of common-user systems as inherently inimical to their interests, the services began to develop sets of dedicated communications networks to meet their own unique, mission-specific requirements.[2] Before long a whole new communications doctrine began to crystallize around the distinction between dedicated and common-user communications systems and technologies.[3]

Despite the apparent decentralizing tendency, the need to create a centralized command structure was also recognized early in the postwar period, at the time the National Security Act of 1947 formally reorganized the defense establishment. The act constituted the Air Force as a separate military department. The secretary of war was replaced by a secretary of defense, who sought to exercise general direction, authority, and control over the three military departments and to serve as the principal assistant to the president in national security matters.[4] The Joint Chiefs of Staff (JCS), established earlier as the supreme military body for directing the Allied war effort, was provided a statutory basis and designated the principal military advisor to the secretary of defense and the president. These new organizations, offices, and departments reflected the war's lessons and insights, preeminently that the advent of revolutionary new weapons had rendered earlier concepts of separate ground, sea, and air warfare obsolete and that future conflicts would involve joint rather than separate operation of forces.[5]

If the National Security Act can be read as a first major attempt to institutionalize the new realities in a more centralized defense management structure, it simultaneously represented an effort to restrain the very centralizing tendencies that it unleashed. It did this by guaranteeing that many of the traditional responsibilities and prerogatives of powerful constituencies within the defense establishment were preserved. The reason for compromise was simple and thoroughly pragmatic: without such concessions, the powerful groups, notably the services, would not support the act. Thus, the National Security Act was less a revolutionary mandate for change than it was a synthesis of the old and the new—a "compromise between the friends and foes of centralization," as one observer phrased it.[6]

The compromise, such as it was, was hardly symmetrical. During the lengthy debates preceding the reorganization, traditionalists frequently held sway over those promoting greater unification, and hence centralization, in defense decision making. While the act gave the secretary of defense formal authority over the defense establishment, the latitude for action was circumscribed by a provision giving the service secretaries the authority to separately administer their respective departments. This provision included, perhaps more importantly, authority over budgetary matters, an arrangement that in practice would decentralize not only day-to-day operational authority to the military departments but practically all true authority as well. The result was that the secretary of defense and his small staff were soon held hostage to the three military services, with their separate secretaries and extensive staffs, which retained the status of individual executive departments. The JCS, lacking a formal chairman and unable to reallocate basic service combat roles and missions—a prerogative of the strongly service-partisan Congress—was powerless as well, able to do little more than attempt to adjudicate interservice conflicts.[7] Nonetheless, this decentralized national military establishment, described later by President Dwight D. Eisenhower as "little more than a weak confederation of sovereign military units,"[8] represented a tentative first step toward greater centralized control of the military.

The 1949 amendments to the National Security Act corrected some of its deficiencies but perpetuated others. The amendments redesignated the national military establishment as the Department of Defense (DOD), over which the secretary of defense was given authority, direction, and control. The Departments of the Army, Air Force, and Navy were downgraded from independent executive status, with their chiefs cabinet-rank officials, to subordinate military departments represented in the cabinet and National Security Council by the secretary of defense alone. The amendments authorized the appointment of a JCS chairman, senior in rank to all other officers, to replace the existing post of chief of staff to the commander in chief. The size of the Joint Staff was more than doubled.[9] It appeared that the forces of centralization were well on the way to achieving ascendancy.

The amendments also introduced into law a series of well-intentioned legal checks and balances against possible abuses of military power that effectively blocked any genuine efforts at unification, and hence centralization of control. The secretary of defense was prohibited from exercising his budgetary power if it interfered with the missions of the military departments. The chairman of the JCS, rhetorically cast as the nation's highest ranking officer, was denied a vote in debates. He could not make decisions in the name of the other chiefs even when the decisions were supported by the secretary, and he lacked even the ability to adjudicate disputes among his separately interested colleagues.[10] In addition, the law limited the size of the Joint Staff, granted the services the right to make appointments thereto, and placed limits on officers' tenures once there. Collectively, these measures limited the continuity and influence of the Joint Chiefs of Staff organization; they had the effect of putting the JCS at a considerable disadvantage vis-à-vis the individual military departments. Despite the changes, then, relationships continued to be bound securely to the earlier system of negotiation.[11]

The consequences were predictable. The services used their best personnel to satisfy their own priority assignments before making assignments to the Joint Staff, which they considered a relatively low priority. Recognizing that a tour with the JCS was out of the service mainstream and thus not career enhancing, the best officers had a major incentive to avoid such an assignment. Those who did receive JCS assignments, including the chiefs themselves, were subjected to the pressures of dual and frequently conflicting loyalties. While in theory joint missions and responsibilities took precedence over the parochial interests of the services, in practice loyalties remained strongly with the services from which officers came and to which they would shortly return. Proposals articulated by the service chiefs tended to come primarily from their own staffs rather than from members of the Joint Staff, with understandable emphasis on service needs and favored positions. Additionally, the fact that the JCS had to make do with officers "who remain in the Pentagon barely long enough to find the cafeteria," and for whom repeat tours of duty were rare, meant that the ability of the joint-service organization to develop the patterns of practices and understandings that constitute organizational memory was severely impeded.[12] In the absence of such memory, team formation was difficult, and it was hard to bring newcomers up to speed regarding the complex issues with which the JCS had to deal. In addition, just when personnel had finally received sufficient exposure to begin to understand, articulate, and advance joint-service concerns, they were rotated out of Joint Staff assignments. Perhaps most important, budgetary control remained with the military departments. This meant that these departments pursued their own political and lobbying agendas with respect to the Congress, from whom they won budgetary approval, and, more generally, with the American public. The resulting structure inevitably was rife with fractiousness, competition, and rivalries that impeded joint-service planning and operations.[13] The National Security Act and its amendments, later characterized by President Eisenhower as "prescribing controversy by law," brought about no genuine unification of forces and did little to advance the cause of greater centralization in defense decision making.[14]

The creation of a series of new organizational entities within the DOD, with their own considerable communications needs, also worked to complicate the picture during the 1950s. Moreover, the communications demands of these actors, perhaps most notably the Strategic Air Command (SAC), were influenced, and continually modified, by the development of a host of new communications technologies. Advances in such areas as ionoscatter and troposcatter transmission techniques, issuing directly from the Semi-Automatic Ground Environment (SAGE), the Air Force's massive air defense effort, had a profound influence on the technologically possible and, by extension, on what was deemed desirable. Such new techniques as pulse code modulation—along with advances in automatic message switching, storage, and retrieval—offered great promise not only for automating the existing communications networks then being developed by the services but also for improving the linkages between them. Many of these advances were conjured into being by new and increasingly sophisticated weapons systems whose use required ever-more-rapid access to accurate weather data, air traffic control information, logistics, and other types of support information.[15]

While there is little doubt that the services were thinking globally (at least in their own terms) as they developed their communications systems, the doctrine of dedicated communications they embraced worked to constrain a truly global capability. Since the services were the ones responsible for developing new communications technologies, the not-so-surprising result was a series of systems that were emphatically service specific. (The Army, for example, operated a large number of dedicated, special purpose, point-to-point communications systems, each with its own terminals and manual cryptographic equipment.) Most of these were incompatible with the others, meaning that however good they might be individually, in the aggregate they constituted no coherent system at all. While it was clearly necessary to do something about this communications "straight jacket," efforts to modernize and automate things were resisted, sometimes quite fiercely, by those with a stake in the status quo.[16] If you change the system, after all, the comfortable bureaucratic world would rapidly devolve toward chaos; careers would be disrupted; authority would slip away. Despite the ever-mounting need, the joint-service philosophy necessary for a comprehensive "systems" approach had yet to take hold. The systems that were developed during the 1950s tended to be vertical, dedicated systems going straight to the top and unable to connect users across different organizational structures. They were ever justified as necessary for the services' unique functions.[17]

Given the prevailing nature of US strategic doctrine—which emphasized deterrence and, in case of a nuclear attack, the ability to launch a devastating reflex counterstrike—these dedicated systems represented no serious national security shortfall. The doctrine of massive retaliation imposed, first, the need for a large nuclear force, and SAC bombers loaded with high-yield nuclear weapons met this need. Second, there was also a need for a sensor system capable of providing early warning of attacks against the United States. A whole array of new warning systems was put into place during these years. These systems included SAGE and such components as the distant early warning line (DEWLINE) of radars, the Ballistic Missile Early Warning System, the undersea Sound Surveillance System, and others. Third and last, all of this hardware called for a centralized command and control structure that would permit the orders of the National Command Authorities (NCA) to be received without a hitch.[18] But under the terms of massive retaliation, the pressures for coordinating efforts were not overwhelming. The services pursued their own preferred ways of contributing to the nation's defense, and there was little reason to fault their separate and generally uncoordinated development of technologies. As a result, an entire generation of complete weapons and the command and control systems appropriate to them were developed during this era, systems which were in almost all important respects entirely independent of one another.[19] In other words, the nature of military doctrine, and by extension the organizations responsible for implementing it, had profound implications for the types of technologies that were conceived and developed during this era.

Yet by no means was it all an issue of technology push, of organization driving technology. For a number of technological changes occurred during this period, many of them involving the strategic nuclear forces, that in turn would have profound implications for change. Such advances as the hardening of land-based missile silos and the later move toward the deployment of a ballistic missile submarine force would soon lead to a reconsideration of American defense strategy. The new strategic doctrine that began to emerge stressed America's ability to react appropriately to the unique exigencies of a broad range of crises—up to and including a Soviet nuclear first strike. As this new brand of strategic thinking began to take hold within the DOD, perceptions of military requirements began to be altered in fundamental ways. The new thinking, which later would acquire the appellation *flexible response*, at first implicitly and later explicitly created the demand for a new generation of weapons more appropriate to the threat that was seen to be emerging. The majority of these technologies would emphasize centralized command and control of US forces to a degree unknown previously, necessitating in turn more sophisticated systems that would permit that control to take place.[20]

Eisenhower's belief that the military departments were the primary obstacle to more effective, centralized defense management was the driving force behind his administration-long effort to reform the Department of Defense. (Shortly after taking office, he moved to centralize decision-making authority by enlarging the Joint Staff, augmenting the JCS chairman's influence by giving him the power to control appointments to the Joint Staff, and substantially expanding the Office of the Secretary of Defense.)[21] His major effort at reform came toward the end of his administration, in April 1958, when he forwarded to Congress a far-reaching proposal described as essential if America were to meet its two "overriding tasks" of ensuring US security through military strength and of working toward a genuine world peace.[22] The proposal began by noting how previous efforts to centralize defense functions had produced predictions of disaster and prompted vigorous opposition. Acknowledging that the desire to protect traditional concepts and prerogatives was sincere and well meaning, Eisenhower then quickly pointed out that it had undercut a fully effective defense. He then issued a resounding call for change: "We must cling no longer to statutory barriers that weaken executive action and civilian authority. We must free ourselves of emotional attachments to service systems of an era that is no more."[23]

Given such a prolegomenon, it was hardly surprising that centralization and the unity it was presumed to ensure were basic to the proposal's two main provisions. The first of these involved giving the joint chiefs operational planning authority over US military forces worldwide. These forces would henceforth be organized into "truly unified commands" instead of the joint-service commands then in place. The unified commands would include personnel from each of the military services coordinated under the operational control of a general or flag-rank officer who would be designated its commander in chief (CINC). Seven such commands were to be established within the DOD, to which all military forces would be assigned. They would be under the auspices of the JCS and independent of the military departments. Their missions would be oriented toward a particular geographic area—the Atlantic, for example, or the Pacific, or Europe. The age of separate ground, sea, and air warfare was gone forever, Eisenhower argued, and what was required was a new conceptual outlook—a whole new philosophy—that took into account the growing emphasis on nuclear weapons and other complex technologies, the fact that these were based at a relatively few fixed sites, and the overarching emphasis on a more static (read *strategic*) version of warfare this implied. Given rapidly improving communications technologies, the time had come, he said, to unify the military services so that during periods of crisis they could function cohesively, as a unified command, responsive to centralized direction. Activities and responsibilities unique to the individual services would, of course, continue, he said, but these would be of secondary rather than primary concern, "the branches, not the central trunk of the national security tree."[24] It was an ordering of authority and priorities that most emphatically did not fall under the current defense organization. It was a resounding call for centralization.

The second proposal involved further enhancing the authority of the secretary of defense, enabling the secretary to function as a fully effective agent of the commander in chief. This clarification of the secretary's role, as the president described it, involved creating a number of several new positions within the DOD and repealing all statutes giving responsibility for military operations to anyone other than the secretary. It would eliminate existing restrictions on the secretary with respect to the transfer, reassignment, abolition, or consolidation of functions within the DOD. It included giving the secretary a direct voice in appointing, assigning, and removing officers in the top two military ranks; the logic being that only those officers who had demonstrated the ability to deal with national security issues objectively—that is, without undue service partisanship—would have their promotions favorably reviewed. Finally and perhaps most importantly, the proposal called for giving the secretary full management authority for directing budgetary expenditures both among and within the military departments. While the secretary already had the authority to place restrictions on the use of funds by the military departments, this amounted in practice to little more than a limited veto power over decisions already made by the services, who actually determined how the funds were to be spent. The president argued that implementing these changes would go far toward unshackling the secretary from legal restrictions derived from the earlier, nonnuclear era.[25]

Opposition to the president's bill was vigorous and immediate. Key members of the Congress asserted that it would destroy the identity of America's armed forces, constitute a complete surrender by Congress of its power over the purse, and concentrate far greater power in the hands of a single individual (the secretary of defense) than was prudent. Carl Vinson of Georgia, the powerful chairman of the House Armed Services Committee, was perhaps the most vituperative. As he sketched out the gloomy scenario of Eisenhower's plan, Vinson noticed that it would turn the traditionally proud and autonomous military departments into little more than supply and service organizations for the new unified and specified commands and make them mere "branch offices" of the secretary of defense. The service secretaries, relieved of responsibility for military operations, would become mere figureheads who would be bypassed in important decision making. Vinson ominously warned of the likely emergence of a centralized, top-heavy defense department decision-making structure in which various assistant secretaries and deputy assistant secretaries would make unilateral decisions and impose them upon the military departments without adequate consultation within the military chain of command. Far from a coherent management structure, he warned, Eisenhower's proposals portended a netherworld of blurred decision making where responsibility was diffuse and lines of accountability were weak. He hinted darkly that one or more of the services might even be abolished altogether.[26]

Vinson and his Armed Services Committee colleagues drafted several key changes to the reorganization bill explicitly intended to counter its centralizing tendencies. These included a provision that while the services would operate under the authority of the secretary of defense, control would continue to be exercised through the secretaries of the military departments. Another change would limit the authority of the secretary of defense to transfer, merge, or abolish important service functions if a member of the Joint Chiefs of Staff objected. Finally, language was introduced to make explicit the right of the individual services to go to Congress on their own initiative to make recommendations or to register complaints. The revised bill, approved unanimously by the 37 members of the House Armed Services Committee, was sent to the White House for review, where the president promptly denounced it as a "bad concept, bad practice, bad influence with the Pentagon."[27] Such vitriolic language, coupled with an unyielding insistence that the offending changes be expunged from the bill, quickly put the president on a collision course with the Democratic-controlled Congress. The Defense Reorganization Act of 1958 had become wholly partisan, an open political test, and the stage was set for a showdown in the House of Representatives.

"In the years that I have served in this body," Carl Vinson intoned before a packed House chamber on 12 June, "I have witnessed many changes in the affairs of our government. But I never thought that the day would come when the duly elected representatives of the people would be asked to appropriate $40 billion to one man and grant him the sole power of determining its expenditure. I never thought that the representatives of the people would be asked to maintain four military services and then surrender to a single man, not elected by the people, the integrity and the justification for the existence of such military services. . . . But that day has come and that is the issue which squarely faces this body today." Vinson opened debate by posing to his colleagues a stark, dichotomous choice: "By your vote you will either wash your hands of your responsibility and abjectly surrender; or you will insist that you have not only the right but the responsibility and duty to have a voice in the defense of this Nation."[28]

Even in a legislative body known for its dramaturgic posturing and rhetorical flourishes, this was heady stuff, and enough of his colleagues ultimately agreed with Vinson to allow the forces of decentralization to carry the field. The Armed Services Committee's amendments to the defense reorganization bill would stay. As this reality became clear, Republicans tried to cut their losses by suggesting the amendments made no real difference anyway; they were merely refinements in language and of no particular consequence. Republicans then quickly threw their support behind the bill, which passed overwhelmingly in both the House and the Senate. Eisenhower signed the bill on 4 August 1958, passing into law a measure he described as "good, but not good enough."[29]

As with earlier attempts at reform, the 1958 defense reorganization bill represented an uneasy compromise between the forces of centralization and decentralization. The expanded role of the secretary of defense and the Joint Chiefs of Staff, coupled with the enactment into law of the unified and specified command structure, established, almost by definition, a requirement for a command and control system capable of meeting the needs of centralized decision makers in Washington. In most important respects, however, the military departments remained independent entities with considerable bureaucratic power. Planning and force structure remained predicated on unilateral service views of priorities and on how a future war might be fought. Views on training, equipping, and supporting forces logically followed, not infrequently at the expense of joint missions and overall combat capability. Each service retained separate responsibility for its own budget and continued to compete vigorously to increase its share of total defense dollars.[30] Under such conditions, any effort to create a command and control system truly responsive to centralized control appeared almost certain to be resisted by the services or subordinated to their unique, mission-specific needs.[31]

This was disconcerting to many both inside and outside the Pentagon, since the existing system of communications appeared inadequate to the requirements of modern warfare and the evolving demands of strategic doctrine. Even though the communications systems of the services were in the broadest of senses quite similar and often worked quite well, the fact that the Army's Strategic Communications System, the Naval Communications System, and the Air Force's Aerospace Communications complex had independently evolved to meet those services' unique mission requirements made them deficient in several key respects. Since research and development efforts were carried out unilaterally by the services, the result was that in many parts of the world there was a duplication of function in the form of a number of separate, essentially identical facilities. Some of these were located literally right next door to one another, where one station could as easily have served all. Such unnecessary redundancy was rightly viewed as a driving force behind escalating costs.

A related area of concern was the lack of interoperability between the services' separate communications systems. Because of their independent evolution, the equipment and procedures employed by the services differed, often considerably so. The result was incompatibilities in such key areas as the modulation systems, frequencies, and message formats employed.[32] Essentially a communications Tower of Babel, the overall "system" produced by this multiplicity of lower level systems appeared to critics to represent considerably less than the sum of its individual parts. There was also a problem of reliability, especially during crises or other conditions of system perturbation and stress. Although much of the time the dedicated circuits of the services' various communications systems were underutilized, they had a tendency to become overloaded during peak usage, when they were unable to handle the increased volume of message traffic. Should something interrupt, damage, or destroy a circuit in a dedicated point-to-point system, there was little possibility for alternative routing of messages; communications between the two points would simply be terminated. Such a network of inflexible, load-sensitive circuits obviously offered little hope for maintaining communications connectivity during major system outages or during periods of degradation that would surely accompany general nuclear war.[33] This separate approach also meant that leasing services from commercial carriers was undertaken in a fragmented manner, disallowing the cost efficiencies of scale that otherwise could be realized. For those taking a broader, defensewide view, things appeared not far short of an organizational disaster.

Many of these problems were identified by an Air Research and Development Command study group, a technical panel of experts assembled at the end of the 1950s to study ways to integrate the separate communications systems then being developed within the DOD. Not surprisingly, the group's vision of the future was technical in orientation, calling for a new computer-based, fully automated command and control system that could serve a large number of users under a wide range of conditions.[34] But with the organizational as well as technological shortcomings of the extant system increasingly apparent, and with a growing level of dissatisfaction with the existing state of affairs both inside and outside of the Pentagon, the need for some form of system consolidation under centralized managerial control also appeared manifest.

Efforts to create precisely such a structure would begin in earnest during the coming decade. Yet, even as the movement toward command and control centralization got under way, it was clear that it faced major challenges. One of these concerned the fact that there was no precedent, no available model on which the centralizing effort could be based.[35] As such, efforts necessarily would have to proceed in ad hoc fashion, an approach that would invariably result in poor decisions and errors. Another challenge lay in the fact that centralization implied a loss of authority for some affected subunits and groups, with corresponding restrictions on their ability to carry out their missions as they saw fit. With resources and thus mission effectiveness at stake, centralization could reasonably be expected to be a source of consternation, tensions, and resistance.

As the decade of the 1960s dawned, the dynamic tension between the forces of centralization and decentralization remained unresolved. Despite an ever-increasing technical capability for rapid global communications, the services—comfortable with their traditional missions, conservative and resistant to change—tended still toward ways of doing business that had proven efficacious in the past. In other words, it was still very much a question whether the best way to proceed was to take a top-down approach, proceed from the bottom up, or to seek some prudent combination of the two.[36] It would fall to Eisenhower's defense secretary, Thomas S. Gates, and to his Kennedy administration successor, Robert S. McNamara, to answer that question.

Notes

1. Geoffrey Cheadle, "Computerized Combat Communications: 'Message Handling Will Never Be the Same,'" *Armed Forces Journal* 108 (19 October 1970): 20.

2. Solis Horwitz, "National Communications for the Nuclear Age," *Signal* 18 (July 1964): 35.

3. Cheadle, 20.

4. S. Rearden, *History of the Office of the Secretary of Defense: The Formative Years, 1947–50*, vol. 1 (Washington, D.C.: Historical Office, Office of the Secretary of Defense, 1984), 24.

5. Dracos Burke, "SAC: The 'Specified' Command," *Air Force JAG Law Review* 10, no. 1 (January–February 1968): 4.

6. Vincent Davis, "Organization and Management," in *American Defense Annual 1987–1988*, ed. Joseph Kruzel, (Lexington, Mass.: Lexington Books, 1987), 179.

7. Allan R. Millett and Peter Maslowski, *For The Common Defense: A Military History of the United States of America* (New York: Free Press, 1984), 480.

8. "Text of President Eisenhower's Message on Reorganization of Defense Department," *New York Times*, 4 April 1958, 6.

9. Rearden, 54.

10. House, *Department of Defense Reorganization Act of 1958*, 85th Cong., 2d sess., 1958, H.R. 1765 (Washington, D.C.: Government Printing Office [hereinafter cited as GPO], 22 May 1958), 27–28.

11. Millett and Maslowski, 480.

12. John M. Collins, *U.S. Defense Planning: A Critique* (Boulder, Colo.: Westview Press, 1982), 58–59.

13. Ibid., 59–60.

14. "Text of President Eisenhower's Message," 6.

15. Albert Mark, "AUTOVON: Inception to Implementation," *Signal* 20 (March 1966): 10.

16. Richard J. Meyer, "Army Support of the AUTOVON/AUTODIN System," *Signal* 19 (December 1964): 7.

17. George Weiss, "Restraining the Data Monster: The Next Step in C^3," *Armed Forces Journal* 108 (5 July 1971): 29.

18. Paul Bracken, *The Command and Control of Nuclear Forces* (New Haven: Yale University Press, 1983), 184.

19. "Electronics in Military Decision Making," *INTERAVIA* 19 (June 1964): 850.

20. J. P. McConnell, "Command and Control," *Sperryscope* 17, no. 2 (third quarter 1965): 1.

21. Millett and Maslowski, 522.

22. Jack Raymond, "Eisenhower Asks Drastic Revision of Defense Set-Up," *New York Times*, 4 April 1958, 1.

23. "Text of President Eisenhower's Message," 6.

24. Ibid.

25. Ibid.

26. House, *Congressional Record*, 85th Cong., 2d sess., vol. 104, pt. 8, 12 June 1958, H. R. 1765, 10891.

27. Russell Baker, "President Scores Congress Version of Pentagon Plan," *New York Times*, 29 May 1958, 1–8.

28. *Congressional Record*, 10894.

29. Russell Baker, "House Approves Bill on Pentagon; President Loses," *New York Times*, 13 June 1958, 1.

30. Alain C. Enthoven and K. Wayne Smith, *How Much Is Enough? Shaping the Defense Program 1961–1969* (New York: Harper & Row, 1971), 10–11.

31. Lawrence E. Adams, "The Evolving Role of C^3 in Crisis Management," *Signal* 30 (August 1976): 60.

32. House Committee on Armed Services, Armed Services Investigating Subcommittee, *Review of Department of Defense Worldwide Communications, Phase I*, 92d Cong., 1st sess. (Washington, D.C.: GPO, 10 May 1971), 18.

33. Albert Mark, "AUTOVON: Inception to Implementation," *Signal*, March 1966, 10.

34. I. B. Holley Jr., "Command, Control and Technology," *Defense Analysis* 4, no. 3 (1988): 276.

35. Brooke Nihart, "The Ungarbled Future: DCA," *Armed Forces Journal* 107 (9 May 1970): 20.

36. Kenneth Allard, "History, Technology, and the Structure of Command," *Military Review* 61 (November 1981): 6.

Chapter 2

Defense Communications Agency and System

On 12 May 1960 Defense Secretary Thomas S. Gates put defense communications under centralized management control. He issued Department of Defense Directive (DODD) 4600.2, which directed that all long-distance, point-to-point, government-owned and -leased defense communications services be merged into a single, common-user Defense Communications System (DCS).[1] In DODD 5105.19, issued simultaneously, Gates created a new agency—the Defense Communications Agency (DCA)—to manage the new system, in the process turning down an Army bid to become the Defense Department's single manager for communications. As the directive described it, DCA's purpose was to ensure that the new system would be "so planned, engineered, established, improved, and operated as to effectively, efficiently and economically meet the long-haul, point-to-point telecommunications requirements" of the Department of Defense.[2]

Gates's authority to establish the new DCA and the DCS derived from the National Security Act of 1947, as amended under the 1958 defense reorganization. As Gates and his boss in the White House saw it, consolidating relevant communications facilities, personnel, and technologies into a single worldwide complex under centralized management would go far toward eliminating duplicate facilities, reducing manpower requirements, and realizing significant reductions in cost through economics of scale. In this spirit of hope the DCA and DCS were conceived, intended as centralizing forces that would couple more tightly the disparate and often contradictory communications elements and efforts of the military departments—creating in the process a more effective system for the command and control of American military forces around the globe. Given the vastness and complexity of the DOD, the ambitiousness of this arrangement can scarcely be overstated.

As with many other major initiatives, an air of urgency surrounded this creation. An announcement accompanying Gates's release of his two directives indicated that DCA would assume its functions on a phased-in schedule over the following 10 months, a highly abbreviated period for setting up a fully elaborated defensewide communications system. Not by coincidence, this arrangement allowed just sufficient time to have the system in place before a new administration arrived in Washington the following January. Promoting this sort of dispatch within the Pentagon bureaucracy would require serious bureaucratic clout. Directive 5105.19 provided this clout by specifying that the DCA director would be a military officer of flag or general rank directly responsible to the secretary of defense by way of the Joint Chiefs of Staff.[3] Keeping the bureaucratic feet to the fire, the implementation schedule set forth in the directive called for the appointment of the first DCA director to take place within a month, a requirement that was met when RADM William D. Irvin was named to the post on 7 June 1960.

Irvin had his work cut out for him. According to the implementation timetable, he was to submit an organizational staffing plan to the secretary within a month. Working closely with members of J-6 of the Joint Staff (communications-electronics), Irvin met this target date, and his plan called for a headquarters organization consisting of some two hundred military and civilian personnel. Of these, military staffers would outnumber their civilian counterparts by a ratio of three to one, a proportion, according to Irvin, deliberately contrived to ensure that military perspectives would predominate within the new organization. Gates approved the staffing plan on 21 July, and the new agency was given office space in the Naval Services Center in Arlington, Virginia, the site of the old Radio Arlington. Irvin dryly remarked that this facility had not been "platinum plated" in anticipation of his arrival.[4]

Although the Defense Communications Agency was intended to be the single management focus for the Defense Communications System, a major problem was that nobody yet knew just what that system would include. One of DCA's first tasks, then, was to prepare for Gates's approval a planning document identifying DCS's constituent elements. In response, some 79 major relay stations scattered around the globe were designated as system assets, in addition to a variety of radio, landline, and undersea cable communications circuits. At that time they represented a total plant investment of $2 billion. DCA assigned responsibility for these elements to the appropriate military service or defense agency. Secretary Gates approved the DCA plan in early 1961, just before the arrival of the new Kennedy administration. According to the implementation schedule, the Defense Communications System would be set up by 7 March. With the organizational wheels turning, that target date was met. In addition, on 6 March 1961, the Defense National Communications Control Center became operational, initiating limited DCA control over the newly identified assets of the DCS.[5] The era of defense common-user communications systems had begun.

In the Defense Communications Agency and the system it managed, the compromise between the forces of tradition and change represented by the 1958 reorganization bill found its uneasy expression. The creation of the DCA was a milestone in the effort to centralize authority within the Office of the Secretary of Defense (OSD). Consider that DCA's charter gave the agency's director operational direction over the DCS, which was defined as the responsibility for assigning tasks to the system's operating elements, establishing a set of standards, practices, methods, and procedures for the performance and operation of those tasks, and conducting ongoing analyses of system performance. The director would exercise this authority not through a specific military service, but through the JCS under the authority of the secretary. This meant that, for the first time, the chiefs were given day-to-day operations responsibility, a major departure from the earlier notion that they should serve solely as a planning and consultative body.

DCA's charter additionally specified that the director would exercise managerial control over those communication assets of the military services, the unified and specified commands, and various defense agencies that directly supported the Defense Communications System. Managerial control was defined as the authority to directly supervise, coordinate, and review those organizational activities and subunits that were relevant to DCS operations, including such things as engineering and programming, prescribing technical standards and procedures, planning, and research and development. Highlighting these points several years later, Defense Secretary

Robert S. McNamara remarked how DCA's goal in developing the DCS had been to achieve a network that makes use of all available circuitry for meeting the priority needs of users. To create such a capability, he said, required the consolidation of existing manually switched communications resources, while moving toward their replacement by a high-speed, automatically switched network. "Only by expeditious pursuit of this goal will it be possible to satisfy the nation's requirements for capacity, reliability, security and survivability of communications at a cost we can afford," he concluded.[6]

On the surface McNamara's statement appeared to be a strikingly broad mandate for centralization and change, and yet from the very outset, it was clear that Gates's directives were not intended to establish a new communications network separate from those of the services, but rather to provide for the more effective coordination of existing service assets under the overall direction of the DCA. Consequently, a number of caveats were built into DCA's organizational authority that would severely circumscribe its ability to exercise control over the system it had been created to manage. Consider, as the military services promptly did, that the "operational direction" and "managerial control" specified in DCA's charter actually constituted a general coordinating role in the development of the Defense Communications System. Unfortunately for the centralizers' ambitious communications designs, such a coordinating function was conspicuously lacking in the bureaucratic muscle necessary for the new agency to enact its agenda and enforce its will over the opposition of other major actors. The services, jealously guarding their independent communications assets as fundamental to the performance of their military missions, saw the new agency and unified system as inimical to those missions.[7] They voiced their serious reservations, and, in various ways were able to erect bureaucratic impediments to limit Gates's mandate.

The first impediment involved the types of communications and associated facilities to be included under DCA's administrative purview. To overcome the objections of the services, in particular the Navy, to centralize functions which had formerly been their exclusive province and to build a coalition supportive of the new DCS and its managing agency, Gates had found it necessary to exclude a number of key communications elements from his unification order. Excluded for this reason were post, base, and local area communication systems. Also excluded were tactical communications systems operated by field commanders, including the Navy's fleet broadcasts and ship-to-shore circuits, Air Force ground-to-air and air-to-air communications, and certain types of Army tactical communications.[8] Thus DCA's "operational direction" and "management control" did not preclude the services from planning for, even operating, their own individual communications systems. By design, then, the Defense Communications System was not intended to include all DOD communications assets. But it was by the services' design, not DCA's.

Another impediment involved existing organizational structures within the services relating to communications that were almost completely unaffected by the arrival of the new Defense Communications System and Defense Communications Agency. Since the services continued to exercise complete control over those communications assets that had not been specifically designated as part of the DCS, their communications branches, developed to meet their specific mission requirements, remained essentially intact.[9] Thus, when DCA was established, no service communications divisions or functions were disbanded. The reason for this apparent organizational lunacy was again bureaucratic pragmatism, an effort by Gates to minimize the services' resistance to the establishment of the new system and agency. Unfortunately, this arrangement also contributed greatly to a condition of perpetual tension between DCA and the communications branches of the services, with the services almost always holding the superior position.

DCA's authority was additionally circumscribed by a provision in its charter explicitly stating that the secretaries of the military departments would remain responsible for the facilities and resources that related to or supported the Defense Communications System.[10] DCA's "direction" of operating elements thus did not extend even to the staffing or command of actual DCS facilities, since the services retained responsibility for training and assigning personnel who manned the facilities. They supplied and maintained those facilities, and they were responsible for the operational activities necessary to provide communications.

The services also were responsible for engineering, procuring, and other activities necessary for expanding and improving the DCS, with DCA merely to provide supervision, coordination, and review.[11] Not only were no personnel eliminated under these arrangements, but the services soon found it necessary to expand, establishing entirely new organizational entities to coordinate their functions with DCA. Accordingly, each military department promptly organized a separate communications command in the continental United States, with field elements located in each commander in chief's geographical area of responsibility—the Atlantic, Pacific, European, and so on. Headed by a flag or general officer, each communications command reported through the chief of the service to the secretary of the department. DCA's "operational direction" was thus severely limited. Not officially a part of the chain of command, it had little true authority over the personnel and facilities that made up the system.[12]

Perhaps the most important way in which DCA was limited was in its ability to make and enforce budgetary decisions. As in other areas, it was the services that furnished the funds to secure personnel and procure equipment. In fact, the services retained such complete financial control that it was possible for funds programmed for the Defense Communications System to be reprogrammed unilaterally by one of the military services without approval from DCA.[13]

For these reasons, the Defense Communications Agency's arrival on the scene by no means represented the military departments' capitulation to the juggernaut of centralization. Responsible for assigning and training personnel, operating facilities, and making major budgetary decisions, the services continued to exercise almost complete control over DOD communications assets. "We actually do not operate the communications which comprise the Defense Communications System," DCA director Richard P. Klocko remarked several years later. "There are three Military Department Operating Commands that actually run the communications. They have the people who are sitting at the consoles and the communications instruments throughout the entire system. Our role is in the management, the operational direction and control, the planning for the communications systems, getting new systems in, and then monitoring the operation of the Defense Communications System after it is in place."[14] The DCS was thus essentially a collection of pieces made available by the military departments, with considerable restrictions and under some duress, over which DCA could exert little in the way of true authority.[15] The system's performance, and perceived effectiveness, would follow directly from this central organizational reality.

The services' resistance to the centralization of authority represented by the DCA and DSC was abetted by a Congress that for both practical and ideological reasons viewed any centralizing effort with considerable skepticism. On the practical side, the belief was that increasing centralization in DOD decision making would reduce efficiency and produce an indecisive, "no decision" attitude conducive to mediocrity among all personnel except for a select few decision makers at the very top. Doubts were also expressed that the top-heavy, centralized defense agencies could be made sufficiently flexible and responsive to function under conditions of stress, something that could seriously endanger America's national security during crises.[16] Other congressmen expounded the ideological thesis that Carl Vinson had articulated at the time of the 1958 defense reorganization. He believed the trend toward centralization represented by the recent rise of DCA and other defense agencies laid the groundwork for a diminution of the role of the military departments, even their possible dissolution, and the ultimate adoption of a single, monolithic "defense concept." It was a condition they had tried repeatedly in the past to prevent, and presumably could be counted on to resist in the years to come.

In fact, congressional concerns over centralizing authority in the OSD ran so high that in March 1962, Vinson, still House Armed Services Committee chair, appointed a special subcommittee to investigate a number of defense agencies that had been created in the recent past, including DCA. The subcommittee's report, released in August 1962, identified as its overarching concern the creation of a vast centralized bureaucracy within the OSD, one rapidly devolving beyond proper control by the Congress. To regain control, the subcommittee recommended that congressional review be mandatory both for expanding existing agencies and establishing new ones. So that there would be no question about where real authority resided, the subcommittee recommended amending the National Security Act once again so that no activities or functions then being performed by the military departments could be transferred, consolidated, or assigned to any defense agency without specific congressional approval.[17] Although these recommendations never were formally enacted into law, the informal consequence of such powerful opposition was a congressional predisposition to withhold from agencies such as DCA the level of authority necessary to perform their functions adequately.

Nonetheless, the responsibilities of DCA were substantial. Key among these was the establishment of three common-user, defensewide networks that would be known as AUTOVON (Automatic Voice Network), AUTODIN (Automatic Digital Network), and AUTOSEVOCOM (Automatic Secure Voice Communications Network). For each, DCA sought to determine its overall system configuration and prepare the technical specifications necessary for the equipment for switching centers, interconnecting transmission media, and subscriber terminals. DCA was responsible for monitoring the procurements as they took place, developing test and acceptance criteria, and performing the testing. It was also responsible for technical management of fabrication, installation, and checkout.[18] Equally important, with the arrival of the space age, DCA would be designated as the "strong focal point" for development, integration, and operation of the space and ground elements of a number of satellite-based communications initiatives. The most important of these would be DSCS (pronounced "discus"), the DCA-managed Defense Satellite Communications System.[19] With these new responsibilities, the influence wielded by DCA was, in theory at least, vast. It extended into the territory of the unified and specified commands, the military departments, and numerous other defense agencies. Its responsibilities increasing, DCA saw a parallel increase in the size of the staff at its headquarters. Less than two years after Admiral Irvin and his two hundred staffers moved into their Arlington facility, they noted that the size of the DCA staff had nearly tripled.[20]

To successfully exploit the diverse group of communications assets now subsumed beneath the expanding umbrella of the Defense Communications System, DCA planning documents described an operational control hierarchy called the DCS Operations Control Complex. Its design contained three distinct levels of reporting responsibilities: national, regional, and area. The first and most important of these, called the National Defense Communications Control Center (NDCCC), was dedicated in early 1961 by Admiral Irvin. This facility would receive status reports on circuit readiness, message traffic backlogs, and other problems from DCS's various operating elements. Data from the status reports would be entered into the center's Philco 2000 computer, where they would update the database and automatically display the current status of system elements as good, marginal, or poor. When problems arose, NDCCC supervisory personnel could send instructions either by telephone or teletype to the relevant facilities to initiate corrective actions. Operated on an around-the-clock basis by DCA personnel, the NDCCC and its automatic data processing (ADP) equipment would keep the Defense Communications System up and running to the maximum extent possible.[21] Irvin and others had argued strongly for locating the NDCCC in a survivable, hardened facility, but primarily for reasons of cost, the decision was made to collocate the center with DCA headquarters near Washington.

The second level of the DCA Operations Control Complex involved three major geographic areas of operations: Europe, the Pacific, and the continental United States. Within these three general areas, four Defense Area Communications Control Centers (DACCC) were established to decentralize major spheres of communications operations to important geographic areas of the world. The European Area Center, initially located in Paris, was moved to Vaihingen, West Germany, near Stuttgart, following France's withdrawal from the North Atlantic Treaty Organization's (NATO) command structure. In the vast Pacific area, there were to be two DACCCs. The first, called the Pacific Center, was located at the headquarters of the commander in chief of Pacific forces on the Hawaiian island of Oahu. The second, the Alaskan Area Center, was based at Elmendorf Air Force Base (AFB) in Anchorage. In addition, an area center for the continental United States was established at North American Air Defense Command headquarters in Colorado. Provided with substantial ADP capabilities, the centers were responsible for assessing the status of circuits and message flow within their geographic areas of responsibility, restoring circuitry during outages, and reallocating and redirecting the flow of message traffic to reduce backlogs.[22]

Finally, six Defense Regional Communications Control Centers (DRCCC) were established, decentralizing communications control even further to more specific geographic regions of concern worldwide.[23] To coordinate all of this, DCA designed a formal system called 55-1 for reporting the status of DCS elements and for making management reports. This system included several basic types of reports, distinguished primarily by the speed with which they were delivered; near-real-time reports, periodic reports made every four hours, and end-of-shift reports intended to provide data for subsequent computer analysis of DCS performance. Depending upon the nature of the problem, certain types of reports might be made by way of telephone using AUTOVON or AUTOSEVOCOM, each of which DCA was in the process of implementing. AUTOVON was, in effect, the Defense Department's own telephone system, a vast proprietary communications network serving most major military installations in the continental United States and abroad. It consisted of many elements, including the circuits over which messages travel, automatic switching centers for routing those messages, a primary control facility serving as the network's focal point of control, and the system's subscribers and the equipment they used. AUTOSEVOCOM, a DCA-managed common-user secure voice system that operates over AUTOVON circuits, is perhaps more properly viewed as a secure subsystem of AUTOVON than as an separate network in its own right. Far smaller than its parent network in its number of subscribers, AUTOSEVOCOM is "larger" in the sense that users are not restricted to the DOD, including, for example, the White House, Central Intelligence Agency (CIA), State Department, and a range of defense agencies and offices.

Other reports and database file updates were designed to be exchanged automatically by way of AUTODIN, DCA's Automatic Digital Network, which was also in the process of being implemented.[24] Like AUTOVON, its sister network, AUTODIN consisted of a number of major system elements, including switching centers, transmission trunks connecting the switches, access lines leading into those trunks, and at the very end, individual subscriber terminals. From its inception intended as a central element of the Defense Communications System, AUTODIN sought to make available to DOD and a host of other government users fully automatic, high-speed, high-volume, secure data, and teletypewriter (record message) service.

All in all, it appears a classic example of rational bureaucratic design, involving the centralization of policy-making authority and strategic planning, plus the decentralization of operational authority, where operational direction and problem solving are accomplished at the lowest possible level of the hierarchy.[25] Within DCS, if there was a problem at a single switch, tech control, or other facility that did not impact on the rest of the system, problem solving would take place through the channels of the operating and maintenance agency involved, usually one of the military services. But if the problem implicated other system elements, the regional centers would be brought in. In case of circuit outages or other conditions of system degradation, corrective actions would normally be initiated at that organizational level, generally involving one or two of the stations providing the circuit. Only those message-routing problems that could not be resolved at the regional level or ones that possessed some special interest would be passed up the chain of command to the appropriate area center. And only a subset of these would be of sufficient seriousness or interest to warrant bringing in the national center.[26] This was the shape of DCA and the system it managed during the early 1960s, and despite numerous caveats, proponents of centralization had cause to be sanguine.

Of course, what constitutes sufficiency, and by extension effectiveness, might well turn on whether the system is called upon to perform during crisis, war, or peacetime. Despite all the talk of redundant communications assets, the Defense

Communications System, as it was then being developed, represented essentially a peacetime system that would likely not fare well under conditions of stress. Disturbed by this trend, a House Armed Services subcommittee noted prophetically in 1962 how communications assets that appear duplicative or underutilized during times of peace might prove vastly inadequate under crisis conditions. While the subcommittee acknowledged that many of DCA's initiatives then underway would certainly help bring about a better utilization of existing communication assets, the hope was expressed that those assets would be tested under "maximum-use conditions" to determine how well they might work in a crisis or war.[27] Sagacious advice, perhaps, but in a real world of budgetary constraints, it pointed out a dilemma that would confront DCA throughout the years and decades to come; ensuring system effectiveness under one set of conditions would often mean having to compromise it under other conditions.

Analogously, the subcommittee pointed out the somewhat schizophrenic organizational mission of the Defense Communications Agency. On the one hand, the agency had been chartered to engineer and improve DCS to meet the needs of its various user communities to the maximum possible extent. This obviously suggested a drive toward maximum system capacity and capabilities. On the other hand, DCA was tasked by its charter to emphasize cost-effectiveness in everything it did, to "obtain the maximum economy and efficiency in the allocation and management of DOD communications resources."[28] In practice this often meant having to make some hard choices, and coming up with an appropriate balance between user requirements and available funding would prove no easy task for DCA. This was especially true since, lacking the extensive marketing capabilities of its commercial counterparts, DCA had to depend on the services and other system users to provide requirements and indicators of trends. This obviously implied that users knew what they wanted, both then and in the future, and it required that they transmit this knowledge so DCA could make appropriate communications services available. But reliable indicators were difficult to come by even for the highly market-oriented commercial telecommunications companies. To expect the Army or Marine Corps, the Air

Force or the Navy—entities with many other things on their organizational minds beyond determining communications growth trends—to do a better job was, to say the least, optimistic. DCA officials warned repeatedly that DCS users should not assume the system would have the capabilities to meet their needs, but that was, in effect, precisely what the services and other key-user communities always assumed. This assumption set the stage for a series of problems in the development of the major Defense Communications System common-user elements, AUTOVON, AUTOSEVOCOM, and AUTODIN, part of the communications infrastructure upon which the World Wide Military Command and Control System would later depend.

Notes

1. "Army Loses Bid for Single Manager Defense Signal Task," *Army, Navy, Air Force Journal* 97 (27 February 1960): 3.

2. William D. Irvin, "A Report on the Defense Communications Agency," *Signal* 15 (November 1960): 49.

3. "Adm Irvin Named Communications Agency Chief," *Signal* 15 (July 1960): 16.

4. Irvin, 50.

5. William D. Irvin, "Defense Communications Agency, A Progress Report," *Signal* 16 (November 1961): 8.

6. Albert Mark, "AUTOVON: Inception to Implementation," *Signal* 20 (March 1966): 10–11.

7. "Message Control Unified by Gates," *New York Times*, 13 May 1960, 14.

8. House, Committee on Appropriations, *Department of Defense Appropriations for 1970: Part 2, Operations and Maintenance*, 91st Cong., 1st sess. (Washington, D.C.: Government Printing Office [GPO], 1969), 1005.

9. House, Committee on Armed Services, *Report of Special Subcommittee on Defense Agencies of the Committee on Armed Services*, 87th Cong., 2d sess. (Washington, D.C.: GPO, 13 August 1962), 6601.

10. Ibid., 6603.

11. Brooke Nihart, "The Ungarbled Future, DCA," *Armed Forces Journal*, 107 (9 May 1970): 20.

12. House, Committee on Armed Services, Armed Services Investigating Subcommittee, *Review of Department of Defense Worldwide Communications, Phase I*, 92d Cong., 1st sess. (Washington, D.C.: GPO, 10 May 1971), 18–20, 22–23. (Hereafter cited as *Phase I Report*.)

13. Ibid., 23.

14. Senate, *Department of Defense Appropriations for Fiscal Year 1969*, 90th Cong., 2d sess. (Washington, D.C.: GPO, 1968), 2045–46.
15. Herbert A. Schulke Jr., "The DCS (Defense Communications System) Today and How It Operates," *Signal* 25 (July 1971): 22–24.
16. House, *Report of Special Subcommittee*, 6631–34.
17. Ibid., 6633–34.
18. House, *DOD Appropriations for 1970*, 999.
19. Nihart, 20.
20. House, *Report of Special Subcommittee*, 6597.
21. "Defense Communications Control Center Opened," *National Defense Transportation Journal*, May/June 1961, 11.
22. House, *Report of Special Subcommittee*, 6598.
23. House, Committee on Armed Services, *Hearings before Special Subcommittee on Defense Agencies*, 87th Cong., 2d sess. (Washington, D.C.: GPO, 1963), 6760.
24. House, *DOD Appropriations for 1970*, 1009.
25. Philip S. Kronenberg, "Command and Control as a Theory of Interorganizational Design," *Defense Analysis* 4, no. 3 (1988): 232–33.
26. Schulke, 24.
27. House, *Report of Special Subcommittee*, 6623.
28. *Phase I Report*, 22.

Chapter 3

National Military Command System

The creation of the Defense Communications System represented in substantial measure an effort to achieve greater economies by integrating the services' separate long-haul communications systems and placing them under a single management authority. Yet almost from the outset, DCS was itself considered to be but one element, albeit a central one, in a larger worldwide system whose scope went well beyond DCS. Crises, along with the accompanying perception that the existing command and control structure was ineffective in dealing with them, provided the impetus for this more extensive and ambitious system, both at the time of its birth and through a number of subsequent developmental stages.

The process of crafting this larger structure began almost as soon as the Kennedy administration arrived in Washington in early 1961. The contradictory political impulses of this administration shifted between "hardheaded military pragmatism and liberal humanism."[1] Within a week of taking office, Defense Secretary McNamara had become a convert to counterforce, the idea that nuclear targeting could be employed selectively and flexibly to limit damage to population centers and other civilian targets.[2] McNamara's White House boss, President John F. Kennedy, underwent a similar sort of conversion shortly thereafter. The president's conversion was the result of a briefing in which advisors described to him the probable human consequences of the full-scale nuclear spasm attack called for in the Single Integrated Operational Plan (SIOP). Repulsed, Kennedy publicly renounced the doctrine of massive retaliation, vowing that America would never be the first to strike with nuclear weapons. The search for options was on, a search that would lead to a thorough review of the nation's command and control systems, to a host of new initiatives, and, ironically, that would make nuclear war more thinkable and rational.

The Partridge Report

The Bay of Pigs incident, coming less than two months after Kennedy's inauguration, was, to the thinking of many newly arrived administration officials, exemplary of the problems that resulted from an excessively decentralized military command and control structure. Intelligence and communications difficulties had plagued the aborted effort to overthrow the Castro regime from the outset; the new president and his chief advisors were unable to keep track of troops and events as they unfolded in the swamps of southern Cuba, and events rapidly devolved beyond anyone's control. The embarrassing and costly debacle provided, in the minds of a host of administration officials from the president on down, a striking example of the need for better communications and for a more centralized, coherent, integrated, and effective structure for managing military operations.

The need having been established, in May 1961 Kennedy called for the creation of a command and control system that, although located within the Department of Defense, would be responsive to the needs of central decision makers and remain under ultimate civilian control at all times. The system must be maximally survivable, he said, and offer protection both from the effects of nuclear weapons and from electronic interference. "We propose to see to it," the president declared, "that our military forces operate at all times under continuous, responsible command and control from the national authorities all the way downward, and we mean to see that this control is exercised before, during, and after any initiation of hostilities."[3] Survivability, interconnectivity, and endurance were the essential criteria being called for, a distinct departure from the past. Kennedy acknowledged that developing this kind of system would involve a major effort on the part of the United States, but that effort, he said, "vital to the existence" of the nation, was worth it.[4]

In line with the president's vision, Defense Secretary McNamara directed that a study be conducted to assess the ways in which this more centrally responsive command and control system might be achieved most readily. The research group that was assembled, called the National Command and

Control Task Force, included a number of military personnel, some of them from the recently established Defense Communications Agency. Civilian participants included members of the White House staff and a number of outside consultants from such interested organizations as the RAND and MITRE Corporations. Earle E. Partridge, a retired Air Force general and former commander in chief of the Air Defense Command, was recalled to active duty to head the study team. The task force promptly set to work and received considerable impetus to its efforts in August 1961, when the Soviets began the construction of the Berlin Wall. During the Berlin crisis, the national leadership considered a series of possible military responses, ranging from an attack by conventional forces up the autobahn linking Berlin with the rest of West Germany to the firing of nuclear warning bursts over unpopulated areas of the Soviet Union. With nuclear confrontation a distinct possibility, the need for an effective command and control system appeared urgent in the extreme.

The top secret *Final Report of the National Command and Control Task Force*, more familiarly known as the Partridge Report, was completed on 14 November 1961.[5] Concluding at the outset that the capabilities of US weapons systems had outstripped the ability to command and control them, the report provided the administration with a comprehensive blueprint for the integrated, survivable, worldwide system it desired.[6] Although the conventional forces were by no means neglected, the major emphasis of the report focused on the strategic nuclear forces. But despite its expansiveness, the report contained no radically new ideas and proposed few new initiatives or systems. Rather, what it did was more along the lines of describing an intellectual context, establishing a framework for streamlining, modernizing, and centralizing command and control that both reflected the new administration's concerns and gave coherence to a number of military command and control programs then in the planning stages, under development, or already deployed.[7] This framework was called the National Military Command System (NMCS).

Partridge's task force envisioned that action to implement the NMCS should proceed in a series of steps. Of foremost importance was the need for a large, technically sophisticated

National Military Command Center (NMCC) to replace the Joint War Room at the Pentagon. Plans for upgrading the small, overcrowded war room with its extremely limited data-processing capabilities had, in fact, already been drawn up by the Defense Communications Agency. Partridge's team drew heavily from these plans as they considered ways to improve this, the most important node of the proposed National Military Command System.

Recognizing the near certainty that the Pentagon would be destroyed in the opening moments of a nuclear attack, Partridge's team next recommended upgrading an alternate, more survivable command center to which the national leadership could repair during times of crisis. Called the Alternate National Military Command Center (ANMCC), this backup "underground Pentagon" was intended to perform the critical functions of the NMCC if the Pentagon facility was destroyed or rendered inoperable. The ANMCC, the location of which was originally shrouded in secrecy, (code-named "Site R" for security reasons) was hardened against the effects of nuclear detonations by being buried inside Raven Rock, a mountain in southern Pennsylvania. This site was located within the boundaries of the Fort Ritchie military reservation in Maryland, about eight miles from the presidential retreat at Camp David.[8] The idea for the ANMCC did not originate with Partridge's team, however, for the Raven Rock facility already existed. During the 1950s President Eisenhower and his cabinet, on several occasions, had convened there, as well as at other locations, to participate in nuclear war exercises.[9] As with the NMCC, the Partridge team's point was to identify existing resources that would serve as crucial elements of the new National Military Command System.

A third command post whose critical nature the Partridge's task force underscored was the headquarters of the North American Air Defense Command (NORAD). Thinking as the task force did of possible nuclear crises or conflicts, NORAD's missions were indeed critical ones: surveillance, detection, and identification of aircraft operating over or near the North American continent, and, by way of the Semi-Automatic Ground Environment, operational control over US and Canadian air defense forces. With the coming of the space age,

NORAD's mission would be expanded to include surveillance, tracking, and cataloguing of all man-made objects in space and the detection of missile launches and nuclear events around the globe. In all cases, NORAD's most important job was to provide US and Canadian leaders with timely warning that North America was under attack. Given the vital nature of this mission, a major Air Force concern at the time was the construction of a new, survivable NORAD headquarters. Planning for the new facility had in fact commenced years earlier, and by the time the Partridge Report was issued, work on Program 425L, NORAD's new Cheyenne Mountain facility, was already well under way. With NORAD, as with the other command centers, then, the Partridge Report broke no new ground.

Finally, the Partridge Report underscored the need for a survivable airborne command post. In February 1961, just days after the Kennedy administration's arrival in Washington, the Strategic Air Command's first EC-135 "Looking Glass" aircraft, designed to provide control of the strategic nuclear forces after a nuclear attack, took off from Offutt AFB in Nebraska.[10] SAC would keep one such aircraft continuously airborne for the subsequent three decades. Concerned that civilian leaders be accorded an equally survivable airborne command post, McNamara directed SAC to station similar EC-135 aircraft at Andrews AFB near Washington so that the national leadership could rapidly board in case of crises. SAC responded by deploying three of the planes, one of which was kept on continuous ground alert,[11] to Andrews the following year, 1962. Here again, the Partridge Report simply underscored the importance of initiatives already well advanced.

Since the NMCS would be a system linking the national leadership, both civilian and military, to the operating military forces worldwide, it consisted not only of the various command posts but also of the communications media linking them together. As conceived, the NMCS would not be a separate, dedicated communications system within the DOD. Rather, it would constitute a collection of existing resources—some of which had been designed and provided in response to national level requirements—that the NCA could draw upon as necessary. These assets, which included the Defense Communications

System as a central element, ensured that the Defense Communications Agency would be a key player in NMCS design, engineering, technical supervision, and support.

It is worth underscoring Kennedy's point that the NMCS, while a military system, would remain under ultimate civilian control at all times. To that end, the NMCS by necessity would interface not only with a wide range of military forces but also with a number of civilian agencies and offices. The NMCS would provide direct connections to the White House situation room, for example. The White House's own command center, the situation room was headed by a military officer and operated by the CIA. NMCS would also be linked directly to the State Department's Operations Center, an element established in 1961 to deal with international crises. A third key civilian interface would be with CIA headquarters in Langley, Virginia.[12]

The major emphasis of the Partridge Report, its cutting edge, is found in the final word of the phrase "National Military Command *System*." As the phrase denotes, the thing being created was to represent an integrated whole, a body with articulation among its constituent parts. If NORAD and its sensors were the body's eyes and ears, if SAC were its muscles and its fists, the new command system would constitute its brain and central nervous system—the network that permitted all of the parts to operate as a single coherent entity. The ensuing 30-year effort to create a more coherent and centralized, interconnected, and survivable system for military command and control, one with a strong and ever-increasing emphasis on the control of the strategic nuclear forces, can be traced to the criticisms and recommendations contained in the Partridge Report.

Given the fierce resistance to centralization that had been encountered during the 1958 defense reorganization, it was hardly surprising that the Partridge Report and other similar recommendations were greeted in some quarters with skepticism and even hostility. Many experienced military personnel deeply resented the White House-to-foxhole approach to military operations implicit in an increasingly centralized command and control structure. To them, the White House approach smacked of meddling, of a looking-over-your-shoulder sort of micromanagement. Socialized to the merits of a bureaucratic hierarchy as formally expressed in the military chain of command, they considered efforts to bypass or ignore that chain injurious and potentially dangerous. These efforts, they believed, deprived those both at the top and the bottom of necessary advice and guidance. As one officer ominously cautioned, "If the chain of command is not employed, it is not needed. If it is not needed, it should be abolished. If it is abolished, everything will have to be run from the top."[13] Of course, in this judgment and the judgment of many other professionals, attempts to run things from the top would only increase the likelihood of errors and mishaps. In addition, diminishing the importance of the chain of command would deprive those therein of their prerogatives, even their reason for existence. The result was that the NMCS concept was resisted from the outset, often openly, but even more so in secret. "You didn't go to West Point twenty-five years ago and train your whole professional life to have somebody look over your shoulder," after all.[14]

The new administration firmly believed in the merits of centralization, however, and with the Partridge Report in hand as guidance, McNamara set out in late 1961 to establish the operational framework for precisely such a centralized system. The president reinforced the effort through his public statements, asserting how "we propose to see to it . . . that our military forces operate at all times under continuous, responsible command and control from the national authorities all the way downward—and we mean to see that this control is exercised before, during, and after any initiation of hostilities."[15] Noting that America's nuclear monopoly had ended, Kennedy pointed out how this hard reality had forced defense planners to consider new contingencies that in turn required a new emphasis on improved command and control. Thus, the development of the new National Military Command System recommended in the Partridge Report a system that would directly support the National Command Authorities under all conditions of peace, crisis, and war.

Implementing the National Military Command System

On 2 June 1962 Secretary McNamara issued a memorandum directing that the NMCS be put into operation. The memorandum assigned to the Joint Chiefs of Staff the responsibility for working out the user requirements and the functional design for the new system. To meet those responsibilities, the chiefs acted along two fronts. They immediately began using the Joint War Room as a nucleus for the National Military Command Center, designated the NMCS's key node. They then turned to their Joint Command and Control Requirements Group (JCCRG) for help.[16] The JCCRG owed its existence to the 1958 defense reorganization, which had dramatically expanded the role of the joint chiefs and made their role far more detailed and complex. With each passing day, it became increasingly obvious to the joint chiefs that fulfilling their new responsibilities required the development of some sort of joint-service command and control system complex, one that was global in scope and responsive to the needs of the National Command Authorities.[17] By the end of 1959, the need was sufficiently acute that a number of major reviews had been commissioned. (One of these, known as the Winter Study, consisted of two dozen separate panels—some 140 Air Force, industry, and other personnel in all. The study examined key command and control issues, mostly from an Air Force point of view.[18] Another effort was the Navy's Pangloss research into ways to improve communications with the then-emerging ballistic missile submarine force.) The possible forms that a national level command and control system might take were myriad. The joint chiefs needed help sorting through the possibilities, so the JCCRG was set up in January 1960 on an informal, advisory basis. Over the next few years, the JCCRG would expand, both in size and in the number of system strategies it considered.

These efforts were still in progress when the Kennedy administration arrived in Washington. Suspicious of the old Pentagon regime, Defense Secretary McNamara was not about to accept the conclusions of these studies without a second opinion. That was where General Partridge's Task Force came in.

But Partridge's key conclusion—that a National Military Command System be established—proved to be almost identical to the conclusions of the various other studies then under review by the JCCRG. McNamara accordingly directed the joint chiefs to implement the system, and to accomplish this task, the joint chiefs turned to the JCCRG, their in-house experts. Already a substantial if informal bureaucratic presence, the JCCRG was promptly formalized, its bureaucratic clout increased by upgrading its chief to a two-star billet.[19]

The tasks facing the JCCRG were substantial. The first was to inventory available facilities and assess objectives. What exactly was the system in question? Precisely what was it supposed to accomplish? How could various system assets be tied together to do this? As one officer assigned to the JCCRG described the process, "As we look in detail we find we are also looking at the broader view. And, as we work closer and closer with the full worldwide system and its requirements, we are able to more clearly define and refine the requirements of the NMCS."[20] If it all sounds more than a bit ambiguous, this is hardly surprising. From the outset, the thorniest of the problems facing the JCCRG was the identification of requirements. What specific types of information do users require across a spectrum of possible situations?[21] For only when these problems had been specified could the most appropriate communications, automatic data processing, and organizational technologies for meeting them be identified. And only when all of this was done could the best arrangement be devised for linking these elements together so that they could, in fact, function as a coherent system.[22]

Unfortunately, this quest for centralized coherence was imperiled from the very outset by fragmented NMCS management responsibilities. McNamara had specified in his 2 June 1962 memorandum that the joint chiefs would work out the new system's requirements, and his rationale made considerable sense. The secretary understood that many military leaders felt they they did not have sufficient input into the design of earlier command and control systems. These systems were inappropriate, failed to meet their needs, and were consequently rejected by the very users they were intended to serve. By implicating the joint chiefs, McNamara felt, military leaders would have input into the conceptual development of the system, making it more useful to them and lessening their resistance to its implementation.

But the major problem was that nobody knew precisely what was needed, or what the appropriate requirements were. The president had called on the new system to be survivable, interconnected, and endurable; and defense journals were now discussing how these were the key words for NMCS development. But precisely what did such terms mean in the real world of proliferating nuclear weapons and limited budgets? Did they mean redundancy, the use of a wide range of communications frequencies and system approaches, including very low frequency and microwave systems, terrestrial and undersea cables, ionospheric and tropospheric scatter systems, and, when they became available, satellite and rocket communications systems? Did they mean internetting, the ability to communicate between system elements even in case of communications breaks or outages? Did they mean having the ability to ride out a nuclear attack and still have the means to assess what was happening and to pass launch orders to the strategic nuclear forces?[23] But the major problem in virtually every case involved having little real-life experience to go on and having no way to determine what sort of demands would be placed on the NMCS in the event of major crises. The conceptual undertaking was thus ambiguous in the extreme, for as director of defense research and engineering Eugene G. Fubini summed up the problem, "In the final analysis, what we are really dealing with is a system whose configuration depends on answers to the larger question, 'What is the proper texture of a democratic government under stress?'"[24] This was a question for which there was no single or simple answer.

But whatever that answer might eventually prove to be, one thing was apparent to everyone from the outset: the NMCS-to-be would rely heavily on automatic data processing technologies. Along these lines, the first computer designated exclusively for command and control support of the National Command Authorities, an International Business Machines (IBM) 1401, was installed in the fall of 1962 in J-3, the joint chiefs' operations area in the Pentagon, and another simultaneously installed at the Alternate National Military Command

Center's computer facility. Initially, these computers were run by personnel attached to the Defense Atomic Support Agency, but on 1 January 1963, they were transferred to the Defense Communications Agency. At that time the Pentagon computer facility was renamed the National Military Command System Support Center and placed under DCA operational control. At first, these computers were not considered "organic" elements of the NMCS, since, as National Military Command Center director Paul Tibbets noted, the NMCC did not deal with a volume of data sufficient to justify automation. But Tibbets and others knew well that this was soon to change, for in terms of ADP support for national-level command and control, the surface had only been scratched.[25]

The search was on for an optimum computerized system, one capable of providing the ADP support required at every level of the military hierarchy while being able to feed information quickly up the line. Yet while providing unprecedented ability to process data, the introduction of computers did little to help clarify the fundamental issue of requirements. Since nobody was really sure what information was needed, by whom, or when, huge sums of money and countless hours of effort would be invested by a staggering number of commands and agencies (DCA perhaps chief among them), trying to identify and specify the appropriate database that should be available in the various NMCS command centers for decision-making purposes. Nor in the absence of well-defined requirements was it clear with precisely which systems, and under what conditions, the NMCS should connect. It was a circular problem, to be certain, for only when requirements had been specified was it possible to develop specific, appropriate supporting ADP subsystems and the interfaces between them.[26] With the program thus wallowing in conceptual ambiguity, what was happening was that the system was essentially designing itself.[27]

In the absence of a well-elaborated set of user requirements, many officials believed that the best approach to the ADP and interface problems, and by extension the best strategy to pursue with respect to the design of the NMCS, was the basic one of standardization. Since the system required the capability to exchange data among its constituent elements, and since this process by definition involved a large number of interfaces, what was called for was the use of similar types of computer hardware, associated equipment, standard data formats, and standard software applications.[28] One early effort along these lines was the National Military Command System Information Processing System (NIPS). Developed in 1963 to run on the IBM 1410 computer, NIPS sought to achieve a measure of compatibility in the areas of hardware and software, to reduce the (even then) staggering costs of software development, and to promote standardization of information, equipment, and training among the various elements of the NMCS. But since it was being forced down from the top, such compatibility would not be accomplished immediately.[29] In the interim, the system would continue to design itself.

By this time, several other defense organizations already had developed automated command and control systems, and it might seem that these could provide operational models for the NMCS, thereby avoiding much, if not all, of the developmental "ad hocery." SAC and NORAD, for example, had computerized systems that were extremely sophisticated compared to what was then available in the Pentagon. But although some of the technologies developed under these programs would later be drawn upon for use in the NMCS, they were far from appropriate models. Consider that both NORAD and SAC performed relatively clear defense functions. Knowing what had to be done served to foster a climate of aggressive technological innovation, especially in fashioning automatic data processing as an effective tool for command and control. In contrast, nobody knew exactly what functions NMCS was supposed to perform. But even with the relative clarity of their missions, NORAD's and SAC's problems with their automated command and control systems were vast—indeed, the stuff of which bureaucratic legends were made. For these reasons, they were far from perfect models for developing a computer-based National Military Command System. As for the other major commands, since they had absolutely nothing in the way of ADP support for command and control, they provided no examples at all, useful or otherwise.[30]

So it was not the computer technologies themselves that represented the major limiting factor in the birth and subsequent evolution of the National Military Command System.

Rather, the problem was more in conceptualizing their usefulness to decision makers and their supporting organizations. At base, it was a people problem, figuring out what information was required and then having the personnel adjust to the technologies capable of providing it.

As McNamara's memorandum laid out the responsibilities for the new National Military Command System, the joint chiefs and their in-house experts, the JCCRG, working closely with the OSD, would generate policies, concepts, and requirements. The Defense Communications Agency was given responsibility for systems engineering. This engineering function would be performed under the broad policy supervision of the director of defense research and engineering, who would create the necessary NMCS support organization. The military departments, for their part, would program, deploy, and support the operation of the various NMCS subsystems.[31] That was how things were supposed to work. The problem was that, while the secretary's memorandum seemed to clearly establish the basic relationships and responsibilities within the NMCS, the reality was otherwise. Many of the instructions in the memorandum were quite general, and there were numerous caveats, both stated and implied. Given the widespread resistance to anything that smacked of centralization in military operations (resistance, that is, to the very thing exemplified by the NMCS concept), and given the institutional strength of any number of individuals and organizations who "viewed the world from the castle walls on their manor lands and fiefs," the results were predictable.[32] The ambiguities in McNamara's instructions were quickly exploited, interpreted in terms of existing interests, and used as justification for new programs and initiatives where specific guidance was weak or absent.

This gave extraordinary latitude to powerful constituencies to shape the form that the NMCS would take. To make certain that their interests were represented, many of these—in particular the services—established staff sections to monitor the development of the systems they were to use. When it came time to deploy these systems, the services also accepted this responsibility. For example, two mobile command posts, the National Emergency Airborne Command Post and the National Emergency Command Post Afloat, were in large measure service initiatives, actively pushed by the Air Force and Navy, respectively. (The Army was already in charge of the ground-based Alternate National Military Command Center at Fort Ritchie, and these mobile command posts assured that the other two services would not be cut out of the NMCS action.) In other words, some of the most crucial decisions concerning the design, capabilities, and evaluation of the new NMCS were determined by organizations whose interests ran in the direction of keeping things decentralized. And things were made all the easier since the JCCRG performed its advisory function with one eye focused directly on service needs.

As things were supposed to work, once requirements had been worked out by the JCCRG, the joint chiefs would turn them over for approval to the secretary of defense. The approved requirements would then be passed along to the director, Department of Defense Research and Engineering (DDR&E), whose agency was responsible for ensuring that a system would be engineered to meet them. To meet those responsibilities, McNamara instructed the DDR&E (at the time Harold Brown, himself later a secretary of defense) to establish the position of director of NMCS technical support within his office.[33] By separating the functional roles of the Joint Chiefs of Staff and DDR&E in this fashion, McNamara hoped to eliminate the frequent and injurious power struggles that had characterized the relationship between these two organizations, thereby allowing the groups most directly responsible for national-level command and control system development to work together more smoothly. DDR&E took its mandate seriously, and during the system's early years, a substantial percentage of its funds would be utilized to establish the NMCS.[34]

If DDR&E was given an overall supervisory function in the technical implementation of the NMCS, the Defense Communications Agency had a more hands-on role—take identified system requirements and turn them into a detailed set of technical specifications. Specifically, DCA would prepare a detailed technical plan for the NMCS that would include both a system design and a strategy for acquiring the system. It would be responsible for preparing cost estimates, performing technical analyses, and otherwise providing the joint chiefs with the necessary systems engineering and technical support for the system. So that his agency could do these things, the director of DCA was tasked to establish an NMCS technical support element within his agency, analogous to that already established within DDR&E.[35]

The DCA promptly established its NMCS element at its Arlington, Virginia, headquarters. It was headed by John B. Bestic, an Air Force major general, who was given the title deputy director of the NMCS. Staffing for this new NMCS headquarters directorate included a mixture of military and civilian personnel, people whose backgrounds were mainly in the areas of communications, automatic data processing, and engineering, but who also represented, in Bestic's words, "the various disciplines contributing to the solution of command and control problems."[36] In one of those dubious official accounts, (devoid of the black humor that makes stories memorable outside of a self-serving, bureaucratic context) when the freshman staff of the NMCS assembled on its Defense Communications Agency "campus," one exuberant young major was supposed to have exclaimed, "What NMCS means, General, is No More Confused Situations. Right?"[37]

Using the plan drawn up by the JCCRG as its basic framework, Bestic's group soon developed a detailed set of technical specifications for the National Military Command System, effectively a master plan for its form and evolution.[38] The MITRE Corporation was awarded the contract for the technical planning, designing, managing, and integrating of the NMCS project, and its approach to system development was described as "evolutionary." MITRE's reasons for taking this sort of approach seemed eminently reasonable.[39] Several existing major command and control systems (such as those at NORAD and SAC), designed essentially as "turnkey" systems, had proved to be major failures in many respects. As the realization began to sink in that command and control systems might not be amenable to the usual acquisitions (weapons system) approach, and as it became clear that military requirements 10 or 15 years in the future could not be predicted with any certainty, system designers at MITRE decided that a better strategy was to follow the lead of the commercial telephone companies. Instead of "dropping a new system on top of the others," in the words of Esterly Page, DDR&E's technical director for the NMCS, the system would be designed within the state of the art, modified as necessary to meet changing requirements, and be flexible enough to exploit technological innovations as they became available.[40] MITRE's evolutionary approach seemed fully consistent with the intent of McNamara's directive, which emphasized creating command and control systems with steady-state learning and growth bases, that could be continually improved without widespread equipment obsolescence.

To those ends, the approach emphasized a close and continuous interaction between system users—commanders and their staffs—and the engineers and technicians who designed the systems for them. Through this interaction, requirements could be evolved and new systems developed, but in a way that permitted greater all-round understanding. Users would become aware of the practical possibilities offered by new technologies, especially in the area of automatic data processing. They would come to understand their limitations as well. Engineering personnel at DCA and other agencies, for their part, would learn to appreciate the complexity of operational situations—in particular, the frequent lack of complete information and the need to deal with surprise and uncertainty. The technologies that were recommended for the system would presumably reflect this understanding and result in an NMCS that could better cope with real-world situations.[41] While the evolutionary approach sounded positive indeed, it would prove to be the source of a great many problems in the future.

Within a year and a half of McNamara's order that the NMCS be put into operation, more than 40 command and control systems operated by the services, defense agencies, and the unified and specified commands were tied into it with varying degrees of success. Some of these, including such key commands as NORAD and SAC, were rapidly moving toward computer-based, fully automated reporting systems. The system also included such communication networks as AUTOVON and AUTODIN that linked these facilities with the National Command Authorities, the unified and specified commanders, service headquarters, and other designated agencies, including the National Security Agency and the Defense Intelligence Agency.[42] With matters apparently proceeding well, enthusiasts were not shy to declare, with some hyperbole, that before long the NMCS would provide, for the first time, a means by which the National Command Authorities could maintain instantaneous and detailed contact with all levels of US military forces worldwide.[43]

Notes

1. Janne E. Nolan, *Guardians of the Arsenal: The Politics of Nuclear Strategy* (New York: New Republic Books, 1989), 64.
2. Desmond Ball, "The Development of the SIOP, 1960–1983," in *Strategic Nuclear Targeting,* eds. Desmond Ball and Jeffrey Richelson (Ithaca, N.Y.: Cornell University Press, 1986), 62.
3. Lawrence E. Adams, "The Evolving Role of C^3 in Crisis Management," *Signal* 30 (August 1976): 60.
4. "The USA's National Military Command System," *INTERAVIA* 19 (June 1964): 854.
5. *Final Report of the National Command and Control Task Force,* 14 November 1961, attachment 2 to Joint Chief of Staff Publication 2308/64.
6. Lee M. Paschall, "The Command and Control Revolution," *Air Force Magazine* 56 (July 1973): 39.
7. C. W. Borklund, "National Military Command Control: The Problems in Brief," *Armed Forces Management* 9 (July 1963): 20.
8. "The USA's National Military Command System," 855.
9. Ted Gup, "The Doomsday Blueprints," *Time,* 10 August 1992, 34.
10. Kenneth J. Stein, "E-4B Boosts SAC's Communications Net," *Aviation Week & Space Technology,* 16 June 1980, 77.
11. Bruce G. Blair, *Strategic Command and Control: Redefining the Nuclear Threat* (Washington, D.C.: Brookings Institute, 1985), 110.
12. J. H. Wagner, "NMCS: The Command Backup to Counterforce," *Armed Forces Management* 9 (July 1963): 25.
13. William E. Burr, "Centralized Control," *Military Review* 46 (May 1966): 82.
14. Daniel Ford, *The Button: The Pentagon's Command and Control System—Does It Work?* (New York: Simon and Schuster, 1985), 175–76.
15. Lee M. Paschall, "The Role of the Present and Future DCS in Support of the WWMCCS," *Signal* 29 (March 1975): 38.
16. Wagner, 23.
17. "Command and Control Spectrum: JCS to Battlefield," *Armed Forces Management* 14 (July 1968): 37.
18. Peter Pringle and William Arkin, *SIOP: The Secret US Plan for Nuclear War* (New York: Norton, 1983), 210–11.
19. Wagner, 23.
20. "Command and Control Spectrum," 37.
21. Borklund, 20.
22. "Command and Control Spectrum," 37.

23. "Can Vulnerability Menace Command and Control?" *Armed Forces Management* 15 (July 1969): 41.

24. Borklund, 20–21.

25. Wagner, 25.

26. "NMCS: The National Voice of Command," *DATA* 11 (January 1966): 24.

27. Leonard Maley, "NIPS—The System that Invented Itself," *Systems Engineering Forum* (n.d.), 32.

28. Borklund, 21.

29. Maley, 32.

30. Ibid.

31. "The National Voice of Command," 24.

32. Maley, 32.

33. "The Search for Effective Command and Control," *Armed Forces Management* 8 (July 1962): 19–20.

34. Senate, Committee on Appropriations, Subcommittee on Department of Defense, *Department of Defense Appropriations for Fiscal Year 1969: Hearings before the Subcommittee of the Committee on Appropriations,* 90th Cong., 2d sess. (Washington, D.C.: Government Printing Office, 1968), 2349.

35. "The Search for Effective Command and Control," 20.

36. "Bestic's Assignment: Define Command/Control For All Services," *DATA* 9 (October 1964): 32.

37. John B. Bestic, "No More Confused Situations," *Signal* 21 (March 1967), 53.

38. Lee M. Paschall, "WWMCCS: Nerve Center of U.S. C^3," *Air Force Magazine* 58 (July 1975): 54.

39. "The USA's National Military Command System," 854.

40. Borklund, 20.

41. Wagner, 25.

42. Hubert S. Cunningham and William Edward Kenealy, "The Joint Chiefs of Staff and Command and Control," *Signal* 29 (March 1975): 15.

43. J. S. Butz Jr., "USAF and the Computer Revolution," *Air Force Magazine* 47 (March 1964): 34.

Chapter 4

WWMCCS Is Born

The effort to establish the various elements of the National Military Command System had scarcely begun when the October 1962 Cuban missile crisis electrified the nation and the rest of the world. The crisis highlighted a number of command and control problem areas that remained unaddressed. It also provided considerable additional impetus for creating the type of comprehensive system, responsive to the needs of the national leadership, that President Kennedy and his advisors said they required. And not without reason: for a few tense hours, the United States and the Soviet Union stood on the brink of the nuclear precipice, the first and most serious such moment in the annals of the cold war.

A series of hard lessons were forthcoming from those tense and dangerous October days. The first of these lessons underscored what had been abundantly and painfully obvious to the administration during the Bay of Pigs debacle—the flow of intelligence from the field to the national leadership needed substantial improvement. The missile crisis now vividly posed the question, "What price information?"—to which Kennedy's emphatic response was "almost any price."[1] The crisis again made it apparent that in many instances it was not possible for the civilian leadership to exercise effective control over the operating military forces. In part, this was simply because the necessary communications systems were not in place. But even where they existed, administration officials encountered a deeply ingrained military resistance to any effort to exercise centralized direction and oversight of local operations. The civilian-military tension had been especially palpable when administration officials went outside of the usual military chain of command to speak with the commanders of vessels participating in the quarantine operation.[2] To those in the services, professionals steeped in a military culture where hierarchy and decentralization were considered both virtues and necessities, this sort of supervision was anathema.[3]

The Cuban missile crisis's third lesson concerned the need for improved civilian communications. Since the conduct of operations during the crisis and its eventual resolution required the use of both defense and nondefense communications systems, the importance of each for successful crisis management was underscored for the administration. Deficiencies in existing civilian communications systems were especially apparent when the president was unable to inform South American leaders of his intended actions because the State Department's communication system was overloaded. They were apparent again when US ambassadors to a number of Latin American nations were unable to contact their governments because of international communications bottlenecks.[4] These ostensibly nondefense systems clearly performed an important defense function.

A closely related lesson concerned the need for linkage between the communications systems used by the military and civilian agencies of government. The Cuban crisis amply demonstrated the intimate relationship between these systems and made explicit for the first time the need to link them together into a single, integrated system.[5] But linking them together was obviously no small undertaking. As was already apparent to the Defense Communications Agency and other entities then attempting to integrate military communications systems, the problem was that different systems had been developed separately by different constituencies for different purposes. They tended to be technically and procedurally incompatible, and, if connected at all, connection occurred at only a few points.[6] Adding a plethora of civilian communications systems to this mixture would only compound the difficulties.

A final lesson administered by the crisis involved serious deficiencies in communications security. The need to communicate orders and information to a number of US embassies and military facilities abroad regarding the impending blockade of Cuba, coupled with a lack of secure communications circuits by which to do so, had permitted the Russians to intercept many of the messages, thereby gaining advance knowledge of impending US actions. When President Kennedy appeared on national television to inform the American public of his decision to blockade Cuba, he did not realize the Soviet

Union had almost certainly known of the decision for several hours.[7]

The Orrick Committee Report

The missile crisis was a nerve-shattering experience. In its aftermath, the president, his National Security Council, and the Department of Defense engaged in considerable soul-searching regarding the appropriate nature of crisis management and decision making in the nuclear age. Immediately following the crisis, an interagency working group headed by the deputy undersecretary of state for administration, William H. Orrick, was set up to do a postmortem report and to make recommendations for change.[8] Not surprisingly, the Orrick Committee report focused on the overall effectiveness of worldwide US government communications. It concluded that a need existed for a flexible communications system to give the president and other elements of the NCA control over the nation's total governmental communications facilities, both military and civilian. Such a system must be highly capable, the committee argued, with the ability to provide users with fast, continuous, and reliable services. In addition, it should be able to function during periods of high tension—even periods of nuclear conflict—without suffering serious degradation, if the prevailing counterforce doctrine of selectively targeting Soviet military facilities rather than cities were to be anything more than a theoretical promulgation.[9] Emphatically so if, as McNamara's Pentagon was now promising, the United States would be able to terminate a nuclear war on favorable terms by threatening further attack, implicit in which was the ability to communicate both with one's own forces and with the enemy.[10] Accordingly, Orrick recommended that this new system emphasize physical hardening, mobility of assets, and circuit redundancy, ensuring the ability to transmit and receive message traffic under all conceivable circumstances.[11]

With the Orrick Committee's recommendations as a call to action, McNamara issued DODD S-5100.30, *Concept of Operations of the Worldwide Military Command and Control Systems*, before the month of October was out. The secret directive formally identified two distinct yet related sets of requirements, one military and the other civilian. At the same time, McNamara assigned the director for operations of the Joint Staff, Maj Gen Ferdinand T. Unger, the task of establishing a national level command and control framework that was fully in accord with both. From McNamara's point of view, the need for a system incorporating all government communications was obvious, and his authority to establish it apparent. As secretary of defense, he was, after all, ultimately responsible for military operations. Ordering those operations required adequate communications. Since he was ultimately responsible for these as well, McNamara was determined to direct them in the manner he deemed appropriate. The result was a decades-long effort to create a national level command and control framework, one that would include, but that went considerably beyond the mandate of the Defense Communications System and the National Military Command System.[12] This framework was referred to as the National Communications System (NCS), and the director of DCA was designated system manager.

Theoretically, the NCS would provide the NCA with the crisis management capabilities they desired, but the problem was that the system came into being without a guiding conceptual rationale. Was there a need for all of the communications systems then in place? Which were duplicative? Of these, which provided necessary redundancy and which were superfluous? Who should make these judgments? What sort of organization was most appropriate to manage this vast metasystem?[13] There were no precedents to draw upon, and no analyses had been conducted to answer questions such as these.

The World Wide Military Command and Control System

The NCS's military component, its dominant and by far most important part, took its name from a slight variant of DODD S-5100.30's title. It was called WWMCCS (World Wide Military Command and Control System), and something new and considerably more complex than just another defense system was being created here. Throughout this time, a number of systems useful for command and control purposes already had made their appearance or were then under development.

Included were such command facilities as the National Military Command Center, the Alternate National Military Command Center, NORAD headquarters, SAC's command post in Nebraska, and the airborne command posts. There were sensor systems such as the Ballistic Missile Early Warning System. There were vast, automated systems such as SAGE and the then-under-development SACCS, SAC's Automated Command Control System. Communications systems such as SAC's Primary Alerting System and the Joint Chiefs Alerting Network had also made their debut. These were the tools of command and control, to be certain, but for the most part, they were viewed as unrelated military capabilities. If the DCS had begun to change all of that, moving things in the direction of larger scale and greater connectivity, the WWMCCS concept would do so even more emphatically. With WWMCCS, no longer would these systems be viewed in isolation. Instead, they would be viewed as integral parts not only of a new defense-wide metasystem but of an entirely new military discipline and science—that of command and control.[14]

In many respects, and for a number of years, the World Wide Military Command and Control System, like the NCS of which it was formerly a part, would remain essentially a bureaucratic fiction, an organizational concept rather than a hard commitment of funds, hardware, personnel, and managerial authority. On a theoretical level, at any rate, such an organizing principle made perfect sense. There was a myriad of command and control assets then in existence throughout the defense establishment, and with proper organization, these assets could provide the connectivity the administration desired. The logic was described by an Air Force officer who noted at the time how low-level tactical command and control assets were properly viewed as subsystems of a higher echelon system. These higher systems, in turn, were themselves subsystems of the national level system serving the president and other members of the NCA.[15] Therefore, the assets were there, and the problems seemed to be ones of design, connectivity, and coordination. If those problems could be solved, the president's vision of a system that would permit the national leadership to electronically orchestrate its military responses to crises could be realized.

To provide a centralized focus and coherence to the new World Wide Military Command and Control System, DODD S-5100.30, WWMCCS's founding document, gave the Joint Chiefs of Staff overall responsibility for the planning, implementation, and operation of the system. Like the DCA's mandate regarding the DCS, however, the document would turn out to be an instance of considerable formal responsibility with little real authority. Since the joint chiefs were supposed to devise ways to merge the command and control assets already in existence rather than create a new system from scratch, they received no funding or permanently assigned personnel to accomplish their mission. Control over resources remained where it had always been—overwhelmingly with the military services. Despite their apparently substantial mandate, then, the joint chiefs' ability to integrate the military's disparate command and control assets into a single, centrally responsive entity called WWMCCS was limited from the outset.

The impetus toward centralization, and yet the simultaneous endeavor to strike a balance with decentralized needs, was made manifest during the days that followed in the form of several additional DOD directives. In the area of centralized control, one of these confirmed the National Military Command Center as the military's principal command post. The purpose here, in McNamara's phraseology, was to ensure that during times of tension and crisis, the NMCC would be the "focal point to which the Joint Chiefs of Staff and higher authority turn for an immediate review of the situation and for advice as to the available course of action."[16] There, the NMCC's deputy director for operations and his staff continuously evaluated the political and military situation around the globe, anticipated problem areas prior to their becoming actual crises, and tracked the progress of crises then in progress. They were linked by way of direct circuits to all key operational centers in Washington, including the White House situation room, State Department, CIA, the services, and unified and specified commands.

But at a later press conference, McNamara dismissed the notion that his purpose was to concentrate all decision making at the top level of the defense hierarchy. Rather, he said, what was being called for was a system that could reap maximal benefits from both centralization and decentralization. On the one hand, decisions should be made at the lowest possible level, he said, "taking account of the political ramifications." Since centralized leaders by definition lacked the detailed understanding of local conditions possessed by the operating forces, they believed those forces should be capable of acting on their own initiative without becoming paralyzed by an absence of instructions from the top. More specifically, what this arrangement implied was that routine matters, including the day-to-day conduct of military operations and war, would be left to the commander on the scene, since that was where the situation was most clearly understood.

On the other hand, nonroutine situations, those "untoward circumstances" with the potential for escalation, had to be dealt with differently; they required that command take place at the highest levels.[17] For this to happen, there was a need for a highly effective means by which the president and his staff could receive, review, and respond to the most important subset of the information generated by various operating elements worldwide. That was where the NMCS came in. The directive designated the NMCS the principal subsystem of WWMCCS and the hub of the national level command and control structure. In effect, the National Military Command System was to serve as a command and control "bridge" linking the NCA to the rest of WWMCCS, and in turn to the operating military forces in the field.[18]

A subsequent McNamara directive, issued on 26 October 1963, modified the basic responsibilities of the unified and specified commanders concerning the command and control systems they used. This directive charged the CINCs with establishing their operational requirements and submitting them to the joint chiefs and the secretary of defense for approval.[19] The directive went on to direct the CINCs to participate in formulating plans for engineering, management, procurement, facility construction, and operation to satisfy those requirements. The view held that the CINCs were such a fundamental part of the system that it was necessary to implicate them in the development and operational phases, to bring them into the process in a more central way, to give greater input into the command and control systems that were being acquired for them, and to permit them to make ongoing changes to that system as required.[20]

Throughout this time, McNamara's explicit intention was to move toward a new approach to command and control system acquisition, something that would later become known as the evolutionary approach. To create systems that were flexible enough to keep pace with changes in technology, military doctrine, and the perceived threat, it was necessary, first, to establish a core capability and, second, to improve it incrementally in response to changing conditions, emerging new technologies, and changing military requirements.

While this approach had its shortcomings, it appeared to many in McNamara's Pentagon to be a vast improvement over the dominant weapons system approach that had characterized command and control system development and acquisition through this time. This was the philosophy in which the commander identified a population of system users, specified their responsibilities as best he could, and then identified the information required by each individual at each organizational level under the full range of possible conditions. It meant answering a series of questions a priori: Who were the likely users? Who was responsible for the transmission of various types of information? What quantity and quality of information were necessary? Which frequencies and formats were most appropriate? How would the message be routed? Who was responsible for receiving, processing, displaying, interpreting, and acting on the information once it was received? Using the answers to all of these questions as guidance, a technical agency would then piece together a development plan. From that point on, as one defense journal described it, the command and control system "might as well have been a missile or an airplane."[21] The contractor for the system would take over, often working closely with military experts in such development organizations as the Air Force Systems Command. Performance specifications for the various equipment subsystems would be prepared. For automatic data-processing equipment, for example, such specifications would include processing speed, storage capacity, subsystem availability, reliability, and so on. Only when each of these concerns had been resolved would the actual equipment be procured.

Problems with this approach were many. Although the users for whom the system was being built would provide liaison throughout this process, the arrangement frequently turned out to be far from satisfactory. Commanders tended to have the performance of their military mission foremost on their minds, meaning, of course, that they would seldom appoint their best people as liaison officers. The chain of events resulting from this was thoroughly predictable. Officers of lesser caliber and expertise were appointed. These persons were, by definition, less able to represent their organization's command and control needs competently. Their voices muted, commanders had little real input into the system being developed for them, and they quickly became detached from the development process.[22] Problems of authority and a lack of clear lines of responsibility resulted. Under the weapons system approach, problems would also result if the user's requirements changed while the system was under development. This eventuality was not at all unlikely, given the rapid and accelerating pace of technological advance.

When systems were planned as a single (albeit relatively inflexible) package, they at least tended to have coherence. They also had the attraction of being developed as a unity by an outside contractor and then turned over to the user, turnkey fashion, on a given day. But this approach had not worked at all well when applied to command and control systems. Given the complexity of such systems and the continuous appearance of new technologies, especially in the area of automatic data processing, almost any set of requirements was quickly rendered outdated, often before the system was even brought on line. Changes in requirements obviously ensued, but not infrequently, these changes could be met only by design changes that reduced or eliminated the coherence that had been the primary attraction of the system in the first place. Higher costs, impaired performance, and not infrequently both, were the result. Two major examples of this process in action were Program 425L, NORAD's Combat Operations Center, and Program 465L, SAC's Automated Command and Control System—"everyone's example of how not to develop a command and control system."[23] And there was little alternative to making the necessary changes, given the size and cost of programs such as these: these systems could not simply be ripped out and replaced every time technologies, military doctrine, or the nature of the threat changed.

What was called for was a new philosophy for planning, developing, and acquiring command and control systems, an approach that would recognize that defining user requirements was extremely difficult (not infrequently resulting in the overstating of needs known as "gold plating") and that would permit the continuous evolution of systems in response to direct and ongoing user participation. Most users don't know precisely what their needs are, after all, especially in a context of rapid change where methods for improving a system often become available even before it reaches operational status. In other words, it appeared to McNamara that the only requirement that could be predicted with any certainty was that of greater flexibility in system design at all levels.[24] And so under his regime, each unified and specified commander would be responsible for coming up with specific proposals for improving the effectiveness of his own command and control systems in an evolutionary fashion. These proposals would be submitted to the Joint Chiefs of Staff, reviewed by the Joint Command and Control Requirements Group, and forwarded to the secretary of defense for final approval.

Although the idea was sound, the problem once again revolved around the issue of authority. Because the CINCs were charged with developing and deploying their own command and control systems, it was hardly surprising that the systems that resulted addressed their specific needs as a first order of business and were only secondarily concerned with the larger issue of integration across systems.[25] This development might not have been so bad had those systems truly been designed for joint-service operations where integration was implicit, but this was hardly the case. From the outset, it was the services, not the CINCs or the joint chiefs, that controlled funding decisions. Money talks, and as a consequence, the services made the key decisions with respect to how the CINCs' new systems would be designed, what assets would be acquired, and how those assets would be deployed. The CINCs would review the designs and specifications proposed by the services, but they could not veto them nor initiate new programs.

Whatever comments they might have had were to be submitted to the service secretaries, individuals whose first loyalty was always to their service branch, not to some new, rather diffuse, and frequently suspect concept of unification.[26] In short, the CINCs' role was essentially one of direction, guidance, and validation. This way, no sooner had the parameters of a more centralized command and control system been defined by the creation of WWMCCS than centrifugal forces came into play. Their effect was to decentralize decision-making authority back toward powerful organizational subunits.

First Fruits: The RB-66

The benefits or liabilities of the new World Wide Military Command and Control System were quickly put to the test. On 10 March 1964 an Air Force RB-66 reconnaissance aircraft, attached to the 19th Tactical Reconnaissance Squadron at Toul-Rosieres Air Base, France, experienced what was later described as "navigational trouble" and intruded into East German airspace. Allied and US radars in West Germany had observed the RB-66 entering the air defense identification zone (ADIZ) near the Berlin center air corridor, the Air Force later said, and attempts had been made to warn the plane, telling it to reverse course. The RB-66 did not respond to the warnings, however, despite standing regulations specifying that aircraft within the ADIZ (within 50 miles of the East German frontier) identify themselves. It continued its flight into East Germany. American and allied radar operators watched as Soviet MiG fighters were sent aloft to intercept the intruder. And they watched as the intruder disappeared from their radarscopes in the vicinity of Gardelegen, East Germany.[27] The time in Washington was 9:06 A.M.

Notification of the probable downing of a "U.S. or friendly aircraft" was immediately flashed to the Pentagon, where it was received in the National Military Command Center at 9:10 A.M. The on-duty watch team promptly informed Secretary McNamara and the White House, the State Department, and the CIA. The team also informed DOD's own International Security Affairs and Public Affairs offices. Within the next minute, the commander in chief, Europe (CINCEUR), telephoned the chairman of the joint chiefs at the NMCC, confirming that a shootdown had taken place. The identity of the downed plane had not yet been established, CINCEUR said, but the Air Force had intercepted a radio report by a commercial aircraft operating in the area saying that three people had been sighted parachuting from an aircraft in distress.[28] In short, within six minutes of the shootdown, all the Washington principals knew that the aircraft had been intercepted, fired upon, and that three people had bailed out, apparently safely.

A few minutes later, the president's military assistant informed the NMCC that President Lyndon B. Johnson had been advised of the incident, wished to know the identity of the downed plane as soon as possible, and expected a full report by noon. The NMCC relayed this demand to CINCEUR, whose completed report was received in the NMCC by 11:15 A.M. While all of this was going on, the NMCC representatives of the State Department, CIA, National Security Agency (NSA), and other agencies were reporting the facts to their agencies as they became available. The US Military Liaison Mission at Potsdam commenced efforts to recover any surviving crew members.

At 12:30 P.M. the White House issued the first official press release on the incident, stating that the Soviets had shot down a US plane in distress. According to the official account, the RB-66, which was admitted to be a reconnaissance aircraft, had been on a low-level navigator training exercise. During the flight, the navigator was said to have become disoriented and had accidentally penetrated East German airspace. Official protests were subsequently filed with the Russians. Yet, however poorly events worked out that day for the RB-66 and its crew, everyone agreed that the National Military Command System and the new World Wide Military Command and Control System of which it was a part had performed admirably throughout the affair.

Automatic Data Processing

Incidents like the RB-66 shootdown created an insatiable demand for information. In terms of sheer quantity of data, it was a demand that would be met soon, indeed with a vengeance, by the dramatically improved sensor and communications technologies that were now coming on line. But if the quantity of data available to decision makers was increasing each year, quality was a different issue entirely. Raw data tend to be meaningless and without value until they have been sorted, processed, compared, interpreted, summarized, and displayed. In short, they have no value until they have been transformed into interpretable information. The problem for the new WWMCCS, and the defense establishment generally, was that the volume of data flowing into command centers was becoming too large to handle. A solution to this problem would be difficult, indeed impossible, without the development of sophisticated high-speed automatic data-processing technologies—the computer hardware, software, and associated peripherals.[29] Many rightly viewed computers as the solution to a wide range of military problems, calling them a panacea for command and control, and it is no overstatement to say that automatic data processing ranks equally with nuclear energy and the rocket engine as the major revolutionary technologies of the postwar period.

Concern for WWMCCS's automatic data-processing capabilities began when the system was formally established. At that time, the Pentagon solicited the help of Herbert Goertzel, a computer scientist who would later earn the affectionate appellation "Mr. WWMCCS" for his spirited advocacy of the joint-service command and control system. He had worked with UNIVAC 1, the first commercial computer, while on the staff of the Atomic Energy Commission. Afterwards, he had participated in the SAGE effort to integrate the US air defense and missile systems. Recognized as one of the few people with both the vision and the technical knowledge necessary to develop the Pentagon's envisioned global command and control network, Goertzel was invited to Washington and offered the job of chief of the Information Systems and Standards Division at the Joint Chiefs of Staff. Such were the inauspicious beginnings of WWMCCS automatic data processing.[30]

Goertzel's first task was to develop the automatic data-processing requirements for WWMCCS's key component, the National Military Command System. Drawing upon plans already developed by Gen Earle Partridge's National Command

and Control Task Force and working closely with the Joint Command and Control Requirements Group and the Defense Communications Agency, Goertzel and his colleague, Mal Billings, laid out the ADP support requirements for the Pentagon's National Military Command Center, for its backup site at Fort Ritchie, and for the alternate airborne and sea-based command posts. This was no small task, for one of the major problems confronting the new WWMCCS, and the DOD as a whole at that time, was the vast array of incompatible computing equipment then in operation. As things stood, most of the major WWMCCS headquarters had already introduced computers to support their command and control functions. Describing the existing state of affairs, Goertzel said it consisted of essentially a "collection of autonomous subsystems which provided little or no potential for fulfilling the command and control requirements of the National Command Authority and the Joint Chiefs of Staff."[31]

If this was the case with the headquarters, it applied even more strongly to the various lower-level WWMCCS elements which, working within the acquisition framework of the past, had separately determined their own ADP requirements, developed the systems to meet them, and then deployed and operated those systems. The Department of Defense was still a strongly ramified and decentralized organization, after all, and acquiring technologies that dovetailed with existing decentralized organizational structures made perfect sense. And so all government agencies, military and civilian, contracted with outside companies to meet their specific data-processing needs with little if any consideration as to how information might be exchanged with the other services, agencies, and commands. It was simply how things were done.

By the time WWMCCS was established, each of the services and defense agencies had its own automatic data-processing system up and running. Since these systems had been developed to meet individual needs and mission requirements, incompatibilities were commonplace. Each of the services had developed, or was developing, its own software programming language. The problem was that software designed for one computer type would not function with computers by another manufacturer, and there were no common standards for data input, storage, or output.[32] Most of the computer hardware then in use had similarly been tailored to specific service missions.[33] Since the machines themselves were essentially custom made, their parts were not interchangeable. Operating personnel trained in the use of one system required extensive retraining to achieve proficiency in another. The situation was analogous to people trying to communicate by telephone who not only used technically incompatible telephones but who also spoke different languages, a veritable communications Tower of Babel.[34] Goertzel's task was to try to get these elements to play effectively together with common equipment, computerized data formats, and other common user elements, creating a system in fact as well as in name.

Goertzel's initial standardization effort was called NIPS, for National Military Command System Information Processing System. Developed in 1963 and designed to run on the IBM 1410 computer, NIPS descended directly from the Navy's Intelligence Data Handling System. It sought to achieve a measure of compatibility in hardware and software, to reduce the staggering costs (even then) of software development, and to promote standardization of information, equipment, and training among the various elements of the NMCS.[35]

Despite these early efforts, things were not well with WWMCCS ADP, and it was not long before they began to reach crisis proportions. By the end of 1965, the Office of the Secretary of Defense and the JCS recognized that the automatic data-processing systems then in place or in the pipeline did not provide adequate support for national level requirements. They lacked growth potential. The systems employed incompatible hardware, software, and database structures, and could not transfer data and information efficiently between the various WWMCCS sites. They also lacked the ability to provide multilevel security access for users with different security clearances. Finally, these systems were costing the Pentagon far too much because of the piecemeal way in which ADP assets were being pursued and because of redundant, replicative software development by members of the WWMCCS community. These uncoordinated efforts also resulted in an inability to get discount prices for hardware through consolidated purchases.[36]

Automatic data processing also presented a number of human problems that extended well beyond the technical issues of compatibility and considerations of cost. Computers were anything but familiar to many of the commanders who were then being persuaded to acquire and take responsibility for them.[37] Technical expertise was not abundant, and what existed was spread thinly. When these mysterious machines arrived at a command center, civilian technicians frequently accompanied them, creating a situation not at all to the liking of many commanders. Such programs as the Air Force's back-to-the-cockpit program that was responding to the escalating war in Vietnam further exacerbated the situation, sending the none-too-subtle message to officers that computers should not figure too prominently in their career plans. These and other influences had the effect of inducing considerable skepticism among military officers regarding computers and conservatism with respect to their deployment and use. With users unwilling to take risks, the introduction of the new automatic data-processing technologies was slowed at least temporarily.

In short, with the individual members of the WWMCCS community pursuing ADP technologies more or less as they always had in support of their own unique needs and specific missions, there was still no organizational center of gravity with sufficient force and authority to cohere these disparate pieces into a single coordinated system.

Problems of Definition

Perhaps the most basic problem with the new WWMCCS was definitional, reflecting not only simple semantic differences among involved constituencies but also more profound differences of interpretation and philosophy. Defense Secretary McNamara and others had identified a number of major system effectiveness criteria to be actively pursued in WWMCCS's development—in particular, survivability, flexibility, standardization, and economy—but the problem was that nobody could agree on exactly what any of these terms meant. Yet, agreement was essential to define what the system was and how it should develop.[38] To the Pentagon leadership, then, reaching a common understanding of what was in fact a part of WWMCCS and what was not appeared urgent in the extreme.

The search for answers was on, and that search was conducted in several ways. On 31 March1964 Deputy Secretary of Defense Cyrus Vance established a group under the leadership of General Bestic, DCA's deputy director for the NMCS, and charged it with the responsibility of coming up with a definition of command and control.[39] (The group's major task was to review the assets and programs of the services and defense agencies to try to determine which of these items related to the command and control function.) Similarly, the Joint Command and Control Requirements Group was involved with the ongoing refinement of the WWMCCS concept throughout this time. It was trying to determine the assets involved, how these should operate, what interfaces should exist between various WWMCCS elements, and how the system should be developed over time. Finally, in a related move, Defense Secretary McNamara established an annual Consolidated Command, Control, and Communications Program Review Panel, including representatives from the services and a number of defense agencies. Its major task was to consider all proposals for program changes within WWMCCS and to make appropriate recommendations.[40] The point of all of this was to answer the nagging question: "What exactly is WWMCCS?" Like the NCS of which it was a part, WWMCCS was an ambiguous entity that had arrived on the scene without a coherent conceptual rationale for its existence.

Rhetorically at least, things were improving daily as the decade progressed. In his 18 January 1965 defense message to Congress, President Johnson described how the past several years had witnessed "dramatic improvements" in the ability to communicate with our forces. A national system for commanding and controlling US military forces around the globe had been established, he said, employing the "most advanced electronics and communications equipment, to gather and present military information necessary for top-level management of crises and to assure continuity of control through all levels of command."[41]

Johnson should know. One of the most voracious consumers of information the Oval Office has ever known, he had four telephones and two teletypes in his office. So that he would not be out of touch while moving from place to place, and decades before car phones became commonplace, two additional telephones were in his car. All of these were in constant use. Of the approximately 120 independent nations in the world at that time, Johnson was said to have had real-time communications with about two-thirds of them.[42] During the 1965 Dominican crisis, Johnson had, and used, a direct telephone line to the US ambassador. As the president described things at a subsequent press conference, even while the two of them spoke, the ambassador was talking from under a desk while bullets crashed through the window. Johnson said the ambassador had with him a thousand American men, women, and children "who were pleading with their President for help to preserve their lives."[43] It sounded very much as if the White House-to-foxhole variety of communications that had been lacking during the Bay of Pigs and Cuban missile crisis had finally been achieved.

Although in fact such a capability had not yet been achieved, the impetus toward centralized command and control was pronounced during the Johnson administration.[44] Johnson dramatically expanded the communications capability of the White House, indeed, to such an extent that one defense journal noted how "a very real problem now appears to be how much more communications equipment can be squeezed into the basement of 1600 Pennsylvania Avenue." Across the Potomac at the Pentagon, things proceeded similarly. Given the increasing seriousness of crises, the nature of US operations in Vietnam and their potential for escalation, the trend was directed necessarily toward circumscription of decision-making latitude for the commanders on the scene. It was a logical and necessary move, according to Defense Secretary McNamara. (This statement stood in contrast to his earlier claim that the president and the secretary of defense had never usurped the role of the military commanders.) According to McNamara, the ultimate command and control system would provide a "standardized, highly survivable, non-interruptible command capability for a wide range of possible situations, and will provide the national authorities with a number of alternatives through which they may exercise their command responsibilities."[45] The question centered on whether the World Wide Military Command and Control System that McNamara had done so much to establish and advance would be that ultimate system.

Notes

1. F. Boonham, "The Impact of Computers on Military Management and Command," *Journal of the Royal United Services Institute for Defense Studies* 117 (June 1972): 33.
2. Kurt Gottfried and Bruce G. Blair, eds., *Crisis Stability and Nuclear War* (New York: Oxford University Press, 1988), 183.
3. Bruce G. Blair, *Strategic Command and Control: Redefining the Nuclear Threat* (Washington, D.C.: Brookings Institute, 1985), 73–74.
4. Paul Bracken, *The Command and Control of Nuclear Forces* (New Haven: Yale University Press, 1983), 207.
5. "Are the Assets of the NCS Responsive to National Needs?" *Armed Forces Management* 12 (April 1966): 58.
6. Alfred D. Starbird, "NCS Today and Tomorrow," *Signal* 21 (May 1967): 63.
7. "Assets of the NCS," 58.
8. Starbird, 63.
9. "For DCA, A Strengthened Role," *Armed Forces Management* 14 (July 1968): 75.
10. William W. Kaufman, *The McNamara Strategy* (New York: Harper & Row, 1964), 74.
11. Solis Horwitz, "National Communications for the Nuclear Age," *Signal* 18 (July 1964): 34–35.
12. Paul W. Tibbets, "About Our Working National Military Command System," *Armed Forces Management* 10 (July 1964): 26.
13. "For DCA," 76.
14. Irving Luckom, "Overview of WWMCCS Architecture," *Signal* 30 (August 1976): 62.
15. Clifton L. Nicholson, "Command and Control and the Decision-Making Process," *Air University Review* 15 (November/December 1963): 77–78.
16. Tibbets, 26.
17. "Effectiveness, Responsiveness of National Command System Vital to U.S. Security," *Armed Forces Management*, July 1966, 43.
18. Hubert S. Cunningham and William E. Kenealy, "The Joint Chiefs of Staff and Command and Control," *Signal* 29 (March 1975): 15.
19. "Effectiveness, Responsiveness," 45.
20. "How Not to Build C&C Systems Is Still an Unanswered Question in Defense," *Armed Forces Management* 12 (July 1966): 109.
21. J. J. Cahill, "Resource Management: A New Slant on C&C," *Armed Forces Management* 9 (July 1963): 71.

22. "How Not to Build C&C Systems," 109.
23. Ibid.
24. J. S. Butz Jr., "USAF and the Computer Revolution," *Air Force Magazine* 47 (March 1964): 35.
25. Gordon T. Gould Jr., "Computers and Communications in the Information Age," *Air University Review* 21 (May/June 1970): 9.
26. Lee M. Paschall, "WWMCCS: Nerve Center of U.S. C^3," *Air Force Magazine* 58 (July 1975): 54.
27. Arthur J. Olsen, "U.S. Reconnaissance Jet Downed in East Germany," *New York Times*, 11 March 1964, 1.
28. Tibbets, 28.
29. Butz, 33.
30. Carol Hamilton, "Worldwide C^2 System Networks Strategic Data for Joint Chiefs," *Defense Electronics* 20 (June 1988): 50–51.
31. Herbert B. Goertzel and James R. Miller, "WWMCCS ADP: A Promise Fulfilled," *Signal* 30 (May/June 1976): 57.
32. "Electronics in Military Decision Making: The Need for Automatic Aids," *INTERAVIA* 19 (June 1964): 851.
33. "NMCS: The National Voice of Command," *DATA* 11 (January 1966): 22.
34. Edgar Ulsamer, "The Military Decision-Makers' Top Tool," *Air Force Magazine* 54 (July 1971): 44.
35. Leonard Maley, "NIPS—The System that Invented Itself," *Systems Engineering Forum*, n.d., 32.
36. General Accounting Office, *The World Wide Military Command and Control System—Major Changes Needed in its Automated Data Processing Management and Direction, Report to the Congress*, 14 December 1979, ii–iii.
37. "How Not to Build C&C Systems," 109.
38. C. Kenneth Allard, *Command, Control, and the Common Defense* (New Haven: Yale University Press, 1990), 135.
39. Ibid., 33.
40. "NMCS," 18.
41. John B. Bestic, "No More Confused Situations," *Signal* 21 (March 1967): 53.
42. "Evolution and Compatibility: 1965's Key Words in Tactical C&C," *Armed Forces Management* 11 (July 1965): 45.
43. "Transcript of the President's News Conference on Foreign and Domestic Affairs," *New York Times*, 18 June 1965, 14.
44. Alexander Cockburn, "Press Clips," *Village Voice*, 8 November 1983, 10–11.
45. "Evolution and Compatibility," 46–47.

Chapter 5

Three WWMCCS Failures

Despite problems of definition, standardization, and authority, the development of the World Wide Military Command and Control System was accompanied by high expectations and considerable fanfare from the proponents of centralization. But during the decade's final several years, three serious and highly visible incidents cast considerable doubt upon the new system's capabilities and its effectiveness. Each of these incidents involved intelligence activities, each resulted in loss of life to American servicemen, and all were subsequently attributed to communications breakdowns and delays in systems that were designated as part of WWMCCS. A cacophony of criticism was raised as a result, leading to calls for increased standardization of system assets, greater centralization of managerial authority, and a more formalized approach to developing a command and control capability responsive to the needs of the national leadership.

Failure One: USS *Liberty*

The first of the three WWMCCS failures took place in June 1967, the time of the Six Day War between Israel and the United Arab Republic. It involved the American intelligence vessel USS *Liberty*, part of a worldwide fleet of electronic intelligence ships operated by the National Security Agency and designed to intercept communications and other types of electronic transmissions. With tensions between Israel and Egypt steadily building during the spring of 1967, *Liberty* was ordered to sail for Rota, Spain, where she would take on supplies and prepare to proceed to the Eastern Mediterranean area off Port Said, at the mouth of the Suez Canal.[1] While in Rota, *Liberty* received orders that she should approach no closer than 12.5 nautical miles to the coast of Egypt and no closer than 6.5 nautical miles to the coast of Israel, a position that would permit the maximum collection of signals intelligence.

The shooting war between Israel and Egypt began on 5 June, while *Liberty* was steaming toward the Eastern Mediterranean.[2] At 2015Z on 6 June, the commander of the US Sixth Fleet, Vice Adm William I. Martin, instructed all surface and air units under his command to stand at least one hundred nautical miles off the belligerents' coasts. But *Liberty* never received this message, and no subsequent action was taken by Martin to ensure that the vessel complied with his one-hundred-mile standoff order. During the afternoon of 7 June, officials at the NSA also decided to reposition *Liberty* farther away from the coast, and they sent a high-precedence flash message to this effect to the Joint Chiefs of Staff by way of AUTODIN. (Four separate precedence levels had been established for AUTODIN: flash, immediate, priority, and routine. The JCS's criteria called for a flash message to be transmitted within 10 minutes, an immediate message within 30 minutes, a priority message within three hours, and a routine message within six hours.)[3] The JCS responded by preparing a message to *Liberty* that instructed it to approach no closer than 20 nautical miles off the Egyptian coast and no closer than 15 nautical miles off the Israeli coast. The JCS issued this message at 2230Z and directed it to the commander in chief, Europe, for action. Information copies of the message were to be sent to a number of relevant parties, among them Adm John S. McCain Jr., the commander in chief of US Naval Forces in Europe (CINCUSNAVEUR); the commander of the US Sixth Fleet; and *Liberty*. Eleven minutes after the JCS's message was issued, it was given to the Army Communications Station at the Pentagon for transmission by way of AUTODIN. The message was assigned a priority precedence and scheduled for delivery to its addressees within three hours.

It was then that the problems began. Since there was a large volume of higher-priority messages swamping the Army Communications Station, operators did not get around to transmitting the action copy of the message to CINCEUR for more than 14 hours. The information copies of the message, including the one for *Liberty*, were transmitted even later. To make matters worse, the information copies were incorrectly routed to the Navy Communications Station in the Philippines. From there, they were sent to the Navy Communications Station in

Asmara, where they were placed on fleet broadcast a full 23 hours after they had been issued by the joint chiefs. (This was far too late to do *Liberty* any good, for the vessel had been attacked nine and one-half hours earlier.)

This failure did not turn out to be of great importance, however, because long before the message was placed on fleet broadcast, it already had been canceled by a subsequent JCS message. During the hour following the release of their original message, the NSA and the joint chiefs became increasingly concerned over the repositioning of the *Liberty*. It was decided that the ship was still too close to the coast, and so one hour and 20 minutes after the joint chiefs issued their original message, a JCS duty officer used AUTOSEVOCOM, the Automatic Secure Voice Communications Network, to telephone CINCUSNAVEUR headquarters in London. The duty officer there was given a verbal directive to order the *Liberty* to operate no closer than one hundred nautical miles to the Egyptian and Israeli coasts, and he was told that a written message formalizing the verbal directive would arrive shortly by way of AUTODIN. A message ordering *Liberty* to operate at least one hundred nautical miles off the belligerents' coasts was promptly prepared for transmission to Admiral Martin, Sixth Fleet commander, who in turn would pass it on to *Liberty*. Despite the urgency implicit in the JCS verbal directive, the CINCUSNAVEUR staff did not release the message for transmission until the formal written notification was received from Washington.[4] Ever bureaucratically cautious, they wanted written proof that the instruction to move *Liberty* had originated with someone other than the relatively junior staff officer who had placed the call at the JCS's Joint Reconnaissance Center.[5]

The JCS released the written confirmation about one hour after the telephone call to CINCUSNAVEUR headquarters. This delay itself should not have been of consequence, since the joint chiefs had every reason to believe that prompt action would be taken in response to their verbal directive.[6] The written message canceled the earlier (and unbeknownst to the JCS, hopelessly misrouted) message ordering *Liberty* to stand at least 15 nautical miles off the belligerents' coasts, and confirmed that the vessel should remain at least one hundred nautical miles offshore. This message was given an immediate precedence suggesting a heightened level of concern by the joint chiefs and requiring a transmission time of not more than one-half hour. As before, following the chain of command, the action copy of the message was addressed to the commander in chief, Europe, with information copies going to CINCUSNAVEUR, the commander of the Sixth Fleet, and *Liberty*.

The message was released by the JCS to the Army Communications Station in the Pentagon for transmission by way of AUTODIN. But despite the message's precedence level, the Army Station took 44 minutes to transmit the action copy to CINCEUR. The message's information copies fared even worse, with a delay of two hours and 23 minutes before they were transmitted. The only valid explanation for such a delay would be that messages of equal or higher precedence were awaiting transmission; but a congressional subcommittee later pointed out that the Pentagon was unable to furnish any evidence that this had been the case.[7]

The initial delay represented only part of the problem. Army Communications Station personnel assigned an erroneous routing indicator to the information copy of the message intended for *Liberty*, which misrouted it to the Naval Communications Station in the Philippines. There the error was recognized and, within an hour, the message was retransmitted to the Naval Communications Station in Morocco, from which it was to go directly to *Liberty*. This should have solved the problem, except on the way to Morocco the message had the misfortune to be routed so that it again passed through the Army Communications Station in the Pentagon. Rather than routing the message on to Morocco, as should have been done, it was sent instead to National Security Agency headquarters at Fort Meade, Maryland, where it was filed without further action. The explanation offered for the error was that the clerks in the Pentagon had misread the message's routing indicator.

The action copy of the message had made it to the headquarters of the commander in chief, Europe, however, arriving at 0212Z on 8 June. CINCEUR took action a little over an hour later, telephoning Admiral McCain, commander in chief of US Naval Forces in Europe, to take immediate action on the JCS message. CINCEUR followed up its verbal instructions to CINCUSNAVEUR with a formal written directive, the action copy of which was directed to McCain, with information copies going to a number of parties, including the commander of the Sixth Fleet and *Liberty*. The problem was that the written directive was not released for transmission by CINCEUR until 0625Z, more than three hours after the immediate message had been received from the Pentagon. An additional delay of 40 minutes occurred in the CINCEUR message center before the message was finally transmitted.

As before, the initial delays represented only the beginning of the problems. To ensure that the message would get through to its addressees, CINCEUR transmitted it concurrently over two separate relay paths. A good thing, too, since one of the messages was promptly lost at the first station on its transmission path, the AUTODIN relay at Parmesans, Germany. The explanation later offered was that the station was experiencing a heavy volume of communications traffic at the time and that the number of qualified personnel at the station was inadequate to ensure error-free processing of communications traffic.[8] The second relay path worked considerably better for getting the message to CINCUSNAVEUR and the commander of the Sixth Fleet, although not to *Liberty*, which never received it.

The information copy directed to *Liberty* itself did not fare at all well, following a meandering route that took it through a number of intermediate stations. One of these stations was the Army Communications Station at Asmara, where a delay of more than two and one-half hours occurred before the message was passed on to a Navy Communications Station located less than one mile away. A garbled message was placed on fleet broadcast at 1059Z, saying only that *Liberty* was to act on other messages previously received from the joint chiefs—messages which, of course, had yet to be received. By the time the complete message was finally placed on fleet broadcast at 1646Z, more than nine hours had elapsed since the message had been transmitted from CINCEUR. By then, the attack on *Liberty* already had taken place.

One imagines that Admiral McCain's staff at CINCUSNAVEUR should have been concerned about the messages they were receiving. They should have acted to contact *Liberty* in response. After all, CINCUSNAVEUR had received the AUTOSEVOCOM call from the joint chiefs regarding the need to reposition *Liberty*.

Also received was the information copy of the message ordering the vessel to stand off from 15 to 20 nautical miles from the belligerent nations' coasts. Finally, there was the telephone call from CINCEUR concerning *Liberty*. It was only following this final call, at 0325Z on 8 June, that CINCUSNAVEUR established a direct teletype conference circuit with the commander of the Sixth Fleet, who was instructed to take action on the JCS message. This directive was followed by a formal message confirming the specifics of the teletype order.

After receiving these messages, Admiral Martin, the Sixth Fleet commander, should have gotten word to *Liberty*; yet, inexplicably, he did not do so with any dispatch. Despite the concern evidenced by the higher commands, more than four hours would elapse before Martin released his action message for transmission to *Liberty*. To make matters worse, he chose not to contact *Liberty* directly by way of radio but rather to use normal communications procedures. What this meant in practice was that the action message was passed to the communications center on board the Sixth Fleet flagship, USS *Little Rock*, for transmission, where an additional delay of more than an hour ensued. The reason given for the delay was that there were one flash and seven immediate messages being prepared for transmission, and the message from CINCUSNAVEUR was simply put at the end of the queue. It was finally transmitted at 1035Z on 8 June.

This message arrived at the Army DCS station at Asmara at approximately 1200Z. Instead of delivering it to the nearby Navy Communications Station for fleet broadcast, the Army DCS station sent it to the Navy Communications Station in Greece. Aware that an error had taken place, Navy duty personnel there returned the message to the Army station. After additional delays, the message was finally delivered to the Navy Communications Station, arriving there at 1510Z, over six hours after its release by Admiral Martin. The message was put on fleet broadcast 15 minutes later, at 1525Z, more than three hours after the attack on *Liberty* had taken place.

As all of these electronic vagaries were taking place, the going was getting tough on board USS *Liberty*. Throughout the night of 7 June and the morning of the 8th, the ship had been reconnoitered by Israeli fighters and reconnaissance aircraft.

Flying so low that the pilots were readily visible to ship personnel, the aircraft crews could easily see the American flag being flown on board the ship, according to *Liberty's* crew, and they had been overheard reporting the nationality of the ship to ground headquarters. Members of the crew apparently found the close surveillance reassuring. Israel was an ally, after all, and dominated the sky. The nationality of *Liberty* was unmistakable. Under these conditions, an Israeli attack on the ship seemed unthinkable.[9]

This sense of assurance was rent asunder shortly after 1400Z on 8 June, when *Liberty* came under sustained attack by Israeli Mirage jets and torpedo boats. When the attack began, *Liberty's* crew desperately worked to break through Israeli jamming to transmit an urgent request for assistance to the carrier *Saratoga*, operating some five hundred miles away. The ship's commander, Capt Joseph Tully, relayed the message to the Sixth Fleet commander over the Primary Tactical Maneuvering Circuit, duplicated the message by teletype, and sent information copies to Navy headquarters in Washington and London. Admiral Martin promptly directed the carriers *Saratoga* and *America* to launch aircraft to defend *Liberty*, but to little avail. *America* had apparently relaxed from an alert posture and did not respond. *Saratoga* launched planes, but these were quickly recalled by Rear Adm Lawrence Geis, commander of the carrier task force, who had received orders from the Pentagon that the planes should not engage in action until permission was received from the White House. By the time authorization was finally received and the aircraft launched, more than an hour and one-half had elapsed.

Alone and unarmed except for a few machine guns, *Liberty* found things going badly. Israeli jets began strafing runs and bombarded the ship with napalm. The attack was soon joined by Israeli torpedo boats. Several torpedoes barely missed, but one hit the ship in its cryptologic spaces, blasting a 40-foot hole in her hull and killing 25 NSA personnel. According to crew members, the torpedo boat then sat nearby the crippled ship for the next 40 minutes, machine-gunning any personnel who tried to fight the fires or help the wounded. The Israelis shot up life rafts that were launched to save the crew in the water.[10] In all, 34 Americans were killed in the attack, another

75 were wounded, and the ship itself was damaged so badly that it subsequently had to be scrapped.[11]

Did the Israelis know they were attacking an American ship? The evidence is conflicting. A House investigating subcommittee concluded in 1971 that the attack was wholly deliberate on Israel's part. In his book *Assault on the Liberty*, Deck Officer James Ennes seconds this conclusion, arguing that there was no possible mistake regarding the ship's identity. But why would Israel attack a vessel belonging to its benefactor and strongest ally? Perhaps, as Ennes suggests, the motive was to keep the United States from learning of Israel's intention to invade Syria. Only two weeks before, President Johnson had informed Foreign Minister Abba Eban that the United States would not tolerate such a move if it were initiated by Israel. With *Liberty* out of the way, the hostilities could be blamed on the Syrians.[12] On the other side of the interpretive fence are those who say the attack was all a gigantic mistake, the result of a lengthy series of erroneous reports that indicated the ship was non-American and hostile.[13] Whatever the truth of the matter, there is little doubt that WWMCCS performed poorly throughout the entire affair. The incredible odyssey of messages that might have saved *Liberty* as they were lost, misrouted, and delayed, constituted one of the most serious failures of command and control to date.

Failure Two: USS *Pueblo*

Only seven months had passed since the attack on *Liberty* when the second WWMCCS failure took place, this time involving the USS *Pueblo*—also part of NSA's worldwide fleet of electronic intelligence ships. Of the three incidents here recounted, the case of the *Pueblo* is by far the muddiest and most ambiguous. The versions of events recounted later by crew members were often utterly at variance, the official Navy inquiry conducted afterwards was limited, and the entire incident was rapidly shrouded with a veil of secrecy so impenetrable that virtually nobody, likely not even top DOD officials or the president himself, knew what had actually taken place.

In late January 1968 Cmdr Lloyd Bucher and his crew departed for an area off the North Korean coast near the mouth of Yonghung Bay, upon which is located the city of Wonsan and its naval base. *Pueblo* apparently took up a position with-in North Korean territorial waters at the entrance to the bay but hidden from coastal radars by Yo Island. Bucher's mission, which he repeatedly claimed to have carried out to the letter, called for him to invite harassment by the Koreans by putting himself in harm's way. After enemy ships spotted Bucher, he was to remain stationary rather than departing the area. Throughout the harassment, onboard NSA personnel would be intercepting communications to determine whether the Koreans were getting instructions from the Chinese or the Soviets, both of whom regarded North Korea as a client state at the time.

On the morning of 23 January 1968, *Pueblo* transmitted a number of routine messages to the Naval Security Group facility at Kamiseya, Japan. At noon that day (0300Z) a North Korean subchaser came from the direction of Wonsan and circled the ship. *Pueblo* was asked by way of flag hoist signals to identify its nationality, and Bucher responded by raising the biggest ensign he had on board. NSA personnel promptly intercepted a radio report from the subchaser to its base giving *Pueblo's* number, GER-2, and identifying it as American. The North Korean ship then hoisted signals informing *Pueblo* "heave to or I will fire," to which Bucher's apparent response was "I am hydrographic," and "intend to remain in area until tomorrow." From Bucher's view, it was not yet an emergency; harassment of US intelligence ships was common, and his orders were to provoke the North Koreans, then monitor their reaction. Nothing happened during the next three-quarters of an hour. Several messages of a routine nature were sent to the NSA facility at Kamiseya during this period—routine except for noting that there was "company outside."[14]

Three North Korean torpedo boats were then spotted heading toward *Pueblo*. At 0350Z Bucher released the first of two messages designated Pinnacle, identifying the message as of great significance and requiring immediate delivery to the National Command Authorities. The message reported the encounter with the subchaser and the exchange of flag-hoist signals and affirmed Bucher's intention to remain in the area, if feasible. The message was given a flash precedence, presumably guaranteeing extremely rapid delivery to its addressees.[15]

The NSA station at Kamiseya received the message almost instantaneously, and, within six minutes, relayed it to the commander of naval forces, Japan, where it was logged in at 0413Z. And there it sat for the next 47 minutes, despite its Pinnacle designation and flash precedence. The explanation later offered by the assistant chief of naval operations for communications and electronics for this lengthy delay was that much time was required for decision making, part of which involved restructuring the message so it could be taken out of dedicated NSA intelligence channels and transmitted over the general-use AUTODIN network.[16]

Although the message left the commander of naval forces, Japan, at 0500Z, things did not go well. The message was forwarded to the Naval Communications Station, Japan, where it was transmitted to its Navy addressees by way of the Navy Command Operational Network, a communications system used exclusively for Navy operational orders. This system was not instantaneous, requiring one-half hour for all Navy addressees to receive their copies, that is, an hour and three-quarters after its transmission by *Pueblo*. For all other addressees AUTODIN was used, and there things went even worse. The Naval Communications Station in Japan introduced the message into AUTODIN at 0508Z, but it did not reach the first of its addressees until 0600Z, almost an hour later. The message did not reach the Joint Chiefs of Staff until 0624Z, an hour and 16 minutes later, and more than two and one-half hours after its transmission by *Pueblo*.[17]

Aboard *Pueblo*, things were increasingly serious. Shortly after the first Pinnacle message at 0350Z, and with the torpedo boats approaching, Bucher ordered his engines started. The subchaser responded by lowering its "heave to" flags, instructing *Pueblo* to "follow me—have pilot aboard." Bucher ignored this, as well as the frantic signals of a sailor on board the subchaser ordering him to follow to Wonsan. The torpedo boats arrived and formed a circle around *Pueblo*, and two MiG fighters began making passes overhead. At 0415Z soldiers aboard the subchaser were seen transferring to one of the torpedo boats, which then began backing up toward *Pueblo's* stern in preparation for a boarding attempt. The faces of the boarding party could easily be discerned and the cocking of their weapons was clearly audible.

Bucher decided to make a run for it to the east, the direction of the open sea. But even as he did so, he transmitted a second Pinnacle message. This transmission took place at 0418Z, some 28 minutes after the first Pinnacle, and like the first, it was given a flash precedence. In that message, Bucher reported that the torpedo boats had joined the subchaser, that he had been ordered to follow the subchaser, and that instead he was departing the area "under escort." The gravity of the situation should have been apparent now to those receiving this message, since the Navy considered a second Pinnacle to be a "trigger message" that required the most urgent attention.[18]

As before, this second critic from *Pueblo* was received almost instantaneously at Kamiseya, and within six minutes (by 0424Z), it had been relayed to the commander of naval forces, Japan. There the delays were again substantial, and 39 minutes elapsed before the message was passed to the communications station for further transmission. At 0510Z copies of the message designated for Navy addressees were again transmitted by way of the Naval Command Operational Network, and all arrived at their destinations within 15 minutes. But things again went far from smoothly for those messages destined for AUTODIN transmission. For reasons that could never be determined, those messages languished for 18 minutes at the Naval Communications Station in Japan before they were introduced into AUTODIN, and almost another hour passed before they reached all of their addressees. At the Pentagon, the joint chiefs received the AUTODIN message at 0557Z, a full hour and 39 minutes after its transmission by *Pueblo*.[19]

In one sense, the Defense Communications System's torpid performance hardly mattered, since the second Pinnacle message had also been twice transmitted to the joint chiefs by way of NSA's dedicated Critical Intelligence Communications (CRITICOM) intelligence network. Recognizing the seriousness of the situation upon receipt of *Pueblo's* second Pinnacle message at 0418Z, the NSA facility at Kamiseya gave it a critic format, introduced it into the CRITICOM network at 0440Z, and dispatched it to the Pentagon, where it was received at the

National Military Command Center six minutes later. Similarly, the commander of naval forces in Japan gave the second Pinnacle a critic format and sent it out at 0435Z; it was received in the NMCC four minutes later. While from the JCS's perspective AUTODIN's poor performance might not seem to matter much because the intelligence channels performed so well; there were other important addressees that did not have the advantage of CRITICOM. For example, it took more than an hour for *Pueblo's* second Pinnacle message to reach the two Pacific commands that might have been able to send assistance to the beleaguered ship.[20]

Back on board *Pueblo,* things had gone from bad to worse. At 0426Z, 11 minutes after Bucher decided to make a run for it, the subchaser, which had not immediately taken up the pursuit, received orders to fire on the intruder to make it stop. The subchaser rapidly closed the distance between itself and the American ship, lowering its "follow me" signal and replacing it with "heave to or I will fire." Communications operators in the secure NSA portion of the ship promptly sent a message to Kamiseya, repeated three times: "They plan to open fire on us now." This was followed by another transmission of the second Pinnacle message, which was interrupted to repeat, "North Korean war vessels plan to open fire. . . ." Two minutes later, at 0430Z, *Pueblo* transmitted five times, "We are being boarded," which was a bit premature. It was followed by "SOS" and repeated more than 30 times. Throughout these tense minutes, emergency destruction of intelligence materials was under way. Bucher apparently made one last desperate effort to evade the subchaser, which then opened fire at *Pueblo's* antennas, both to bring the ship to a halt and to shut down its radios. Realizing his position was hopeless, Bucher stopped his ship, which was boarded by 0437Z. At 0445Z an NSA operator still at his station in the secure intelligence portion of the ship transmitted to Kamiseya, "We are being escorted into prob Wonson repeat Wonson [*sic*]."[21]

The escort, according to one analysis, was by the Red Chinese. In his book, *The Pueblo Surrender,* Robert A. Liston argues that it was a contingent of Chinese soldiers, not North Koreans, who actually boarded the ship. The Americans were quickly subdued by the soldiers and their AK-47 automatic weapons. The soldiers herded the Americans below decks and blindfolded them. The boarding party broke out the ammunition for *Pueblo's* several machine guns for protection (the ship had been limited to one hundred rounds for each gun), and got the ship under way toward Wonsan as rapidly as possible. Why? Liston contends it was because a Soviet warship was fast approaching from the east, something several of *Pueblo's* crew later reported having seen. Wanting to get hold of *Pueblo* as much as the Chinese did, the Soviet ship fired upon it at about 0500Z, killing one American seaman and wounding three others. Recognizing that the Soviets might slaughter his crew, Bucher somehow managed to bring his vessel to a halt despite the Chinese infantryman at his side. At 0532Z *Pueblo* was boarded for a second time—this time by North Koreans acting on the orders of the Russians. It was some six hours later that the ship finally sailed into Wonsan, where the crew was held in captivity until the following December.[22]

Liston argues that the *Pueblo* surrender, unbeknownst to its captain or crew, was engineered by two high-level NSA operatives to get a "rigged" cryptographic machine into North Korean hands. Both the Koreans and the North Vietnamese used codes devised by the Soviets, and the hope was that when the Koreans used the rigged machine, it would help break the North Vietnamese codes, allowing the United States to gain information about the military buildup then under way. Indeed, in a few days this buildup would result in the Tet offensive.

How then was the NSA to make certain that the North Koreans would seize the ship and its machine? Liston argues that NSA leaked information to both the Russians and the Chinese that *Pueblo* carried a highly sensitive piece of information of interest to both countries regarding Soviet preparations to attack China. Thus, the desperate acts of both nations to board and search *Pueblo* and their desire to have the world think the seizure was the work of the North Koreans alone. If so, it was a bold and brilliant operation: Liston tells us that NSA estimates were that three Americans would be killed, but if necessary the agency was willing to sacrifice the entire crew to save thousands of American lives in Vietnam.

Whatever the merits of these speculations regarding motive, the performance of WWMCCS during the *Pueblo* affair was hardly what many in the Pentagon, or the Congress, believed was required for crisis management. A naval court of inquiry later described the delays in relaying *Pueblo's* Pinnacle messages as "grossly excessive," saying they were at least partially responsible for the failure of US forces to rescue the ship.[23] The House Armed Services Committee concluded that while it was essential for WWMCCS and other systems to be linked together effectively, this obviously had clearly not been the case: Donald T. Poe wrote that "because of the vastness of the military structure with its complex divisions and the multiple layers of command and the failure of responsible authorities at the Seat of Government . . . our military command structure is now simply unable to meet emergency criteria."[24] The *Pueblo* incident had clearly shown that military facilities and command centers lacked the necessary connectivity to give and respond to top-level orders and requests for information.

Failure Three: EC-121

The third WWMCCS failure came in April 1969, when the North Koreans shot down a Navy EC-121 reconnaissance plane over the Sea of Japan. According to the North Korean account, the American plane had penetrated deeply into their airspace, their interceptor fighters were scrambled in response, and one of them had "scored the brilliant battle success" of downing the intruder with a single shot. Officials of the new Nixon administration flatly denied the intrusion charge, saying that at no time during its mission did the EC-121 actually enter North Korean airspace.[25] Whatever the truth of the matter, all of the EC-121's 31 crew members, 30 Navy personnel and one Marine, died in the attack.

The EC-121 was a big, four-engine, propeller-driven Lockheed Super Constellation modified for military use. The EC-121's first electronic defense role back in the 1950s had been as a radar picket in SAGE, where the mission of these "pregnant geese"—flying high-altitude patrols hundreds of miles offshore and working in conjunction with radar picket ships—was to scan the sky ceaselessly to detect the approach of Soviet bombers.[26] But as the years passed and far more powerful ground-based radars came on line, the radar pickets, both airborne and sea-based, were phased out. Their mission an anachronism but their airframes still viable, the EC-121 fleet was again modified during the 1960s, this time to serve as platforms for the collection of electronic intelligence. In its late 1960s incarnation, the EC-121 remained a sight to behold: two large radomes protruding "like goiters" from the top and bottom of the fuselage, which was itself stuffed with six tons of state-of-the-art electronic equipment.[27]

While the EC-121 was a newcomer to the air reconnaissance game, the game itself was of far longer standing. Every year since the dawn of the cold war, thousands of reconnaissance missions had been flown by US aircraft just off the state borders of Communist nations to collect intelligence. The Sea of Japan was of particular interest, for there the borders of North Korea, the People's Republic of China, and the Soviet Union drew together. During the first three months of 1969, almost two hundred such missions had been flown in the area by US military aircraft, without incident.[28] But incidents were in fact far from rare, and the one involving the EC-121 was by no means an anomaly, except perhaps in the large number of American servicemen who lost their lives. Dozens of American aircraft and crews had been shot down over the years in what one author described as a "bloody electronic air war."[29] Why so many, why so bloody, when reconnaissance planes are unarmed? The answer turns on the very nature of airborne reconnaissance. Of particular interest is information concerning a target nation's electronic order of battle—the ways in which radar, communications, and other electronic assets are employed under conditions of stress. Such information is of critical importance to military planners during time of war, and the point is to gather as much of this type of intelligence as possible.

The problem is that the targets of these intelligence efforts tend to be unwilling to accommodate an adversary's curiosity by turning on their electronic equipment so its performance can be assessed, and it is precisely at that point that the passive nature of reconnaissance flights ends. For it is then necessary to create situations of actual emergency, including border penetrations to induce the adversary to react in a way that reveals information of importance.[30] This explains the prodigious propensity of reconnaissance flights to stray from course, and the sometimes deadly reactions their intrusions provoke.

This occasionally bloody cat-and-mouse game reached its zenith with the downing of the EC-121 in 1969, another tragedy that might have been averted had the World Wide Military Command and Control System performed as advertised.

Responsibility for the EC-121's mission was apparently straightforward. The plane was formally under the operational control of Fleet Air Reconnaissance Squadron I, itself attached to the Seventh Fleet, which was under the control of the commander in chief, Pacific (CINCPAC). But final authority for the flight did not rest with CINCPAC, for all such missions had to be reviewed by the Joint Chiefs of Staff and approved by the Office of the Secretary of Defense. Throughout, however, NSA was really in charge of these operations. When the missions took place, they were monitored continuously at the NMCC by relevant civilian and military personnel attached to the Joint Reconnaissance Center, generally with the secretary and the joint chiefs listed as "parties on the line" to receive data or advisories of events and emergencies. In the case of the EC-121, communications up and down the chain of command would be far less clear-cut.

The EC-121's mission was under the operational control of the Naval Security Group, the Navy's in-house intelligence service, based at Kamiseya, Japan, which was in turn operating the flight under the direction of NSA. As with all reconnaissance missions, that of the EC-121 followed a carefully prepared script. After it took off from Japan, the EC-121 was to take a northwest heading until it arrived at a point off North Korea's Musu Peninsula, near the city of Chongjin. The pilot, Navy Lt Cmdr James H. Overstreet, was then to begin flying a series of slow elliptical orbits, each about 120 miles long, that would first take the plane northeast toward the borders of China and the Soviet Union, then back southwest toward North Korea. Throughout the flight, onboard linguists fluent in Korean and Russian would be eavesdropping on military radio communications. At the same time, the plane's electronic equipment would be busy intercepting and recording hostile radar activity and a host of other electronic intelligence.[31]

Thus far, this description makes it sound as if the EC-121 was a completely autonomous operator in the Far Eastern skies, but that was hardly the case. The plane was followed closely throughout its flight by a number of US facilities in the area, the most important of which were the Naval Security Group station at Kamiseya and an Army Security Agency facility in South Korea. Following standard procedures, all monitoring facilities received an advisory indicating the time of the flight and its intended course. As the time approached, the stations began gearing up for the activity that was to follow. Linguists prepared themselves to eavesdrop on radio frequencies used by the target nation's radar stations as they began to track the plane, a procedure known as grid plotting. Other intelligence technicians prepared themselves to monitor the frequencies used by the radars themselves, with still other technicians standing by to pinpoint those radars' precise locations using high-frequency radio-direction-finding techniques.[32] The plan worked. As President Richard M. Nixon noted following the shootdown, there was never any uncertainty regarding the location of the EC-121. "We knew this, based on our radar," Nixon said. "We know what their radar showed. We, incidentally, know what the Russian radar showed. All three radars showed exactly the same thing."[33]

The flight began at dawn on 14 April at 0950Z, local time, when the EC-121 and its crew took off from Atsugi Air Base (AB) near Yokohama, Japan. Eighteen minutes into the flight, the EC-121 transmitted a routine radio voice message saying it had reached its cruising altitude and was en route to its area of operations. Given the relatively slow speed of the prop-plane, nothing more would happen for several hours, except that the various US ground-based intelligence stations involved in the mission would become increasingly alert. On 15 April at 0334Z, the Army Security Agency station detected North Korean aircraft activity, presumably in reaction to the presence of the EC-121, although the North Korean planes were quite a distance away. American personnel continued to track the planes, Soviet-made MiG fighters, for almost an hour, until 0422Z, when they lost their track. This loss caused some concern, perhaps, yet there was no sense of crisis, as evidenced by a stream of routine communications between the EC-121 and ground stations throughout this period.[34]

Fifteen minutes after their radar track was lost, the North Korean MiGs were again picked up on radar. Concerned with the surveillance, the Army Security Agency duty personnel issued a spot report to the Naval Security Group station at Kamiseya. The report came in at 0445Z. It noted simply that North Korean air activity had been observed and that such activity was probably in reaction to the EC-121. In Kamiseya, the message was reformatted, addressed to the JCS and three other addressees, assigned an immediate precedence level, and handed over to Defense Communications System personnel at 0454Z for transmission by way of AUTODIN. Despite the fact that DCS criteria specify that messages with an immediate precedence level should be transmitted within 30 minutes, the spot report was not received by the Joint Reconnaissance Center in the Pentagon until one hour and 16 minutes later.[35]

Even as the spot report was being issued, it was becoming increasingly apparent to the American military watchers in South Korea that the North Korean MiGs were definitely approaching the EC-121. Communications between the ground station and the reconnaissance aircraft were now ongoing. Commander Overstreet was warned of the approach of the MiGs, and, following standard procedures, he immediately aborted the mission, changed course, and headed back to base. But the far faster MiG fighters closed rapidly on the lumbering Constellation. The attack came at 0447Z. The Navy plane disappeared from radar screens about two minutes later.[36]

The fact that a shootdown had taken place was not immediately apparent to American intelligence. Presumably because of the urgency of the situation he faced and a desire to maintain radio silence, Overstreet issued no radio call that he was under attack. Nor was the fact that the plane had disappeared from radar screen itself conclusive. A warning had been issued to the plane, and many officials monitoring the flight believed that it had simply dropped beneath the radar horizon to hide from enemy fighters—a standard practice for planes aborting a mission. Extensive efforts were made during the ensuing minutes to determine the EC-121's location and status, and, given the prevailing atmosphere of uncertainty, Washington was not immediately notified.[37]

Although the fact of the shootdown was not yet known, the situation was nonetheless one of increasing seriousness. The Army facility in South Korea soon issued a follow-up to its spot report, coming at 0503Z and informing Kamiseya and other addressees that MiGs were approaching the EC-121. As with the spot report itself, the follow-up message was received at Kamiseya, reformatted, addressed to the joint chiefs and other recipients, assigned an immediate precedence, and given to DCS personnel for AUTODIN transmission. Although the exact time the message was transmitted is not known with certainty, it most likely took place within 5 to 10 minutes. The Joint Reconnaissance Center at the Pentagon did not receive this message until more than three hours later.

Confusion was now rapidly setting in. Since the EC-121 was on a mission for the NSA, Fleet Air Reconnaissance Squadron I (also known as VQ-1), the group to which the EC-121 was nominally attached, was out of the operational picture altogether. It was not included as an addressee on any of the messages concerning the plane's situation. Nonetheless, in a world of serendipity, VQ-1 personnel had somehow managed to intercept and copy the spot report, giving them their first indication that there was a problem with the flight.[38] Concerned that the plane might be in danger, the VQ-1 duty officer alerted the squadron commander. Within minutes, VQ-1 had intercepted the Army Security Agency's follow-up to the spot report, although again it was not addressed to them. The squadron commander's concern increased, and at 0501Z he scrambled a protective combat air patrol of two F-102s, which was airborne within a matter of minutes. Well aware that standing procedures called for the EC-121's mission to be aborted in case of hostile activity, the squadron commander also initiated a series of calls to determine whether any abort messages had been received from mission aircraft. No such messages had been copied, he was told.[39]

When half an hour had passed since the EC-121's disappearance, and yet no word from it had been received, the Army Security Agency personnel in South Korea issued a second follow-up report saying that the EC-121 had disappeared from radar. This message came at 0520Z and was given a flash precedence. The seriousness of the situation was now also being felt in Kamiseya, where the message was copied, reformatted for AUTODIN transmission, and retransmitted six minutes later. It promptly disappeared into the labyrinthine channels of the Defense Communications System, and would not emerge at the Joint Reconnaissance Center at the Pentagon until 0558Z, some 38 minutes later.

Having failed to reach the EC-121 with repeated radio calls, and with no sign of the plane on their radar screens, Army Security Agency personnel were finally coming to the realization that something serious had taken place. Although they still had no conclusive evidence of a shootdown, when an hour had passed since the plane's disappearance, the decision was made to release a critic message. This message was transmitted by the South Korean station at 0544Z. Kamiseya received it immediately, and, given its precedence level, the message was promptly reformatted and retransmitted. The message arrived at the White House six minutes later and at the National Military Command Center four minutes after that. When we recall that the initial spot report would not reach Washington until 0615Z, the first follow-up until 0807Z, and the second follow-up, with its flash priority, until 0558Z, the critic message represented the National Command Authorities' first indication that a probable shootdown had occurred. This was more than an hour after the EC-121 had in fact been shot down.

When the critic message was received by the Defense Communication System's Pentagon station, it was automatically distributed to its six recipients simultaneously, including the White House and NMCC. A Pentagon duty officer telephoned National Security Advisor Henry A. Kissinger with news of the critic, and Kissinger immediately went to his West Wing basement office in the White House to assemble a crisis-working group. Two hours later Kissinger telephoned President Nixon with the news concerning the EC-121. Nixon's recommendation: "No immediate action is required" against the North Koreans.[40]

Action would be taken concerning further reconnaissance flights over the Sea of Japan; however, they were halted immediately following the shootdown. Yet as had been the case with the *Pueblo,* within a week the flights were resumed. And this time they were resumed on the direct order of President Nixon. His intention was to avoid a direct confrontation over the shootdown, yet at the same time not yield to the pressure to cease reconnaissance flights in the area. Headlines in the press described Nixon's handling of the incident as an "exercise in restraint."[41]

The crisis had passed, and yet the poor performance of defense communications during the incident would continue to generate comment and criticism for years to come. Precisely where had the communications tie-ups occurred? As the DCA director pointed out, the ambiguity of the situation at the time of the shootdown was such that it might reasonably account for the South Korean facility's delay in transmitting its four reports regarding the EC-121. But this in no way explained why the first three of these messages, once having entered the Defense Communications System, required one hour and 16 minutes, three hours and four minutes, and 38 minutes, respectively, to arrive at the Pentagon. The problem, apparently, had to do with the various facilities through which the messages passed. At each point, officials had to consider the message and its implications before deciding whether to pass it up the chain of command. Had a shootdown actually taken place? Apparently, nobody wanted to take responsibility for saying so while any doubt remained, leading to the delays.

The *Liberty*, *Pueblo*, and EC-121 incidents revealed a number of serious inadequacies with the World Wide Military Command and Control System as it had developed during the 1960s in particular, confusion within the system and a lack of clear-cut responsibility. Perhaps more than anything else, the three failures created a climate of criticism that in the decade to come would result in a formal effort to transform WWMCCS into a more coordinated, coherent system.

Notes

1. James M. Ennes Jr., *Assault on the Liberty: The True Story of the Israeli Attack on an American Intelligence Ship* (New York: Random House, 1979), 17.
2. Greenwich Mean Time will be used throughout this discussion.
3. John E. Sobraske, "Check Your Communication Pollution," *Army Digest*, August 1970, 33.
4. House, Committee on Armed Services, Armed Services Investigating Subcommittee, *Review of Department of Defense Worldwide Communications, Phase I*, 92d Cong., 1st sess. (Washington, D.C.: Government Printing Office [GPO], 10 May 1971), 7. (Hereinafter cited as *Phase I Report*.)
5. Ennes, 33–34.
6. *Phase I Report*, 7.

7. Ibid., 8.

8. Ibid., 9.

9. James M. Ennes Jr., "Israeli Attack on U.S. Ship Reveals Failure of C^3," *Defense Electronics* 13 (October 1981): 61.

10. Ibid.

11. *Phase I Report*, 11.

12. Ennes, "Israeli Attack on U.S. Ship," 64.

13. Hirsh Goodman and Zeev Schiff, "The Attack on the Liberty," *The Atlantic*, September 1984, 61.

14. Robert A. Liston, *The Pueblo Surrender: A Covert Action by the National Security Agency* (New York: Bantam Books, 1988), 248–49, 352.

15. *Phase I Report*, 11.

16. Ibid., 11–12.

17. Ibid., 12.

18. Ibid.

19. Ibid., 12–13.

20. Ibid., 13.

21. Liston, 256–57, 355–57.

22. Ibid., 263–67.

23. *Phase I Report*, 14.

24. Donald T. Poe, "Command and Control: Changeless—Yet Changing," US Naval Institute *Proceedings* 100 (October 1974): 24.

25. "An Exercise in Restraint," *Newsweek*, 28 April 1969, 28.

26. John M. Carroll, *Secrets of Electronic Espionage* (New York: Dutton, 1966), 115.

27. "The Varieties of ELINT," *Newsweek*, 28 April 1969, 30.

28. "The EC-121 Incident," *Signal* 24 (December 1969): 34.

29. James Bamford, *The Puzzle Palace: Inside the National Security Agency—A Report on America's Most Secret Agency* (New York: Penguin, 1983), 239.

30. Edgar Ulsamer, "A Strategic Blueprint for the '80s," *Air Force Magazine* 61 (September 1978): 48.

31. Bamford, 239.

32. Ibid., 240.

33. *Newsweek*, "The Varieties of ELINT," 30.

34. *Signal*, "The EC-121 Incident," 34.

35. *Phase I Report*, 14.

36. House, Committee on Appropriations, *Department of Defense Appropriations for 1970: Pt. 2, Operations and Maintenance*, 91st Cong., 1st sess. (Washington, D.C.: GPO, 1969), 1033.

37. *Signal*, "The EC-121 Incident," 36.

38. Ibid., 36.

39. Ibid., 36–37.

40. *Newsweek*, "An Exercise in Restraint," 28.

41. Ibid., 29.

PART II

Formalization

Chapter 6

WWMCCS Automatic Data Processing Upgrade

Whatever else WWMCCS might require, one thing was certain: it needed computers that were responsive and secure, compatible and capable of exchanging information between sites, cost effective, and yet with growth potential. In short, WWMCCS needed computers that could adequately support national level command and control requirements. The problem was that the WWMCCS ADP assets in operation during the latter half of the 1960s could do none of these things effectively because of the uncoordinated, piecemeal way in which they had been developed and acquired.

Even before the three major WWMCCS failures focused a harsh critical light on the system, a series of informal discussions concerning WWMCCS's automatic data-processing problems began between the Joint Chiefs of Staff and the Office of the Secretary of Defense. These talks represented a tentative first step in what would eventually become a comprehensive effort to update WWMCCS automatic data processing. By the second half of the decade, WWMCCS ADP consisted of some 158 separate computer systems, employing 16 different makes of computers, in operation at 81 separate locations. Thirty-two different program languages were then in use throughout the system. Given the haphazard fashion in which these assets had been acquired, and given the myriad problems this situation had occasioned, the words on everyone's lips as they considered a computer upgrade were *compatibility*, *interoperability*, and *standardization*. These concerns were all important new criteria by which effectiveness would henceforth be assessed.[1] To many officials, developing standard automatic data-processing systems that could meet these criteria represented the most challenging issue then facing the Pentagon.

The process of acquiring new computers for WWMCCS was formally begun in January 1966 when Defense Secretary McNamara issued a memorandum directing the joint chiefs to assess the feasibility and desirability of a single, multiyear buy of standard computer hardware for WWMCCS.[2] The chiefs, in turn, delegated this task to their in-house command and control experts, the Joint Command and Control Requirements Group. The JCCRG's chief at that time, Maj Gen J. R. Russ, described his group's methodology for improving WWMCCS ADP as a "straightforward process," the first step of which was to identify user requirements. Once the information required by decision makers had been identified, he said, the next step would be to develop performance specifications, that is, to determine what the system was actually supposed to do. Technical specifications plans drawn up by a service or DCA would flow from this point and describe how to satisfy the operational requirements. These plans would then be validated by higher authority and ultimately submitted to the secretary of defense for approval. Only then would hardware and software be developed, acquired, and installed.

The first step in developing a common set of defensewide ADP standards was to specify what constituted WWMCCS, which, in turn, called for thinking in broader system terms. After all, the World Wide Military Command and Control System, in its broadest conceptualization, could be seen to include everything from the White House to the foxhole. It could encompass the individual automatic data-processing and communications systems of virtually every echelon of military command, the services, and defense agencies—indeed, everything. In practice, however, the system was considerably more restricted. Certain elements were more central and important to WWMCCS than others. The participation of some system elements was substantial, while for others it was minimal. Some elements were WWMCCS on a full-time basis, while others were emphatically part-timers, considered elements of the system only at certain times, such as during crises.

So in thinking through the problem of automatic data processing for the massive and complex system that was WWMCCS, it soon became clear that the capabilities acquired for its various elements should not be the same. Given the diverse missions of the military services, for example, or of the CINCs and defense agencies, what seemed necessary was a system that would permit the free flow of information between users and yet be flexible enough to allow a single user to conduct his own unique mission.[3] In other words, it was not essential that each computer be identical, nor that its database contain everything, irrespective of its utility for a specific commander and his mission. Substantial areas of overlap, however, would obviously be required.[4] So the idea that increasingly received favor was that while both hardware and software would be permitted to vary from site to site in response to specific user requirements, the range of that variation would be tightly circumscribed to a limited menu of hardware and software options.

If the general point that not all WWMCCS elements required the same information, and hence the same automatic data-processing capabilities, this by no means answered the question of what in fact each one needed. And, defining user requirements had always been one of the most slippery issues confronting command and control system designers. "When DOD decides what they want, we'll stack the blocks the way they should be stacked," one industry computer expert had remarked back in 1962. But at the moment, "They're all over the table."[5] To a significant degree they still were in 1966. What was necessary before proceeding with any major WWMCCS automatic data-processing upgrade, then, was a comprehensive plan, one that defined system elements, goals, and specific standards to allow those goals to be achieved. The problem was that no such road map for arriving at a cohesive, centrally responsive system of this sort then existed.[6]

But if no road map was then available from national-level decision makers, the military services were only too happy to provide maps congenial to their interests. For example, for some time the Air Force had been actively developing its own computer systems for command and control purposes, which it wanted the other services to adopt. If they would only do so, Air Force experts argued, the sort of reporting-response capability the JCS had in mind for WWMCCS would take place. The Air Force's voice in these matters was especially eloquent and powerful, its case articulated by its own experts and by officials from the MITRE and RAND Corporations. In the bureaucratic battle that ensued, the joint chiefs, represented by Herbert Goertzel, fought for, and eventually won, a larger mandate. The technical specifications for the new computer system would be developed by the Joint Technical Specifications Group (JTSG), not by the Air Force. That group would be physically located in the Defense Communications Agency but would work under the direct authority of the Joint Staff.[7] Authority to solicit bids would also come from the joint agencies, rather than from one of the military services. Goertzel's success in keeping system design and development within the joint service community has been described as a major victory for WWMCCS.[8]

As things were initially conceived, the WWMCCS procurement would come in stages. Phase I would involve the purchase of computer hardware for software development. This hardware would not have the capacity or speed to satisfy operational requirements after the development phase was concluded. The DCA originally allocated some $3 million of its fiscal year 1968 budget for this hardware, money that was contingent upon receipt of system specifications from the JTSG. The JTSG's first draft of the WWMCCS road map was completed in late 1967. But under heavy pressure to move forward expeditiously with the program, the JTSG then modified the Phase I design, deciding that the fastest and most economical approach would be to purchase computers that would serve both the Phase I requirements for software development and actual operational needs during the subsequent stage of the program, designated Phase II.[9]

The JTSG's specifications were based on four different workload models, each involving different data-processing tasks and requiring its own distinct type, or combination of types, of automatic data processing equipment. The first model was called the Force Control System. It was designed to control combat forces under changing operational situations. Examples of this system then in operation included NORAD's Combat Operations Center and SAC Automatic Command and Control System (SACCS). The second model was called the Scientific System. It came with a large-scale mathematical computational ability to support surveillance, data analysis and reduction, war gaming, and other similar functions. An example of a scientific system then operating was NORAD's Space Defense Center. The third and fourth models were General Staff Support Systems that would provide a wide range of computer support to a headquarters staff. These specifications represented a major milestone in the development of WWMCCS ADP, and the joint chiefs gave approval to proceed with equipment determination.[10]

The draft specifications for WWMCCS's anticipated automatic data-processing system were sent to computer manufacturers for comment in 1967. Accompanying documentation made it clear that the WWMCCS buy would be a unique one, the first time that so large a computer purchase would be made from a single source. Industry immediately began gearing up for the competition.[11] The potential future payoffs were considered to be so substantial that, despite a series of subsequent delays, a number of computer firms would spend as much as $100,000 a month to hold their project teams together.[12]

Another program milestone was reached in August 1968, when the joint chiefs and the director of defense research and engineering jointly submitted to the secretary of defense a development concept proposal for the WWMCCS ADP program. For hardware, the proposal outlined a plan for purchasing a family of compatible computers for WWMCCS. For software, it described a policy of centralized and standardized development and maintenance. Considering the magnitude of the undertaking, the proposal also outlined an evolutionary approach for system development under whose terms new capabilities would be added as user requirements were clarified and experience with the system was gained. That is, both hardware and software would develop incrementally, over the course of several years, resulting, it was hoped, in the coordinated and integrated system desired by the Pentagon. To aid in realizing the goals of system compatibility and standardization, a WWMCCS ADP program project manager was appointed. Finally, the proposal provided a series of ideas for managing the new system once it was in place.[13]

Concerning the actual WWMCCS computers themselves, the development concept proposal reaffirmed the JTSG's call for four standard, generic computer types. For purposes of economy, it was recommended that they be commercially available, general purpose machines. They would be located at such key sites as Strategic Air Command headquarters, the National Military Command Center at the Pentagon, NORAD headquarters, at various elements of the ballistic missile defense system, and elsewhere. They would replace the obsolescent systems at those locations, since those systems required extensive maintenance and could no longer cope with the increasing workload.[14] The recommended number of computers to be purchased by the Pentagon was set at approximately one hundred, and preliminary specifications for the new computing hardware were released to industry in 1968. The following year saw the release of typical benchmark problems.[15] As the 1960s ended, events had advanced sufficiently so that it was finally possible to make a formal announcement on the purchase of WWMCCS ADP hardware.

Software was another issue altogether. An area inherently more complex and confused than hardware, software was always the most important issue, the real pacing item in system development. For major applications, it would often take years of painstaking effort to debug the software, as the experience with SACCS had shown. Two major consequences of this were system utility and cost. As to utility, by the time the software was finally debugged, users' requirements generally had changed, meaning that it was time to start all over again. As to cost, the overall price tag for an automatic data-processing system could easily double as expensive professionals tried to figure out where the software errors lay. And because of the software problems, costs might also increase because more capable hardware would often have to be purchased to allow the software to be written with less care.[16]

The purchase of a family of commercial general purpose computers for WWMCCS appeared especially attractive in addressing the software problem. Bought essentially off the shelf, these general-purpose computers, like all general systems, would be useful for a broad range of applications but not for any specific one. In order that a broad range of users could use them, manufacturers furnished these computers with nonfunctional software—that is, a basic operating system and means of organizing the computer's memory. Users naturally tended to have quite specialized interests, and these oftentimes required additional mission-specific functional software applications. To run on the computer, however, functional software applications had to employ the same data elements and codes as the nonfunctional software. Thus, if lack of software standardization were a problem, one sure way to address it was to do precisely what the development concept proposal was suggesting—standardize the hardware.[17]

Standardization had other benefits as well. Given the military context prevailing at the end of the 1960s—that of an increasingly large and accurate Soviet ballistic missile threat—common equipment and personnel training procedures meant that if some facilities were damaged or destroyed, the surviving elements could pick up the pieces. Equipment from damaged facilities could be cannibalized, surviving personnel formed into new working teams, and the most essential functions restored. In theory, then, by allowing for the reconstitution of the system, standardization contributed to its endurance, a term that would later loom large as a criterion of WWMCCS effectiveness. Looking to the future, standard hardware and software held forth the promise that some day the computers might be linked into an intercomputer network to share information and processing power. Work then being conducted by the Advanced Research Project Agency (ARPA) in its ARPANET project already suggested that users in different locations could in fact share information and processing capacity.[18]

In the organizational realm, it was hoped that the WWMCCS purchase would help solve some of the persistent personnel problems plaguing the system. Training and logistical support were ongoing headaches in a system using different software applications and employing a multiplicity of machines that were of different types and generations. It was also hoped that the purchase would solve the problem of finding programmers, many of them civilians, who would be willing to work with the older computers then in use throughout the system. The problem was that the best civilian personnel simply did not want to work with older equipment because it restricted their learning in a dynamic industry and their ability to find work on newer computers.[19] While this problem would be gradually reduced through the expanded use of military personnel for ADP functions, it remained a serious concern at the time.

Most of these factors had positive implications for program costs. Consolidating hardware purchases would eliminate the need for a series of separate competitive bids for upgrading the many WWMCCS sites and help to reduce overhead costs. The bulk purchase of computers would provide considerable cost savings over site-by-site modernization, since a volume buy raised the possibility for volume discounts, and hence lower per-unit costs. Using standard software would eliminate the considerable costs associated with converting the wide range of individual software applications then in use at WWMCCS sites.[20] The new machines would result in higher-quality personnel being attracted and retained, and the problems involved in extensive retraining of personnel would be eliminated, saving thousands of man-hours and millions of dollars each year.[21] Standardization, then, was viewed as overwhelmingly positive in training, programming, operating, and maintenance.

The WWMCCS Automatic Data Processing Upgrade Program (its formal name) was intended to address many of these concerns. It would do so in a single dramatic step. This step was establishing the foundation for a coherent national-level command and control system from a disparate group of disconnected or only loosely connected ADP elements. But new standard computers would provide only a foundation, not by themselves create a centrally responsive WWMCCS. The system was simply too large for that, since it involved by this time some 81 computers and 129 separate WWMCCS activities or sites. More important was the continuing need for adequately defining system requirements so that competitors could intelligently bid on an eventual request for proposal. But changes were continually being made to the system design, owing to ongoing deliberation and contentiousness regarding the rather basic questions of precisely what the system was and what it should do. Delays necessarily ensued, and the program timetable slipped.

On 12 November 1969 Deputy Defense Secretary David Packard announced that he had approved a plan to purchase a minimum of 34 standard computers to be used throughout WWMCCS, with an option to buy an additional 35. The computers were to be of medium and large size, Packard said, costing between $1 million and $5 million each, and they would have an expected service life of about six years. Procurement responsibility for hardware and nonfunctional software was given to the Electronic Data Processing Equipment Office at the Air Force's Electronic Systems Division. Once the computers were purchased, the Joint Chiefs of Staff would be responsible for allocating them to appropriate locations. Responsibility for developing functional system software was given to the joint chiefs, who tasked it to the Joint Technical Specifications Group.[22] In a bureaucratic move to enhance system integration, general program oversight responsibility went to the director of defense research and engineering, rather than to the Pentagon comptroller, the normal bureaucratic focal point for computer acquisition.

The rhetoric accompanying Packard's announcement was relentlessly upbeat, with Pentagon spokesmen pointing out how the WWMCCS procurement represented the first time the ADP needs of a diverse community of users would be satisfied by computers acquired from a single source.[23] Equally unique was the nature of the procurement process itself. Whereas in the past many of the Pentagon's major computer contracts had been sole source and without competition (with the sole source generally being IBM), Packard, in announcing the WWMCCS purchase, made a point of saying that in the current procurement, IBM, notorious for its reluctance to discount its prices to the federal government, would be given no special consideration. The competition between major computer manufacturers would be "both extensive and equitable," Packard promised, and the excitement among the industry competitors predictably ran high.[24]

This excitement would undergo a measure of tempering in the days following Packard's announcement. For reasons of budgetary economy, the House Appropriations Committee ordered the Pentagon to reassess a number of its programs then in progress, among them the WWMCCS ADP Update Program, which was conspicuous because of the scale of its computer purchase. With the defense budget falling and with no end to the decline in sight, Pentagon officials got the message. In a March 1970 memorandum to the secretaries of the military departments, the Joint Chiefs of Staff, and the director of the Defense Intelligence Agency, Packard solicited information regarding the number of computers to be procured, the costs involved, and these users' views on whether to proceed with the standardization effort.[25] After reviewing the responses, Packard approved on 4 June a substantially downscaled plan for WWMCCS. Rather than 34 computers with an option for an additional 35, the system would now employ a minimum of 15 computers with an option for an additional 20. They were to be divided among the WWMCCS sites and those parts of the Intelligence Data Handling System that worked intimately with WWMCCS.

The best possible bureaucratic face was put on the reduction, which was portrayed by Pentagon officials as a prudent move to minimize the risks of a large-scale ADP conversion in a time of tight DOD budgets. "With big chops coming throughout the DOD budget," one Pentagon official noted, describing the mood, "we might not have enough budgetary support for our program and we'd end up with egg on our face."[26] Another concern was a predictable bureaucratic desire to limit liability by avoiding overly strong commitment to a program whose future was still in doubt, a program that might be in line for further cuts and revisions. Specifically, the recommendations of a blue ribbon defense panel were expected shortly; nobody was sure how WWMCCS would fare in the panel's report, and in the interim a cautious conservatism prevailed. By some accounts, this slowed the purchase of the WWMCCS computers by more than a year.

The reduction in the size of the WWMCCS procurement and the associated delays sat poorly with industry, which saw its own costs increasing even as the program's profit potential was being reduced. Limiting the scale of the WWMCCS procurement also raised industry concerns over the competitiveness of the procurement process, and the fears were not unfounded. The 15 computing systems that were now slated for purchase were not sufficient for all WWMCCS sites, and because of this, David Packard had announced that he intended to designate the third generation IBM/360 computer as a "second standard" for 16 other sites. Many WWMCCS sites already used leased IBM equipment, and Packard noted that economic considerations would determine whether to purchase this equipment, continue to lease it, or replace it with the new WWMCCS computers.[27] Leasing computer equipment as opposed to buying it for WWMCCS had been an economic consideration for some time. Even when procurement funds are limited, money is generally available for a lease, with an option to buy later. "If you lease," one Pentagon official remarked, "it comes out of operating funds. You're not asking for more money. The guy who buys is in a hole. He's asking for a big new chunk of money."[28] For the industry competitors other than IBM, this was bad news. The worry was that IBM might offer to sell the Pentagon some of its leased equipment at a discounted price, further reducing the number of computers that eventually would be purchased and giving IBM the proverbial foot in the door for future leases and ultimately sales.

At stake was an enormous amount of money beyond the 15 computer systems themselves. The profit to be made by selling the first 15 systems was not at all substantial; indeed, it was hardly sufficient to offset the $3–$5 million that various firms already had invested in developing benchmark WWMCCS software. But from that point on, the money would really start rolling in. By one estimate, the Pentagon would have to spend at least $159 million to convert present WWMCCS sites to a common software standard. In addition, another $30 million or so would be involved in providing software integration for the essential WWMCCS sites. Finally, there was at least $62 million to be made selling additional hardware should the Pentagon exercise its option for the additional 20 computer systems.[29]

While this planning was in progress, the much-publicized *Liberty, Pueblo, and EC-121 incidents were taking place. To make* matters more urgent still, a number of specifically computer-related command and control shortcomings contributed to an already dour mood regarding WWMCCS. Some of these shortcomings concerned the need for additional computerization of critical defense functions. For example, when the Pentagon was planning Operation Rolling Thunder, the bombing campaign over North Vietnam, Defense Secretary McNamara had needed an up-to-date inventory of available bombs. No such inventory was available, and McNamara reportedly had to order 14 general officers in the Pentagon to man the telephones, calling air bases around the world to get the necessary information. (By one account, the generals forgot to check with the National Guard or to consider the materiel in the nation's war reserves.) Another story concerned the daily report prepared by the Air Force Command Post on the worldwide status of its forces. It took the Air Force all night to prepare the report, with errors commonplace and timeliness sacrificed as a small army of personnel scrambled to make the morning deadlines. It was also reported that CINCPAC, the commander in chief, Pacific forces, required some 32 hours to produce an accurate tabulation of the status of forces in the Pacific.[30] The rather obvious moral of these stories was that there existed an urgent need for up-to-date automatic data processing.

If anecdote suggested the value of computerization not yet achieved, other stories concerned existing computers that were inappropriate or ineffective. During the 1969 Tet offensive, incompatible ADP equipment resulted in it taking a full hour for a flash message to travel between two offices within the same facility, CINCPAC headquarters and its alternate command post in the Kunia tunnel. Similarly, the computers in CINCPAC's main WWMCCS facility employed paper tape while other WWMCCS facilities in the Pacific used magnetic tape. The problems that existed were increasingly of this type. They were no longer attributable to a lack of computers, but instead to the opposite problem—a proliferation of different ADP systems that all too often could not function together.[31] Incidents such as these seemed to demonstrate the need for a command and control structure that would permit high-level decisions to be carried out in a timely fashion. Doubly so, critics contended, since unlike the earlier incidents, the next crisis might involve nuclear weapons.[32]

Notes

1. Carol Hamilton, "Worldwide C^2 System Networks Strategic Data for Joint Chiefs," *Defense Electronics* 20 (June 1988): 53.
2. John B. Bestic, "No More Confused Situations," *Signal* 21 (March 1967): 56.
3. Gordon T. Gould Jr., "Computers and Communications in the Information Age," *Air University Review* 21 (May/June 1970): 9.
4. "Man and Machine Must Learn to Talk Together," *Armed Forces Management* 14 (July 1968): 111.
5. "The Search for Effective Command and Control," *Armed Forces Management* 8 (July 1962): 21.
6. *Armed Forces Management*, "Man and Machine," 111–12.
7. C. J. LeVan, "WWMCCS Automation—A Team Effort," *Signal* 32 (November/December 1977): 11.
8. Hamilton, 53.

9. Senate, *Department of Defense Appropriations for Fiscal Year 1969*, 90th Cong., 2d sess. (Washington, D.C.: Government Printing Office, 1986), 2051–52.

10. "Command and Control Spectrum: JCS to Battlefield," *Armed Forces Management* 14 (July 1968): 38.

11. "WWMCCS: Report, One Year Later," *Government Executive* 3 (January 1971): 19.

12. Benjamin F. Schemmer, "DOD's Computer Critics Nearly Unplug Ailing Worldwide Military Command and Control System," *Armed Forces Journal* 108 (21 December 1970): 24.

13. Herbert B. Goertzel and James R. Miller, "WWMCCS ADP: A Promise Fulfilled," *Signal* 30 (May/June 1976): 57.

14. Edgar Ulsamer, "The Military Decision-Makers' Top Tool," *Air Force Magazine* 54 (July 1971): 46.

15. Joseph Volz, "Revamped World-Wide Computer Net Readied: WWMCCS Slimmed Down to 15-System Program," *Armed Forces Journal* 107 (27 June 1970): 21.

16. Edgar Ulsamer, "Command and Control is of Fundamental Importance," *Air Force Magazine* 55 (July 1972): 46.

17. Ulsamer, "The Military Decision-Makers' Top Tool," 46.

18. LeVan, 12.

19. Volz, 21.

20. Schemmer, 40.

21. George Weiss, "Restraining the Data Monster: The Next Step in C^3," *Armed Forces Journal* 21 (5 July 1971): 29.

22. "DOD to Acquire New Family of Computer Systems," *Defense Industry Bulletin* 5 (December 1969): 31.

23. Volz, 21.

24. Schemmer, 25.

25. Comptroller General, letter to George H. Mahon, chairman, Committee on Appropriations, House of Representatives, 29 December 1970, 1.

26. Volz, 20.

27. Schemmer, 24.

28. Volz, 21.

29. Schemmer, 25, 40.

30. Ibid., 23.

31. Ibid.

32. Ulsamer, "Command and Control," 43.

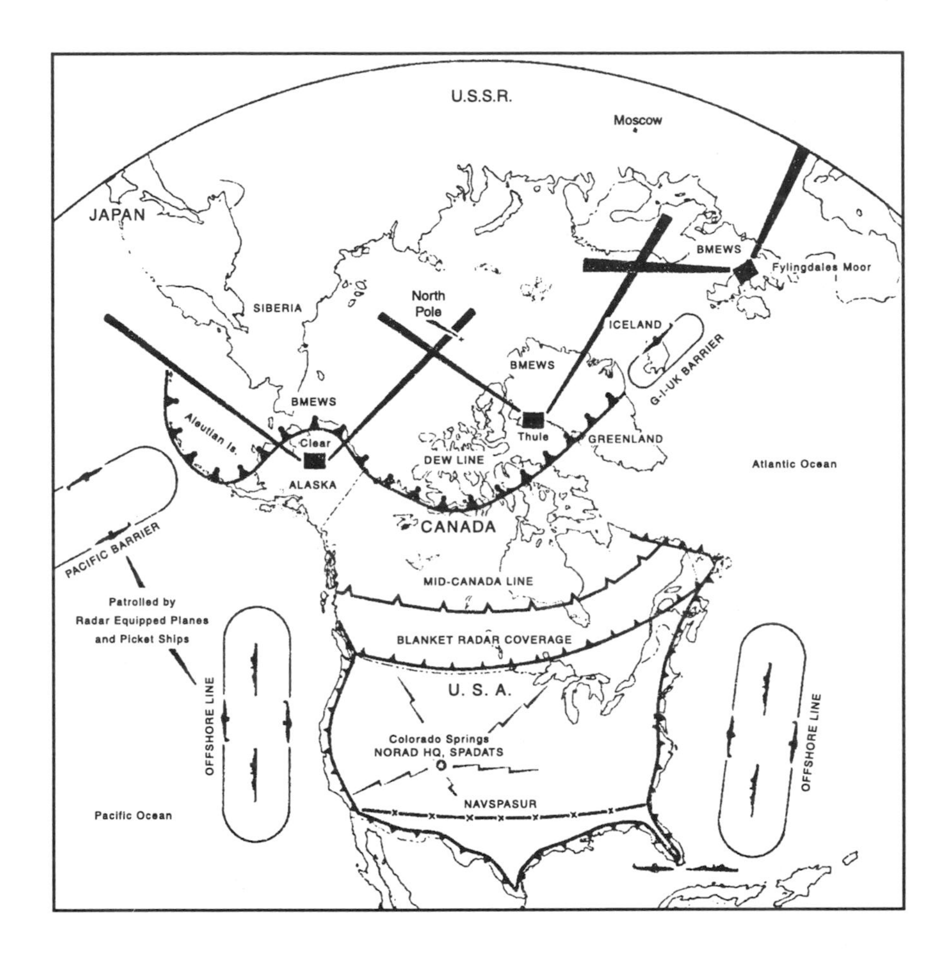

Source: J.P. McConnell, "Command and Control," *Sperryscope,* Third Quarter 1965, 4.

Figure 1. The Early Array of Sensors and Sentinels Forwarding Information to NORAD Headquarters

Source: Charles A. Zraket and Stanley E. Rose, "The Impact of Command, Control, and Communications Technology on Air Warfare," *Air University Review*, November/December 1977, 85.

Figure 2. Cold War Vigilance: Texas Tower Offshore Radar Platform

Source: Signal, March 1961, 29.

Figure 3. SAC's Underground Command Post

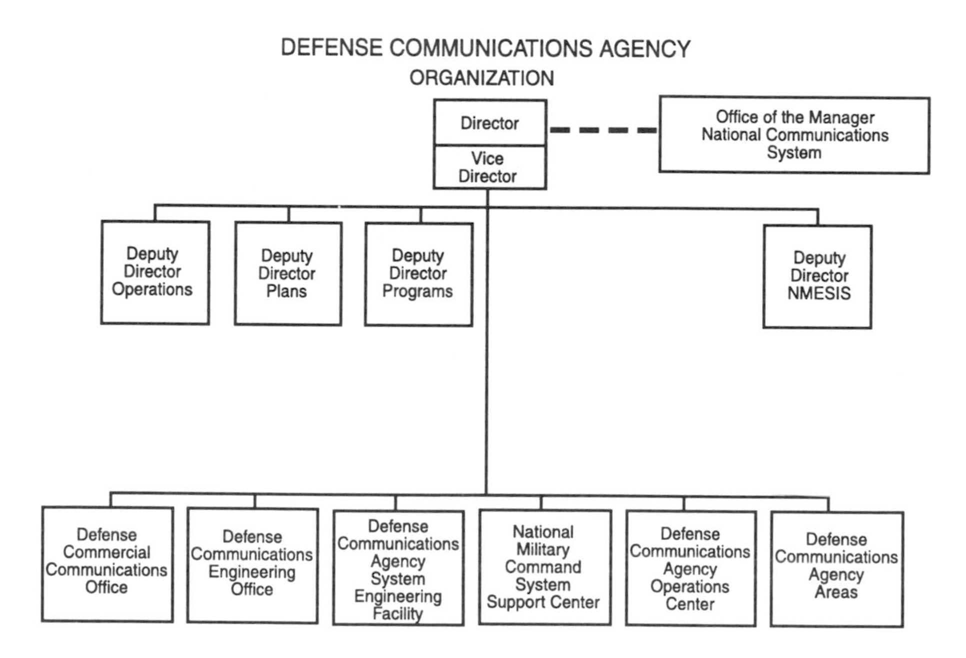

Source: House, *Military Communications—1968* (Washington, D.C.: Government Printing Office, n.d.), 3.

Figure 4. Defense Communications Agency

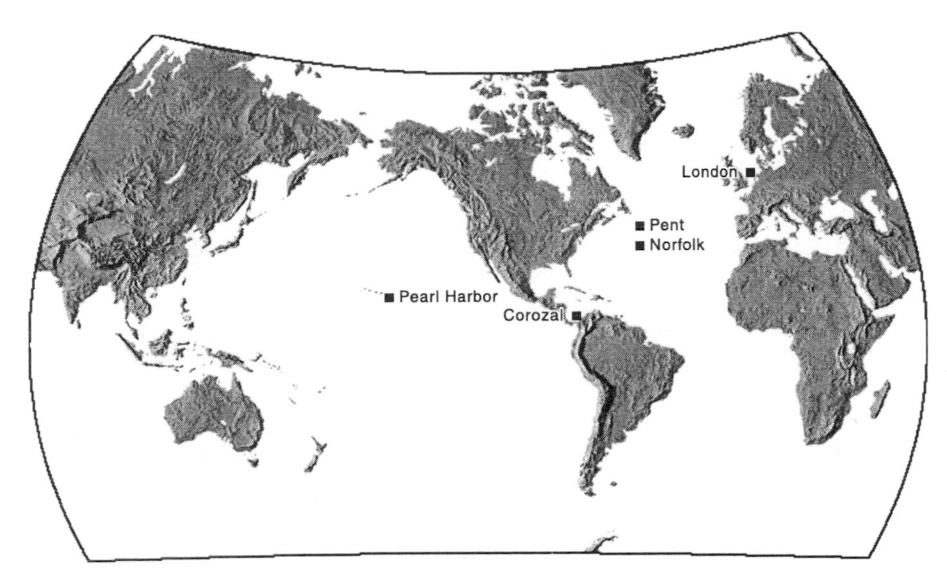

Source: Senate, *Fiscal Year 1978 Authorization for Military Procurement, Research, and Development, and Active Duty, Selected Reserve, and Civilian Personnel Strengths* (Washington, D.C.: Government Printing Office, April 1977), 6822.

Figure 5. AUTOSEVOCOM Switches

Source: "Effectiveness, Responsiveness of National Command System Vital to U.S. Security," *Armed Forces Management,* July 1966, 46.

Figure 6. NMCS Personnel in Action

Source: John B. Bestic, "NMCS Affords U.S. Full Control & Flexible Response," *Data,* January 1967, 29.

Figure 7. USS *Northampton,* National Emergency Command Post Afloat

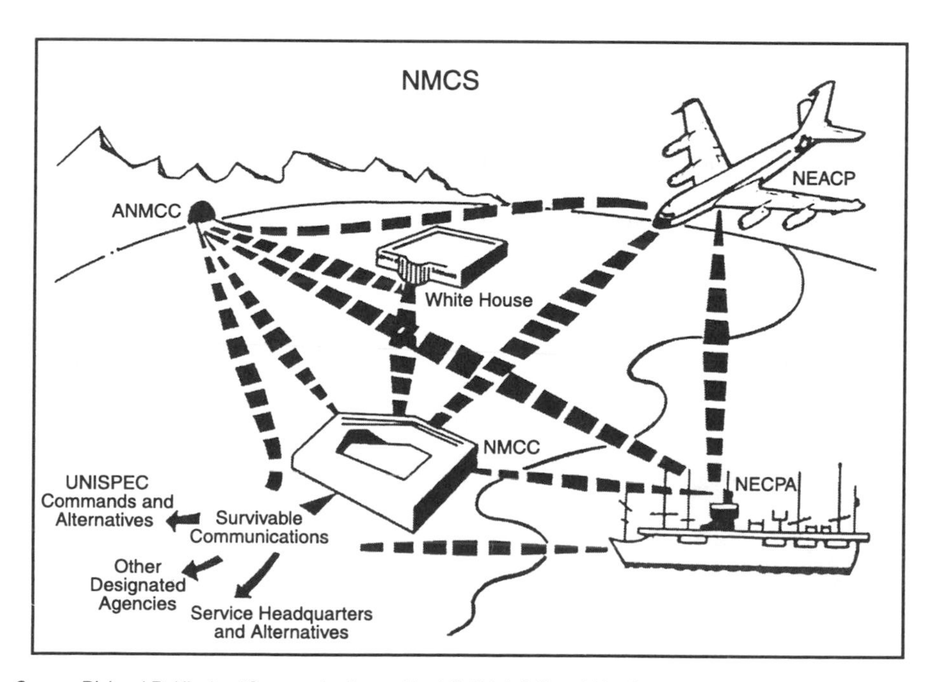

Source: Richard P. Klocko, "Communications—The Vital Link," *Signal,* May/June 1969, 78.

Figure 8. Artist's Conception of the National Military Command System

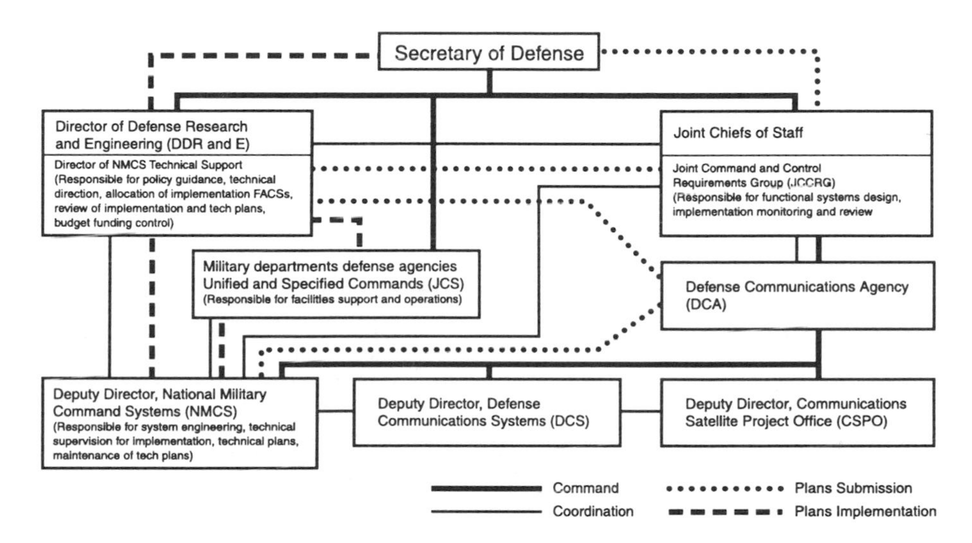

Source: "National Military Command Control: The Problems in Brief," *Armed Forces Management,* July 1963, 21.

Figure 9. Responsibilities for Implementing the National Military Command System

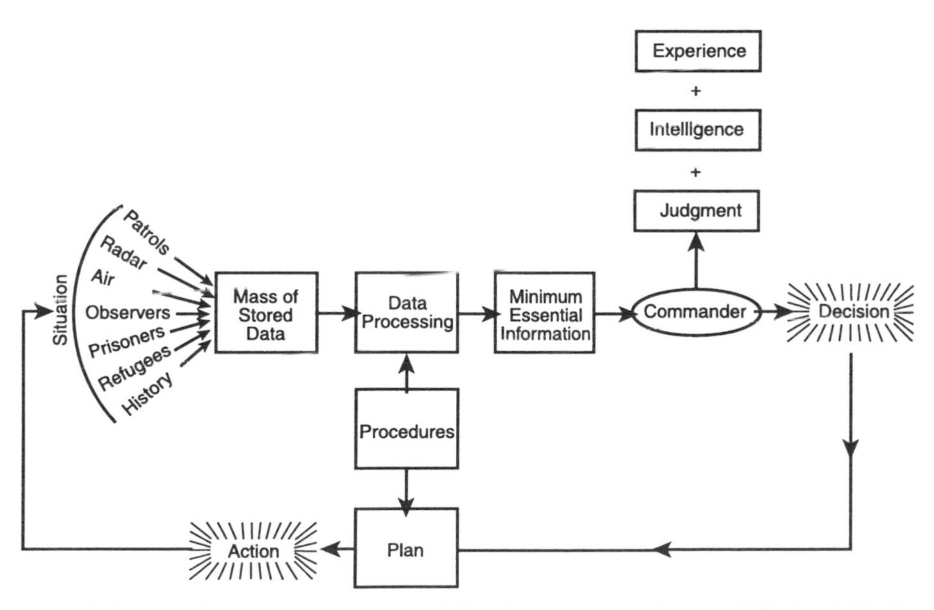

Source: F. Boonham, "The Impact of Computers on Military Management and Control," *RUSI,* June 1972, 32.

Figure 10. Conceptualizing a Command and Control System

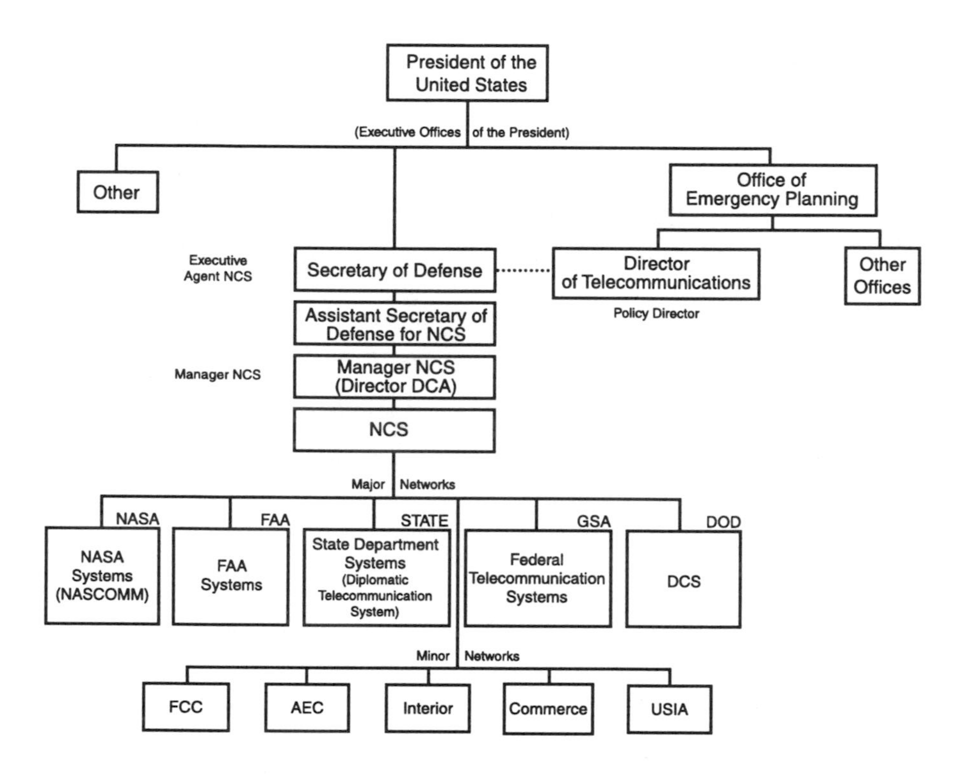

Source: Emmett R. Arnold, "The Commander's Control of His Communications," *Signal,* March 1969, 32.

Figure 11. The National Communications System, Encompassing All Federal Assets

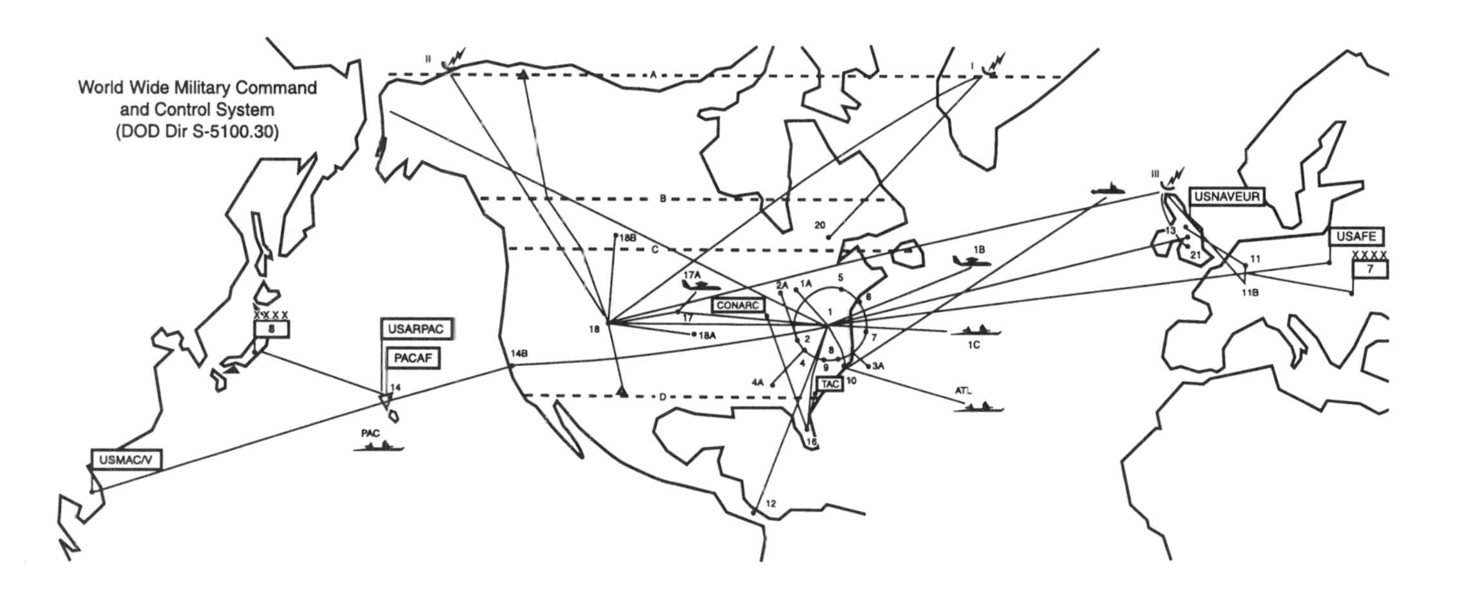

Source: J.H. Wagner, "NMCS: The Command Backup to Counterforce," *Armed Forces Management*, July 1963, 24–25.

Figure 12. World Wide Military Command and Control System as Described by DODD S-5100.30

Chapter 7

Centralizing Communications Management

When the Nixon administration arrived in Washington in early 1969, it discovered that criticism of WWMCCS and of defense communications management generally was broad-based both inside and outside the Pentagon. Not only was there considerable soul-searching by defense officials, but the highly visible nature of the WWMCCS failures had drawn influential new constituencies into the system in the role of evaluators, most importantly the Congress. Their eyes opened to the need for a command and control system responsive to the needs of the national leadership, dismayed that such a system was not in place and concerned with the system's tendency to emphasize the needs of its more influential subunits, these evaluators looked at WWMCCS and judged it ineffective. Addressing these issues both within the Department of Defense and in the federal government more broadly quickly became a priority of the new administration. In addition, a report by the General Accounting Office (GAO) later that year gave added impetus to the desire for change by calling for a major restructuring in government communications management responsibilities.[1] And so, even as David Packard began his energetic efforts to standardize WWMCCS automatic data processing, a series of changes was begun that would affect the organizational and management side of things.

Office of Telecommunications Policy

These changes began in earnest on 9 February 1970, when President Nixon sent to Congress a proposal to establish a new Office of Telecommunications Policy. The plan took the old Office of Telecommunications Management, gave it a new name, and removed it from its home in the Office of Emergency Planning. The old office had become a contentious forum that somehow never managed to find a way to balance conservative desires to make things work with the need to integrate new technologies.[2] As described in Nixon's proposal, the new office would better equip the Executive Branch to deal with the myriad issues arising from recent, dramatic growth in the communications field.[3] The proposed office would be bureaucratically situated in the Executive Office of the President, headed by a director who would be appointed by the president with the advice and consent of the Senate, and elevated in rank so that its director could report directly to the president.[4] The new office would help formulate new policies to better coordinate the myriad defense and nondefense communications systems owned by the government. It also would allow the Executive Branch to speak with a single voice in its policy discussions with industry, other governmental bodies, and the public.[5]

Many experts believed that creating the Office of Telecommunications Policy was overdue. Every president since Harry S. Truman had directed that studies of the communications field be conducted, and each study had recommended the establishment of such an office. A key problem, however, was that communications always tended to be viewed more as a support function than as a critical function in its own right. This meant that policy was formulated essentially on an ad hoc basis rather than as part of a coherent plan to meet the nation's requirements. Vision was required, for without it there could be no "creative shaping" of the telecommunications future.[6]

That shaping process began in earnest under President Nixon. In submitting his proposal to Congress, Nixon argued that the changes being sought were necessary if the government was to respond to the challenges presented by the dizzying pace of change in communications.[7] Congress agreed, Nixon's plan was approved that March, and the Office of Telecommunications Policy was established on 20 April 1970. Clay T. Whitehead was nominated as its director. As described in the Nixon plan and elaborated in Executive Order 11556, which formally established the new telecommunications office, the responsibilities were wide-ranging, and involved communications policy, broadcast and cable media, spectrum management, emergency preparedness, and numerous other areas. This effort to coordinate government-wide telecommunications would soon have its defense-specific analogue.

Assistant to the Secretary of Defense, Telecommunications

Inside the Pentagon, perhaps the most influential of the in-house critics of existing defense communications was Deputy Secretary of Defense Packard. Shortly after taking office, Packard had become aware of serious problems in the way defense communications were being managed. These included a fragmentation of management authority, decentralized control of resources, and the lack of an appropriate means for coordinating the various communications systems of the military services. Equally troubling to Packard's thinking was the fact that no matter how hard he tried, he was unable to determine how much was actually being spent on communications. Packard found this intolerable, but, unlike most critics, he was very much in a position to do something about it.

On 21 May 1970 Packard issued a directive establishing the position of assistant to the secretary of defense for telecommunications, his explicit purpose being to further centralize control of communications policy within the Office of the Secretary of Defense.[8] Through that time, managerial direction and resource control within the OSD had been dispersed among a number of offices. Four assistant secretaries of defense, including those for installation and logistics, systems analysis and administration, comptroller, as well as the director of defense research and engineering, had responsibility for managing various aspects of defense communications. Responsibilities were divided along functional lines, meaning that each office tended to emphasize those communications areas relevant to it. Even the Office of the Assistant Secretary of Defense for Installations and Logistics, the designated staff "point man" for communications, had little authority in many areas. As Packard saw it, there existed no sense of managing the Pentagon's communications assets as an integrated whole. This deficiency resulted in overlapping responsibilities, inefficient use of resources, and the system's inability to perform adequately. What Packard had in mind was a single office with the power to coordinate all of the communications activities carried on by the Defense Communications Agency and the military services. The new assistant to the secretary position was intended to do precisely that, and its first occupant was Louis A. deRosa, a former vice president of the Philco-Ford Corporation.[9]

On paper, deRosa had an exceedingly broad mandate. He had overall coordinating responsibility for the Defense Communications System and the defense-related elements of the National Communications System. He would oversee and coordinate areas of communications used for command and control, including such critical WWMCCS subsystems as MEECN (Minimum Essential Emergency Communications Network). Exempted from this broad slate of responsibilities were only those electronics and communications integral to weapons systems. Packard's clear intent was that deRosa would play a central role in establishing policy and priorities, serving as the focal point for coordinating and reviewing issues related to communications. Under his leadership, deRosa said, the existing "federation of subsystems" would give way to a more unified and coherent system, operating under a centralized management structure.[10]

To combat what Packard and many others believed had been years of loose management practices, the new assistant to the secretary was to account for communications expenditures throughout the DOD; no small order since, as Packard had found, even identifying relevant expenditures could prove an impossible task. Prior to the Nixon administration, no defense-wide cost accounting system had been maintained at all, meaning that even such basic information as total communications resources and costs were unavailable. A formal accounting system for communications costs had been established in 1969, but it would not be equal to the task for some time. One of the key problems was an inability to find properly qualified personnel to staff the relevant resource management sections, a problem with obvious implications for reviewing the Byzantine budgets of the military departments. An inability to easily break down the services' operating and maintenance funds meant that it would be necessary to get the services to conform to new budgeting procedures to accurately ascertain communications costs. Even if the services agreed to the changes, it would take years to determine with any accuracy the amount actually spent on communications. In the absence of firm figures, "approximate levels of expenditures" were provided, figures that varied widely among defense officials.[11] For example, Packard estimated that the communications costs for fiscal years 1970 and 1971 were $2.31 and $2.25 billion, respectively. DOD deputy comptroller Thomas Moran estimated the costs for those same years to be $2.75 and $2.51 billion. Secretary of Defense Melvin R. Laird placed the 1970 figure at $3.06 billion. As Packard pointed out to Congress with some exasperation, the budgeting and financial control system within DOD did not bring together all of the costs relating to communications in any systematic manner. "I can't assure you those [figures] would be within a couple hundred million dollars to be correct," he said.[12]

Even adding Packard's few hundred million dollars estimate to these figures still would likely have resulted in a considerable understatement of the Pentagon's actual communications costs. The figures tended to refer only to costs for "classical communications." They did not include costs for communications systems related to command and control, or those that were integral to weapons systems. Given these exclusions, a more realistic figure for communications of all sorts would probably have been at least double the estimates. A key task of the new telecommunications position, then, would be to bring these figures together and make them visible, an essential preliminary step in improving the management of defense communications resources. But as deRosa told the Congress, it would not be an easy task. His office would have to assess outlays for research and development, procurement, operation, and maintenance for a large number of large systems. These included the Defense Communications System, tactical communications, and command and control communications that were neither DCS nor tactical, including many elements of WWMCCS.[13]

Another major barrier to identifying costs was the inability to find qualified personnel to staff deRosa's resource management sections. Understaffing had obvious implications for reviewing the complex communications budgets of the military services. But even with a properly staffed office, the arcane and frequently idiosyncratic manner in which the services accounted for their operating and maintenance expenditures meant that it would be necessary to get them to conform to new budgeting procedures. Even if the services accepted this sort of conformity, an outcome that was by no means certain, the imprecision of earlier accounting practices meant it would still take years before the amount spent on communications could be determined with any accuracy. Things went similarly with automatic data processing, an area by now inextricably linked to communications. Here, because of exclusion of the cost of depreciation, personnel, and support from cost calculations and reports, the Pentagon could not make appropriate management and budgetary decisions.[14] So urgent did this cost accounting function seem that some members of Congress recommended that the resource management function of the new telecommunications office be staffed as expeditiously as possible to identify total relevant expenditures and to calculate defense communications costs on a year-to-year basis.[15]

The way the defense budgeting process should work with respect to command and control, it was felt, was that the services would first address their priorities across the spectrum of expenditures: weapons, manpower, operations and maintenance, research and development, logistics, command and control systems, and other areas. The services would then submit their appropriations requests to OSD for evaluation. In its evaluation, OSD would strike a balance between the three services and across their various mission areas, with the assistant to the secretary of defense for telecommunications striking a balance in communications. Since service-specific command and control systems had never fared particularly well in the budgetary process, and joint-service command and control tended to fare considerably worse, it was up to the new assistant to the secretary to make certain both did better.[16]

Such were the hopes and the expectations. The very fact of creating the new position represented a clear, highly visible sign of changing priorities within the OSD. Similarly, some of the language in Packard's directive implied that serious bureaucratic muscle would be accorded its occupant. But while the mandate seemed broad and the intention clear, the organizational reality of the situation was less auspicious. The reasons for this were two: the nature of the position itself and the language of the directive that established it.

Consider first the assistant to the secretary position, where the important limitation lay in the single word *to*.[17] In creating the position, Packard was engaging in an artful bureaucratic maneuver, anticipating the release of the president's Blue Ribbon Defense Panel's report, then still two months away. Packard knew full well that the panel intended to recommend the creation of a similar telecommunications post, and to create it at the assistant secretary level. From the beginning, Packard had wanted this new position as well, but an assistant secretary slot required legislative action, whereas an assistant to the secretary position did not. By creating the new telecommunications slot at the lower level, Packard was covering the bureaucratic bases. If Congress favorably received the Blue Ribbon Panel's recommendations, the already existing telecommunications position almost certainly would be upgraded to an assistant secretary slot. No problem there. Indeed, the fact of its existence might well aid in ensuring a warm reception; after all, the Pentagon had clearly identified a need and taken steps to address it. If the congressional reaction to the panel's report was less favorable, however, all was not lost. Packard had at least established the assistant to the secretary position without congressional opposition while the getting was good. So in the interim, Packard was prepared to settle for the assistant to the secretary position, one with substantially less clout than he in fact believed was necessary.

The second problem revolved around the language of the directive, which stated that the responsibilities for management and operational direction of telecommunications resources would remain with the services and defense agencies. In line with many of the Pentagon's unsuccessful centralizing initiatives, the new telecommunications post would serve a coordinating function only. Circumscribing its authority virtually guaranteed the inability of the new post to effect basic system changes.

In many respects, these restrictions were quite intentional. "Stopping a juggernaut" is how one defense journal characterized Defense Secretary Laird's efforts during his first year in office to turn back the clock of centralization that under McNamara's leadership had proceeded far beyond what many desired. Laird immediately began decentralizing decision-making authority from DOD-level organizations, including the director of defense research and engineering and various assistant secretaries, back toward the services. "It is simply foolhardy," Laird said, "not to make maximum use of the great talent, wisdom, and experience available through the Joint Chiefs of Staff and within the services." The role of civilian elements of the Pentagon, he said, himself included, should be one of broad guidance and coordination. On the most important major programs the service secretaries would regularly report to him, he would personally sign off on the programs, and a similar role of oversight and coordination would be assumed by other DOD-level elements. Far from limiting the civilian leadership's control over the military forces, Laird said, this brand of decentralized decision making with periodic centralized oversight would only strengthen civilian control.[18] Making certain that the responsibility for communications management and operational direction remained with the services and defense agencies was thus fully consistent with Laird's desire to stop the centralizing juggernaut. It obviously put him on a collision course with such proponents of centralization as David Packard.

The Blue Ribbon Defense Panel Report

The Blue Ribbon Defense Panel's report was released in July 1970, the month following Packard's downscaling of the WWMCCS computer procurement. The report suggested primarily that the Pentagon effect a major reorganization of its national-level command and control structure. The report went on to blast the Pentagon for a wide range of computer problems and related management shortfalls. The panel pointed out that of the approximately 2,800 computers then in operation throughout DOD, some 36 percent were obsolete first- or second-generation machines running on vacuum tubes or transistors.[19] Far too many of the computer files being maintained were wholly independent and lacked the ability to be interconnected. Computer utilization was poor (they were used less than 16 hours a day), whereas commercial firms had utilization rates that were far better. Finally and predictably, the panel noted the serious lack of ADP uniformity and standardization.

As a result of these problems, the Pentagon was said to be spending at least $500 million more each year than was necessary for its automatic data-processing functions. The Blue Ribbon Panel scored the Pentagon for the lengthy delays that generally occurred between the time a need for new ADP assets was recognized and the time those assets were finally deployed. It also urged that greater management attention be paid to the ADP function.[20] To ensure that attention and to provide a focal point for command and control generally, the panel recommended the creation of an assistant secretary for telecommunications within the OSD.[21]

WWMCCS's Standard Computers

These changes and criticisms naturally provided additional impetus for the purchase of standard computers for WWMCCS. But standardization would not solve every problem, and might actually provoke some. This was especially true in the organizational realm, where one of the program's most serious potential problems involved bureaucratic resistance to changes that almost certainly would cause a loss of discretion and autonomy for major WWMCCS users. It is not that the standardization process completely ignored the specific requirements of WWMCCS's users for hardware configurations, functional software, and peripheral equipment tailored to each user's needs.[22] The point, rather, was that those individual needs, however well met, were to be subordinated to the primary objective of creating a system responsive to the needs of central decision makers. By definition, a big standardization effort like this meant that users would have to give up some independence. Indeed, some grumbling and resistance among system users was considered inevitable.

Inevitable it was, and it was not long before the grumbling reached epidemic proportions. The program timetable for the WWMCCS ADP update called for requests for proposal to be issued to industry during the week of 17 August 1970. On 12 August, however, NORAD's commander in chief, Air Force general Seth J. McKee, wrote to the Air Force chief of staff, Gen John D. Ryan, to outline a series of objections he had to the WWMCCS computers. Principal among these objections was that the program's technical specifications called for equipment that would operate in batch (sequence processing) mode rather than in an on-line interactive mode. Batch processing means that to run a software application, the computer's operating system has to allocate sufficient space in the main memory for the entire program rather than simply the portion actually being executed. Because entire programs are thus constantly being moved in and out of memory, data-processing times are far slower than for computers with circuitry designed for interactive processing. In effect, the batch processing computers "think" one step at a time, in a series of preprogrammed steps, whereas more modern interactive computers can perform many steps simultaneously.[23] While batch processing is normally not a problem during routine operations, the concern was that it could create an internal electronic "traffic jam" during times of high-volume use. By limiting the computer's processing speed and responsiveness at precisely the time they are most needed, such as crises, batch processing represented a serious bottleneck to the flow of essential command and control information.[24]

Batch processing also meant that a commander could not precisely specify the information he wanted. "Say the PLO hijacks a plane and lands it somewhere in the desert," Col Perry Nuhn, the Pentagon's director for information systems and command, control and communications, remarked. "If I've got to provide help, I need to know where the nearest airfields are, how much fuel they have on hand, how long their runways are, and dozens of other support questions." Unfortunately, he noted, computers designed for batch processing cannot answer questions at that level of specificity. "They may have to dump out information about a whole set of nearby countries and all their airfields. And you've got to go through the doggone things by hand."[25] General McKee was deeply troubled by this, considering real-time capabilities to be essential for the performance of NORAD's early warning mission. He requested that the Air Force chief of staff give his personal attention to the matter, and urged that the release of the WWMCCS request for proposal be delayed until the computers it specified met NORAD's operational requirements. Officials at the National Military Command Center articulated similar reservations regarding the ability of the new WWMCCS ADP equipment to satisfy critical mission requirements.

The release of the WWMCCS request for proposal was delayed by these objections, but it was a delay without effect, for the objections voiced by McKee and others were sternly overruled. A response received later that August from the Air Force vice chief of staff pointed out to McKee that the North American Aerospace Defense Command was part of WWMCCS and that WWMCCS hardware and software must therefore be used in NORAD's 427M Program, then being planned as an upgrade to the computational and tracking capabilities at its Cheyenne Mountain headquarters. While some merit was found in McKee's objections to the proposed WWMCCS equipment, and while it was acknowledged that some satisfactory resolution for these shortcomings had to be found, he was informed that whatever the eventual fix might involve, it had to remain within the overall WWMCCS framework.[26]

Requests for proposal (RFP) for the first 15 WWMCCS computers were released to industry on 1 October 1970.[27] Originally, it was intended that RFPs would go to 34 companies, which included all major computer manufacturers. Eventually, this number was cut into half, and 17 RFPs were solicited, although only seven of the companies (IBM, Control Data, RCA, Univac, Honeywell, NCR, and Xerox Data Systems) were considered serious enough contenders that they were expected to submit proposals.[28] As outlined in the RFP, the purchase was to take place in calendar years 1972–73. In what proved to be another unique aspect of the program, one that reflected recent declines in defense spending and David Packard's own cost-cutting emphasis, the request for proposal stipulated that the price tag for the first 15 computer systems should not exceed $46.2 million. If industry failed to abide by the price ceiling, the Pentagon warned, the system's requirements as outlined in the RFP might have to be returned to the drawing board, ensuring additional delays and the need to resolicit proposals because of whatever changes resulted.[29] It was even hinted, darkly and none too subtly, that a failure to adhere to the price limit could result in the program being scrubbed altogether.

Proposals from the competing companies were due on 1 February 1971, and, as one military officer described the situation, "The computer industry waited with bated breath to determine who would get the lucrative contract for computer hardware that will be tied together with such sophisticated software that all the computers will be able to talk with each other."[30] Industry's breath would remain bated for some eight months, a delay that involved more than just the need to properly evaluate the computer vendors' proposals. Given the Blue Ribbon Defense Panel's recommendation that a sweeping reorganization of WWMCCS be effected, it involved also the need to evaluate the larger command and control context within which the WWMCCS computer purchase would take place. Specifically, David Packard wanted to see how the panel's recommendations would play out before making a final commitment to the computer update.

He also wanted time to assess the fallout from a House investigating subcommittee report, released that March, reviewing the effectiveness of worldwide defense communications management, its goals, and the economy and efficiency with which the communications network operated. The report described management as "inefficient and ineffective," pointing the finger of responsibility at the fragmented and overlapping responsibilities that existed within the Department of Defense. The subcommittee proposed a number of specific changes to eliminate the fragmentation and lack of coordination. Leading the list was a recommendation that the secretary of defense centralize DOD communications, including the Defense Communications System and WWMCCS, under the authority of the newly created assistant to the secretary of defense for telecommunications.[31] This office would then be responsible for establishing a centralized accounting system to fully identify communications expenditures.

By the fall of 1971, the shape of the changes-to-come was reasonably clear, and Packard was convinced that the new computers would not be inconsistent with them.[32] The WWMCCS ADP Update Program could proceed, and on 15 October 1971 the Air Force Systems Command's Electronic Systems Division awarded the fixed-price, fixed-quantity contract to Honeywell Information Systems, Inc.

Under the terms of the contract, the Pentagon would purchase 35 computer systems from Honeywell's 6000 series (models H-6060 through H-6080).[33] In fact, the contract involved 76 individual central processing units, since some WWMCCS sites were designed to employ two or more central processing units (CPU) linked together for their enhanced processing capability. The price tag for the 35 computers was right on the Pentagon's target, $46 million, some 35 percent less than the General Services Administration's scheduled price and presumably reflecting a bulk discount by the manufacturer. Installation of all 35 computers was to be completed by the end of 1973, within two years of the contract award. On the surface it seemed like a good deal.

In fact, to many it seemed too good a deal. Skeptics at the GAO viewed the Honeywell bid as unrealistically low, a "buy-in" wherein the computers were intentionally offered below cost so that the company could make its real profits when additions to the system were made.[34] There was ample reason for the skepticism. Consider that the Pentagon's earlier development concept paper had estimated the price of 35 computers purchased in fiscal years 1972–73 to be $91.6 million. Honeywell's contract cost for the same system, however, was much lower.[35] Additional doubt arose because neither the Pentagon nor Honeywell seemed to want it known that Honeywell had been selected as the prime system contractor. That fact only became public when, in a Honeywell parking lot, Sen. Barry Goldwater informally announced that the company had been chosen to provide the WWMCCS computers.[36]

Still worse from the perspective of many system users was that the Honeywell computers already were becoming outdated. The 6000-series computers, first produced in May 1964, were the follow-on to General Electric's GE-600 series. (Honeywell earlier had acquired GE's computer business, and had designed the 6000 series with an eye both to retain GE's customer base and to appeal to new buyers.) The result, according to a computer industry publication, was a "strongly GE-flavored product line that blazed no new technological trails but exploited the current state of the art in a highly cost effective manner."[37] That was in the mid-1960s, and by this time the state of the art had advanced considerably.

Specifically, and as critics had predicted, the problem was that the computers' circuitry was designed for batch processing. Because of this oversight, the hardware architecture—a military version of Honeywell's General Comprehensive Operating Supervisor—was also designed to operate in a batch-processing mode. Batch-processing systems had, of course, been specified in the program's technical specifications, but there was no technical justification for them. When Honeywell was awarded the WWMCCS ADP contract, a number of computer manufacturers, Honeywell among them, already were marketing comparable systems to operate in an interactive mode. But despite the availability of more adequate alternatives and despite the deep reservations of many key WWMCCS users regarding this type of equipment, including NORAD's commander and the JCS, the decision was not reversed. The purchase of the Honeywell 6000s proceeded anyway, and it was a decision whose ramifications would be felt within the WWMCCS community for years to come.

The installation of the WWMCCS computers was completed on schedule. Installation began in March 1972 at SAC headquarters in Nebraska and was completed in December 1973. (An additional site was subsequently added at Taegu, South Korea, which was completed in May 1975.) WWMCCS automatic data processing had become a reality, and it was in many ways a reality without precedent. But because of the novelty of such a large system and having to address and overcome the host of technical and operational problems that inevitably accompanied its implementation, considerable work remained to be done. Training procedures had to be developed for system users so they could understand and rapidly come up to speed on the system. To discuss problems as they occurred and to provide a forum for propounding possible solutions, semiannual conferences were established. User support also was critical. In one particularly amusing example of the support provided by the WWMCCS ADP community, one of the major WWMCCS sites immediately began to experience mysterious and apparently completely random signal interruptions. Analysts were rushed to the scene to examine possible sources of the interference. Several exhausting and frustrating months passed before someone discovered that the interference was caused by helicopters passing through the microwave circuit carrying the signals as they took off and landed.[38]

Not all of the problems with the new WWMCCS computers were resolved as easily, however, and almost as soon as installation began, major difficulties with the Honeywell 6000s began to surface. To get around the problem of limited space in the computers' main memory, for example, additional memory capacity was added wherever needed, the amount depending upon the specific user's data-processing requirements. While this solved the immediate problem of capacity, permitting a single computer to handle larger blocks of data than before, it introduced new difficulties when the computer was called on to exchange information with other WWMCCS sites. Because not all sites had made identical upgrades to their computers' memories, users frequently found that exchanges could not take place unless the receiving site happened to have the same memory capacity as the originating site. To overcome this shortcoming and a host of other problems, Honeywell and other contractors were repeatedly called on to provide upgrades and fixes. Indeed, within the first few years of the new computers' operation, more than 60 changes to the original WWMCCS contract were negotiated.[39] Almost all of these were expensive, and many of them proved to be less than completely successful.

Major problems also surfaced with system software, and not all members of the WWMCCS community were using it. The complexity and limitations of the WWMCCS standard software had led many individual system users to feel, often quite correctly, that their command and control requirements were being inadequately served. Therefore, many users found it necessary to develop their own software applications to work around the limitations of the system. Because this type of software was developed locally by its users, consideration of other users' requirements was limited or nonexistent. Duplication of effort was frequent and excessive costs unavoidable. To make matters worse, because locally developed software was by definition not used throughout WWMCCS, the ability of its users to exchange information with others was often severely impaired. This was precisely the situation that had prevailed during the 1960s, and precisely what the WWMCCS ADP

Update Program had been intended to remedy. Years later that remedy was still not clearly in sight. Thus, at precisely the time that the star of centralized communications management was on the rise, just when national-level command and control was assuming greater importance in defense planning, WWMCCS (its recent upgrade notwithstanding) would continue to develop in ways inimical to the needs of centralized decision makers.

Prototype WWMCCS Information Network

On 7 September 1971, more than a month before the contract for the 6000-series computers was awarded to Honeywell Information Systems, the Joint Chiefs of Staff issued JCS Memorandum 593-71, "Research, Development, Test, and Evaluation Program in Support of the Worldwide Military Command and Control Standard System." By this time, the chiefs and other top-level officials were fully aware that the computing capabilities called for in the WWMCCS ADP Update Program's technical specifications were inadequate. Something more was clearly required, and the joint chiefs' memorandum proposed that a first step in getting it was to develop what they called a Prototype WWMCCS Intercomputer Network (PWIN—pronounced pee-win).[40]

The idea of computer networking was relatively new at the time. Experience with actual network operations based on the packet-switching concept was then being gained through the ARPANET, the first of whose nodes came on line in 1969 with spectacularly successful results.[41] The logic behind packet switching was that a message from an originating host computer would be broken into a number of packets, each containing a maximum of 5,000 bits of information (625 characters). Each packet was then given a header by the computer to identify the message's sender, recipient, and other information. Using complex network control protocols, the computers would then independently route the packets from their point of transmission to a series of network nodes, called packet switches, which were digital computers instead of the manual or electromechanical switches used in other types of communications systems. After arriving at a switch, the packets' headers would be automatically examined, and, after that brief delay, forwarded along any available path to the next node. The technique was known as "fail softness," since if some of the network's circuits or nodes were out of service, messages would be routed another way.[42] This procedure was dramatically different from circuit switching, which involves the use of dedicated circuits, and from store-and-forward switching of the type used in AUTODIN; in the latter case, complete messages had to be accumulated at each switch before retransmission. When the packets arrived at their final destination, they would be reassembled into a complete message by the computer and forwarded to the recipient. At that time, acknowledgment of receipt also would be automatically transmitted to the sender, and, in the event the message was incomplete or otherwise incorrectly received, the sender would be instructed to retransmit.

In principle, these characteristics made for a faster, more capable network, and despite the circuitous routes the packets might travel and their numerous stops at the packet switches, message delays in the ARPANET averaged only about one-quarter of a second.[43] A network employing packet switching also promised far greater communications security than one in which entire messages were transmitted intact. But however promising the technology appeared, additional network experience more directly relevant to the operational demands of WWMCCS was considered essential; after all, the ARPANET linked the computer systems at a number of research institutes, laboratories, and universities, not command and control facilities.[44] Precisely what command and control functions should be supported by networking? How should they be supported? What would the benefits and liabilities be? No one was really certain, and that was where PWIN came in.

Creating PWIN as a test bed to determine the operational benefits of networking for WWMCCS made abundant sense. PWIN could determine the specific characteristics that an intercomputer network would require to support the command and control function, and it could assess a method for applying the technology of computer internetting to WWMCCS—all without a full-scale advance commitment to the networking concept.[45] This development was definitely in keeping with Deputy Secretary of Defense Packard's "fly before you buy" approach to systems acquisition, whereby working prototypes would be built and thoroughly tested prior to the award of production contracts. His was a responsible approach to systems acquisition—one that had, for example, given the Air Force two of its most successful aircraft, the F-16 fighter and the A-10 attack plane. (It was also an approach that would be all but abandoned by the Pentagon after Packard left office.)[46] Following this approach, if networking proved inappropriate for WWMCCS, things had been kept small scale and not much was lost. But if the prototype proved successful, as was hoped, it would constitute a baseline system, a foundation for the operational network that was to follow. It was a plan that would be given considerable impetus when, just three months later, Packard established the modern WWMCCS structure.

Notes

1. Brooke Nihart, "Coordination of C^3 Giving Way to Tight Control?" *Armed Forces Journal* 107 (17 August 1970): 22.

2. W. J. Baird, "The Office of Telecommunications Policy," *Signal* 25 (5 September 1970): 5.

3. George A. Pappas Jr. and T. G. Adcock, "The Role of Satellite Communications in the DOD," *Signal* 27 (September 1972): 27.

4. "U.S. Telecommunications Reorganization," *Signal* 24 (April 1970): 44.

5. Clay Whitehead, "On the New Office of Telecommunications Policy: Where Communicators Communicate," *Signal* 25 (22 February 1971): 20.

6. Charles E. Shepherd, "The Office of Telecommunications Policy—Central Issues," *Signal* 29 (February 1975): 18–19.

7. Ibid., 19.

8. House, Committee on Armed Services, Armed Services Investigating Subcommittee, *Review of Department of Defense Worldwide Communications, Phase I*, 92d Cong., 1st sess. (Washington, D.C.: Government Printing Office [GPO], 10 May 1971), 17. (Hereafter cited as *Phase I Report.*)

9. Ibid., 20.

10. Hubert Summers Cunningham and William Edward Kenealy, "The Joint Chiefs of Staff and Command and Control," *Signal* 29 (March 1975): 16.

11. *Phase I Report*, 22.

12. Ibid., 20.

13. Ibid., 22.

14. General Accounting Office, *Accounting for Automatic Data Processing Costs Needs Improvement* (Washington, D.C.: General Accounting Office [GAO], 7 February 1978), 13.

15. *Phase I Report*, 22.

16. Samuel L. Gravely Jr., "The DCA—A Rock and a Hard Place," *Signal* 34 (April 1980): 9.

17. Nihart, 22.

18. "Mel Laird: Coach, Quarterback, or Both?" *Armed Forces Management* 16 (October 1969): 35.

19. "WWMCCS: Report, One Year Later," *Government Executive* 3 (January 1971): 19.

20. Benjamin F. Schemmer, "DOD's Computer Critics Nearly Unplug Ailing Worldwide Military Command and Control System," *Armed Forces Journal* 108 (21 December 1970): 22.

21. House, Committee on Armed Services, Hearing before Subcommittee on Investigations, *Review of Department of Defense Command, Control and Communications Systems and Facilities*, 94th Cong., 2d sess. (Washington, D.C.: GPO, 1977), 29.

22. Joseph Volz, "Revamped World-Wide Computer Net Readied: WWMCCS Slimmed Down to 15-System Program," *Armed Forces Journal* 107 (27 June 1970): 21.

23. Frank Greve, "Pentagon Calls Super-Computer a 'Disaster,'" *Parameters* 10, no. 1 (March 1980): 96.

24. GAO, *The World Wide Military Command and Control System—Evaluation of Vendor and Department of Defense Comments*, LCD-80-22A (Washington, D.C.: GAO, 1980), 5.

25. William J. Broad, "Computers and the U.S. Military Don't Mix," *Science*, 14 March 1980, 1186.

26. GAO, *NORAD's Information Processing Improvement Program—Will It Enhance Mission Capability?* LCD-78-117 (Washington, D.C.: GAO, 1978), 45–46.

27. Herbert B. Goertzel and James R. Miller, "WWMCCS ADP: A Promise Fulfilled," *Signal* 30 (May/June 1976): 58.

28. Nihart, 23.

29. George Weiss, "Restraining the Data Monster: The Next Step in C^3," *Armed Forces Journal* 108 (5 July 1971): 28–29.

30. Carl J. Weinmeister III, "Command and Control Means Teamwork and Responsibility," *Defense Management Journal* 8 (April 1972): 41.

31. *Phase I Report*, 4.

32. "WWMCCS Report,"19.

33. Goertzel and Miller, 58.

34. Broad, 1186.

35. Goertzel and Miller, 58.

36. Carol Hamilton, "Worldwide C^2 System Networks Strategic Data for Joint Chiefs," *Defense Electronics* 20 (June 1988): 53.

37. GAO, *The World Wide Military Command and Control System—-Major Changes Needed in its Automated Data Processing Management and Direction, Report to the Congress*, LCD 80-22 (Washington, D.C.: GAO, December 1979), 38.

38. Ibid., 57.

39. Goertzel and Miller, 61.

40. *The World Wide Military Command and Control System—Major Changes Needed in its Automated Data Processing Management and Direction*, 81.

41. Stephen T. Walker, "Department of Defense Data Network," *Signal* 37 (October 1982): 42.

42. Edgar Ulsamer, "Machine Intelligence Shapes Global C^3 Nets," *Air Force Magazine*, July 1977, 72.

43. David C. Russell, "C^3 Research at DARPA," *Signal* 31 (March 1977): 39.

44. Richard Hartman, "Internetting and ADP Bolster Strategic C^3," *Defense Electronics* 11 (September 1979): 55.

45. Lee M. Paschall, "Command, Control and Technology," *Countermeasures*, July 1976, 42.

46. Richard A. Stubbing and Richard A. Mendel, "How to Save $50 Billion a Year," *The Atlantic Monthly*, June 1989, 56.

Chapter 8

The WWMCCS Council and the Modern WWMCCS Structure

A prominent aspect of Deputy Defense Secretary Packard's tenure at the Pentagon was his "personal crusade" to improve communications, command, and control throughout the Department of Defense. In his many bureaucratic battles, Packard focused considerable, specific attention on WWMCCS, and he was the driving force behind efforts during the early 1970s to rationalize its management structure. But his was a crusade that continually floundered in a sea of Pentagon resistance to his centralization efforts, and, having run hard upon the shoals of bureaucratic intransigence, Packard abruptly resigned from office in December 1971, improbably citing "strictly personal reasons" for his departure.[1] Immediately prior to his resignation, however, Packard took action. On 2 December 1971 he issued a revision of DOD Directive 5100.30, WWMCCS's founding document. Working closely with JCS chairman Adm Thomas Moorer, Packard sought to clarify responsibilities and centralize authority within the system. In so doing, he defined the modern-day WWMCCS structure. If Herbert Goertzel deserved to be called Mr. WWMCCS during the system's formative years, Packard, more than anyone else, merited the sobriquet during his three years in government, 1969 through 1971.

Packard's directive first redefined the overall mission of WWMCCS. "The World Wide Military Command and Control System," the directive began, was the system that "provides the means for operational direction and technical administrative support involved in the function of command and control of U.S. military forces."[2] The directive delineated the system's major missions and ordered them hierarchically. WWMCCS's primary mission was to support the National Command Authorities. It would provide strategic warning, intelligence, and other pertinent information upon which timely and appropriate decisions could be reached. Once the decision-making process had been completed, WWMCCS would constitute the mechanism through which the decisions were implemented. It would be used for applying resources to the military departments, assigning military missions, and providing direction to the unified and specified commands. To this end, the National Military Command System was formally designated WWMCCS's priority component. The directive went on to specify that the NMCS should be the most responsive, reliable, and survivable system possible, given available resources.[3]

WWMCCS's other purpose, one clearly designated as subordinate to the fulfillment of national-level requirements, was to support the command and control systems of the unified and specified commands, the military services and their individual service commands, and several defense and nondefense agencies—and pretty much in that order. The rationale behind establishing a hierarchy of importance among key organizational subunits, with the National Military Command System heading the list, was simple. As Packard put it, "Instead of the local commanders now having as their first priority to design their command system to meet the requirements of their mission, they first have to have a design to meet the requirements of the national command system."[4] Only then, as a secondary concern and not to interfere with the primary mission, could they design systems to meet their specific mission requirements. Now the emphasis was clearly on WWMCCS as a national level system, particularly concerning control of the strategic nuclear forces.

Shortly after the revised directive was released, Packard elaborated on the need for setting priorities within the overall WWMCCS mission at a meeting of the Aviation Space Writers Association. Reviewing the evolution of WWMCCS over the preceding decade, he explained to the assembled journalists how a series of directives concerned with national-level command and control issued early in the 1960s had contained two major emphases. The first was that the unified and specified commanders should have the authority to build their own command and control systems in ways most responsive to their specific mission requirements. This they had done, and reasonably well, Packard said, particularly the Strategic Air Command, and, to a somewhat lesser extent, the Navy. With regard to the directives' second emphasis, however, the linking of these various elements into a system responsive to the needs of the NCA, things had not worked out nearly so well.

When WWMCCS was established, the hope had been for a coordinated and organized system. The explicit intent in bringing the unified and specified commanders into the development process had been to produce greater commonality of command and control assets among those commands. The plan was to allow the CINCs considerable latitude in developing their command and control systems while at the same time exhorting them to be responsive to the needs of the national-level leadership.[5] The problem with this approach to system development was that it simply did not work. During the first decade of WWMCCS's existence, the forces of decentralization held sway. The unified and specified commanders, responding heavily to the demands of the military services of which their commands were constituted, developed and deployed command and control assets to meet their individual needs. It was a situation of developmental ad hockery in which command and control systems that were tailored to the requirements of system subunits proliferated without adequate consideration for how they might interact, or fail to interact, with other systems. It was a case of subunit optimization at the expense of the national leadership, of individual systems going their own separate ways.[6] During the 1962–71 period, the only element of WWMCCS that seemed truly designed for national-level decision makers was the National Military Command System itself.[7] The overall system that resulted from this subunit-dominated process was variously, and rightly, described as a "command and control federation," a "loosely knit federation,"[8] a "loosely-defined, loosely-gathered federation of subsystems with no clear central purpose,"[9] and similar characterizations.

The inevitable consequences of this confederated approach, according to Packard, were that "communications sometimes didn't work, that the messages generally got mixed up in coming out to [the] field, to the local command, and in some way didn't get into the central communications system, which in fact, works very well."[10] Unscrambling this sentence, what Packard was saying was that the decentralized service-specific systems worked well. Many of the more centralized systems also worked well. The problem was that none of them worked well together, a situation that many officials believed the nation no longer could afford. They felt what was needed was some overarching rationale, some larger structure or architecture, that would guide the development of command and control systems and ensure their interoperability and standardization.

In theory, the Defense Communications Agency was to have provided the organizational focus that was lacking in command and control. But as an earlier congressional study of the Defense Communications System had pointed out, DCA had its own share of problems, ones that appeared to be a microcosm of those found in the broader defense establishment. The problems with the DCS were more organizational than technological, and preeminent among them was inadequate managerial control. Better management, the study concluded, would go far toward addressing the widespread confusion, unnecessary duplication, and fragmented areas of responsibility that afflicted the system. What was needed was a "proper mix" of people, cooperation between participating organizational subunits, and dynamic leadership not only at the DCA but also at the Offices of the Secretary of Defense and the Joint Chiefs of Staff.[11] And better management was precisely what David Packard had in mind in issuing DOD Directive 5100.30. He wanted to take this loosely knit confederation and forge from it a truly responsive capability worthy of the name *World Wide Military Command and Control System.* Whereas before WWMCCS had essentially been an instrument of the unified and specified commanders, who in turn were heavily influenced by the military services, it would now constitute a direct link between the NCA and the operating forces. To effect this change, the directive took WWMCCS, formerly managed by the JCS as a corporate body, and made it the sole responsibility of the JCS chairman.[12]

In light of the reorientation of WWMCCS's mission, one of the most important changes wrought by DOD Directive 5100.30 was definitional. Specifically, the concept of the NCA was redefined so that it now consisted of only the president and the secretary of defense, or their duly deputized alternates or successors. Before this time, the NCA had included the JCS and the unified and specified commanders. The new definition did not change the formal chain of command, however, which ran from the president, through the secretary of defense, to the joint chiefs, then to the commanders of the unified and specified commands, and to the operational forces in the field. Packard's quite deliberate intent in removing the military commanders was to simplify the processes of decision making and order execution.[13] To this end, the new definition created a sort of bureaucratic loophole, a mechanism that could be employed at the highest levels of national decision making to streamline the chain of command by bypassing the joint chiefs and the unified and specified commanders.

Packard's redefinition of the National Command Authorities was not, however, designed to have decisions concerning every minor incident or event made at the top of the military hierarchy. Rather, it simply emphasized that such capability should exist, to be exercised at the discretion of the NCA. With this sort of flexibility, the top leadership could examine the requirements of a situation and decide whether centralized or decentralized decision making was more appropriate. The level of control could then be adjusted, like a light switch.[14] The hope was that the new arrangement would capitalize on the advantages of both centralization and decentralization.

Given the crises and other time-sensitive situations that characterized the nuclear age, this sort of streamlined decision-making approach appeared to make abundant sense. During peacetime, the flow of data to the national leadership would take place in a routine fashion. But consider that military and nonmilitary organizations tend to undergo structural changes when confronted by conditions of uncertainty and stress, and such changes tend to lean toward simplification. When crises occur, central leadership wants to focus directly on the area, requiring that those standing in the way step aside.[15] Packard's directive simply recognized this tendency and codified it. Now, in principle at least, all intervening levels of command between the top levels of government and the operational forces in the field could be eliminated, and central leaders could have the same information as commanders on the scene. (Those who were normally in the chain of command but were cut out of the action in this way also would have the same information available to them, but for standby purposes only.)[16] Indeed, the extreme case permitted under the directive was precisely the White House-to-foxhole communications scenario derided by the military services and their congressional supporters in times past. DOD Directive 5100.30 now made it mandatory that a capability exist to communicate directly with the operating forces, whether conventional forces involved in a crisis in a remote part of the globe or, more importantly, the strategic forces responsible for executing the single integrated operational plan.[17]

All of this substantially blurred earlier distinctions between strategic and tactical command and control. If indeed the entire communications capability was integrated to provide national level leaders with detailed knowledge of the situation and the ability to direct the military forces, existing strategic and tactical command and control systems had to become fully interoperable. If the old thinking held that strategic and tactical systems were completely separate and unrelated functional entities, then the new thinking would remove all technological or organizational barriers between the strategic and tactical worlds.[18] The idea was that these systems had to be standardized and essentially transparent.

David Packard's willingness to undertake what essentially was, in effect, a frontal assault on the military hierarchy arose from his conviction that the chain of command itself was to blame for many of the communications problems of the past. The way things stood, messages moving up or down the military hierarchy necessitated multiple reformattings and retransmissions, making delays unavoidable. Packard and his supporters considered such delays intolerable during the critical, time-sensitive situations that characterized the nuclear age, and they were determined to do something about it.[19]

A series of other substantive changes were also promulgated in this revision of WWMCCS's founding document. Major system responsibilities were allotted and roles were reassigned, resulting in the establishment of the modern-day WWMCCS management structure. The directive divided responsibility for the system among several interested parties. The chairman of the JCS was given responsibility for the operation of the National Military Command System, which included developing and validating requirements for the various elements of the NMCS itself, ascertaining the command and control requirements for the unified and specified commands, and ensuring interoperability among all of these by developing an overall WWMCCS objectives plan. The JCS organization was given responsibility for ongoing evaluation of WWMCCS.[20]

Another star in the revamped WWMCCS management firmament was the newly created post of assistant secretary of defense for telecommunications. Recall that without legislative action, David Packard had created the position of assistant to the secretary of defense for telecommunications, providing a bureaucratic springboard for a full assistant secretary position. As Packard had anticipated, the Blue Ribbon Defense Panel recommended the creation of the new assistant secretary position. Packard then moved swiftly to effect the upgrading, and his efforts bore fruit when Defense Secretary Laird informed Congress that consolidated management of defense communications created "an urgent requirement to upgrade his telecommunications manager to the rank of Assistant Secretary."[21] The Armed Services Committee responded favorably to Laird's appeal. Legislative action approving the change was completed in December 1971, and the new assistant secretary of defense for telecommunications was sworn in the following month. His responsibilities included advising the secretary of defense on all matters of telecommunications systems design, development, procurement, and performance. The only area exempted from his slate of duties was automatic data-processing equipment, for this responsibility was assigned to the DOD comptroller. The new assistant secretary also would be a major player in a brand new organizational entity created by DOD Directive 5100.30: the WWMCCS Council.

The WWMCCS Council

The WWMCCS Council was intended to be a management body that would act in effect as a WWMCCS board of directors.[22] Like any other board, it would be a high-level decision-making body, and, as such, would not be concerned with the details of day-to-day operations. Also, like any other board, its members were heavyweights, consisting of the deputy secretary of defense, the chairman of the Joint Chiefs of Staff, the newly created assistant secretary of defense for telecommunications, and the assistant secretary of defense for intelligence (a position also established by DOD Directive 5100.30, and given responsibility for advising the secretary of defense on matters relating to strategic warning and intelligence). The council chair was given to the new telecommunications assistant secretary to lend greater authority and prestige to his position.

The WWMCCS Council was the brainchild both of Packard and Admiral Moorer, the JCS chairman, and their reasons for creating it were several. First, the council was intended to provide general policy guidance to the JCS regarding the operation and future development of WWMCCS. Second, it was to serve as an agent of centralization quite deliberately intended to "bypass most of the ants on the Pentagon log," cutting through bureaucratic red tape so that improvements could be brought rapidly to the system.[23] Third, it was to serve an adjudicatory function, helping to resolve policy conflicts that might arise. It was also intended that the council would review the results of system testing, a clear sign that effectiveness criteria would now be promulgated at the top rather than at the subunit level. A fourth reason for creating the WWMCCS Council was to facilitate work on a number of projects considered necessary for strategic command and control modernization. Opposition was viewed as inevitable because the shift in emphasis toward strategic concerns necessarily implied a shift in resources, and Packard's intent was to have the council serve as a high-level advocate for these projects in future budgetary wars.[24] In summary, Packard's whole thrust was to implement a centralized, top-down management structure focusing on the needs of the NCA as a priority for crisis management. That structure would replace the existing bottom-up approach in which the majority of initiatives were taken at lower levels by the operating commanders whose projects currently received the lion's share of resources.[25]

The increasingly strategic emphasis of WWMCCS and the need for energetic efforts to improve it followed directly from the strategic emphases of the Nixon administration. As President Nixon had rhetorically queried in a speech to the Congress in early 1970, "Should a President, in the event of a nuclear attack, be left with the single option of ordering the mass destruction of enemy civilians, in the face of certainty that it would be followed by the mass slaughter of Americans?"[26] Elaborating on the implications of this strategic shift, the president noted in a subsequent speech: "We must ensure that we have the forces and procedures that provide us with alternatives appropriate to the nature and level of the provocation. This means having the plans and command and control capabilities necessary to enable us to select and carry out the appropriate response without necessarily having to resort to massive destruction."[27] Such capabilities would be enshrined in doctrine in January 1974 with the issuing of National Security Defense Memorandum 242, which formally dispensed with the earlier doctrine of assured destruction. This doctrine was still in place, although increasingly disregarded since McNamara's days in the Pentagon. Elevated in its place was flexible response, the United States's de facto doctrine. With flexible response, the United States was no longer bound to a massive, reflex reaction to a nuclear provocation. Throughout a range of crisis and wartime conditions, responses would henceforth be calculated, modulated, and precisely controlled. Providing such capabilities was now formally the job of WWMCCS, further underscoring that command and control would now be considerably more than a technical detail.

Deputy Secretary of Defense William P. Clements Jr. arrived in Washington in early 1973, at the beginning of Nixon's second term. A former chairman of the board of Sedco, Inc., an oil-drilling firm, and a member of the Blue Ribbon Defense Panel whose earlier report had called for major changes in command and control, Clements quickly became baffled with the World Wide Military Command and Control System. One of his responsibilities was as chairman of the WWMCCS Council, and in his words, "It took a year to get common understanding of what we were talking about."[28] As his understanding grew, so did his realization that a major system overhaul was required. By the end of that year, Clements and the other council members agreed that WWMCCS, as it was currently being developed, lacked a coherent organizing logic. Its overall goals were vague or undefined. Growth had occurred in a conceptual vacuum, in a context of managerial nebulousness and appropriations ad hockery. As individuals concerned with total system integration and coordination, the WWMCCS Council concluded that WWMCCS failed to provide a responsive command and control capability to the national leadership.

A good deal of the concern emphasized that WWMCCS was not easy to define. What precisely *is* WWMCCS? During the previous decade, two major schools of thought emerged, the first and more general being that WWMCCS was really a "concept"; less a system in its own right than a guiding set of principles for how to use existing command and control assets.[29] Such a definition is in obvious ways naturally genial to proponents of decentralized control.

The second view held that WWMCCS was in fact real, a "system of systems" or metasystem, a vast assembly of national-level capabilities that cut across service boundaries and subsumed beneath its compass a large number of assets. These included many, although not all, of the assets developed and deployed by the military services, commands, and defense agencies.[30] They also included a number of WWMCCS-unique circuits, equipments, and subsystems that permitted the NCA both to communicate with subordinate commands and to execute time-sensitive operations up to and including the single integrated operational plan.[31] In this macrosystem view, WWMCCS was far more than an organizing principle (although it was certainly that as well). It was real, however vast. It extended from the White House to the foxhole, encompassing the individual communications systems of virtually every echelon of military command, and it was devoted—although by no means exclusively—to the command and control of the strategic nuclear forces.[32] For obvious reasons this view has proved attractive to proponents of centralization.

The lack of agreement as to what WWMCCS actually was can be attributed largely to the essential nature of the system, to its size and complexity, and to the myriad technologies and groups that, in whole or in part, fell within its compass. A clear specification of boundaries, of who was in and who was out, obviously was of central relevance for the WWMCCS Council, for without this specification, "one man's internal system turns out to be another man's external system."[33] At first glance it might seem appropriate to include certain defense agencies and groups—for example, the Defense Communications Agency—as WWMCCS subunits. DCA is, of course, a key

WWMCCS actor both through its operation of the Defense Communications System, a significant WWMCCS subsystem, and its responsibility for a variety of WWMCCS support functions. But it is also responsible for a variety of non-WWMCCS programs, systems, and facilities, meaning that it is not synonymous with WWMCCS.[34] Determining where DCS ends and WWMCCS begins is thus far from straightforward.

Or consider the military services, which operate the WWMCCS command centers, technical systems for early warning, data-processing facilities, and other assets. At first glance, it might seem appropriate to include them within the boundaries of the system. Yet the services also have myriad interests, missions, and obligations that have little or nothing to do with WWMCCS. Additionally, the personnel, facilities, and other resources they contributed to WWMCCS frequently were called upon to do double duty, serving service-unique as well as joint-service needs. Because of these characteristics, it is probably most useful to consider the services in much the same way as such defense agencies as DCA—as partially overlapping, rather than synonymous with WWMCCS. In addition, it might be argued that a large number of agencies and organizations of the OSD and the JCS organization concerned with joint-service operations should be designated a part of WWMCCS. But what about defense-related industries, civilian institutions such as research universities and think tanks, and Congress, the group with ultimate power of the purse?

So in answering the question regarding which groups, systems, agencies, and so on, are part of WWMCCS, the answer was "all of them" to some degree, under some circumstances, at some times. The common-user switched networks of the DCS, for example, while intended primarily to meet the routine needs of the operating military forces, could be used for a variety of nonroutine functions also. That is, only part of them was WWMCCS, or they were WWMCCS only part of the time. Other subsystems were full timers yet infrequently used, special purpose elements committed to such unique functions as the various elements of the Minimum Essential Emergency Communications Network. Such was the lack of conceptual clarity confronting the WWMCCS Council that at times it was far from apparent where WWMCCS ended and other systems began. (Indeed, the scale of the system is potentially so vast, its boundaries so all-encompassing, that virtually all Pentagon elements, and myriad other elements as well, are possible candidates for inclusion. WWMCCS rapidly expands to incorporate the world.)

The WWMCCS that confronted Clements and his fellow council members may well have been inevitable given its subunit-dominated character and the evolutionary approach that had governed its development. Many of WWMCCS's varied subsystems had been introduced as a quick response to an increased threat, because of individually perceived subunit needs, or to take advantage of sudden advances in technology. But if serious deficiencies existed with the system's ability to serve the needs of the central leadership, they did not pose a problem from the perspective of influential system subunits, whose needs were often served quite well. If WWMCCS resource acquisition failed to take place in a context of clearly specified system goals, guided by some larger vision or central plan, this is not to say that goals did not exist. Goals there were, and in abundance, but they tended to represent subunit needs and interests. More specifically, they represented the needs and interests of the military services, whose missions were frequently at cross-purposes with those of other subunits and with any broader system requirement for centralized control. Consequently, coordination was problematical, and duplication was commonplace. That it took Clements a year to understand "what we were talking about" is thus hardly surprising.

The demands imposed by a truly functional WWMCCS were considerable. Command centers, the strategic nuclear forces, and the communications channels that linked them together had to be survivable. There was the requirement that communications channels be secure, permitting discussions with various force elements and with our allies as alternative courses of action were considered. Flexible response also implied the ability to change plans rapidly, to retarget weapons as conditions changed, and to withhold weapons for future use. Just as Packard and others had anticipated, this indeed represented a substantial blurring of the boundary between strategic and tactical operations. The purpose of the WWMCCS Council was to ensure that flexible response found expression in command and control technologies and organizational structures, ultimately allowing the National Command Authorities to control escalation, conduct nuclear war, and negotiate termination of hostilities on conditions acceptable to the United States. This meant that earlier emphases on cost effectiveness, primarily by the Defense Communications Agency, would now be downplayed. As DCA's director put it, for those assets implicated in WWMCCS, the criteria of effectiveness would be "survivability, reliability, security, and cost—and in that order."[35]

Given the revised doctrinal context, the WWMCCS Council's concerns focused on a series of projects. One of the most important of these involved improving the National Military Command System. Through this time, the NMCS had consisted of four major command centers: the National Military Command Center in the Pentagon, the Alternate NMCC at Fort Ritchie, the National Emergency Airborne Command Post, and the National Emergency Command Post Afloat (NECPA). Each of these system elements was slated for improvement except for the NECPA, the need for which had been eliminated, first informally by the Navy, which wanted no part of the program, and later by DOD Directive 5100.30.[36]

As a first step, the Pentagon facility was to be substantially upgraded. Past crises had led many in the command and control community to conclude that the existing NMCC was simply too small to accommodate adequate staffing. In addition, major technological deficiencies existed in the equipment used at the center, including equipment being used for automatic data processing, information display, and secure voice and conferencing capabilities. To correct these shortcomings, the council outlined an initial improvement project, to be undertaken by the Air Force, that would effectively double the size of the facility and introduce a large number of technical improvements, including automatic distribution for incoming messages, automated access to the WWMCCS computer database, and televisual display of critical data.[37] When these improvements were completed, a follow-on phase would commence in which a major element of the National Military Intelligence Center (NMIC), operated by the Defense Intelligence Agency, would be moved into the newly expanded facilities. The WWMCCS

Council's point here was to provide direct connections with the NMIC for exchange of data, fusing the operational and intelligence arms of the Department of Defense in support of the NCA. A number of problems were encountered in meeting these initiatives, including the purchase of equipment whose usefulness had not been determined and resistance by the military services to the centralizing tendency inherent in a truly viable NMCC.

The Alternate National Military Command Center at Fort Ritchie was also slated for an upgrade. The major effort here, to be performed by the Army, involved providing the center with a more survivable communications infrastructure.[38] Finally, beneath the rubric of the National Military Command System, the WWMCCS Council gave its priority concern to the Advanced Airborne National Command Post (AABNCP) program. The council wanted to replace the current fleet of three National Emergency Airborne Command Post planes and four SAC Looking Glass planes—the modified Boeing 707s designated EC-135—with seven reconfigured Boeing 747s designated E-4s. Because of its far greater endurance, its ability to carry a larger battle staff and additional electronic equipment, its more powerful communications capabilities, and its hardening against nuclear effects, the AABNCP program was a high priority of the council. In fact, one of the first actions taken by the council was to recommend the acquisition of the new airborne command posts, the first of which was to be delivered in December 1974.[39]

Other WWMCCS-related improvements concerned the council, including the upgrading and expansion of various elements of the Minimal Essential Emergency Communications Network. Relevant programs here included upgrades to the various very low frequency systems for communicating with the ballistic missile submarine force and a more aggressive pursuit of an extremely low frequency (ELF) capability. Of interest were efforts to expand the Air Force's Emergency Rocket Communications System and a program to replace that service's aging Emergency Message Automatic Transmission System (EMATS) with a modern space-based communications system that would later come to be known as the Air Force Satellite Communications System.[40] In addition, there were programs to upgrade the nation's early warning system. Here, the WWMCCS Council supported Air Force efforts to acquire several new phased-array radars for the detection of sea-launched ballistic missiles and for upgrades to the aging Ballistic Missile Early Warning System.

In light of the doctrinal shift then taking place, the WWMCCS Council became concerned with the need to develop a communications satellite capable of surviving a nuclear exchange. Here, the council threw its support behind efforts already under way to harden and otherwise enhance the survivability of the Defense Satellite Communications System. With an eye to the future, attention was then given to developing an entirely new satellite intended from the first to perform in a war-fighting context. A decade later this effort would result in the extraordinarily ambitious Military Strategic and Tactical Relay satellite program. Two experimental satellites were then under development by the Massachusetts Institute of Technology's Lincoln Laboratory that were intended as a technical test bed for such a system (known as LES 8 and 9) which would be orbited in 1976 after considerable delays.[41]

A final council concern was the WWMCCS ADP Update Program, already well under way at the time DOD Directive 5100.30 was reissued, but with the installation of the first computer system, at SAC headquarters, still more than a year away. Here, the council immediately assumed its intended leadership and advocacy roles, facilitating the acquisition of the standard WWMCCS computers and actively promoting the development of standard system software to enhance interoperability. Interoperability of system elements was a concern of long standing, of course, and numerous DOD guidelines had been released to address it.[42] What the WWMCCS Council did was to take these issues and move them to the front burner, making standardization and interoperability active considerations in future system development. The goal was a system in which information and commands originating anywhere in the system could flow unrestricted to any other point. A corollary goal was to ensure that the council appointed a project manager to oversee WWMCCS automatic data processing.

Since computers communicate with one another in digital rather than analog form, the council saw as a closely related concern the need to substantially increase the number of digital communications channels in the DCS to accomplish an all-digital system. A majority of usable DCS channels at that time was allocated for voice use, and many were either nearing the threshold of their capacity to carry data or had already passed their limit. Additionally, many channels were not designed to carry the type of high-speed, high-quality data generated by modern computers, and digital transmission was viewed as the only economical means of increasing the capabilities of WWMCCS. In addition to the higher data rates made possible by digital transmission, there would be improved transmission quality, lower error rates, greater resistance to jamming, and enhanced communications security.[43] Digital transmissions of all types—voice, record, data, and facsimile—were considered inevitable for the communications systems of the future. The WWMCCS Council became the principal advocate for developing AUTODIN II, the planned follow-on to DCA's common-user AUTODIN, and, more importantly, a new WWMCCS Intercomputer Network (WIN) that would make use of the upgraded AUTODIN capabilities.

PWIN Design

Transitioning to the WIN of the future is where the Prototype WWMCCS Intercomputer Network came in. Following the reissue of DODD 5100.30, the joint chiefs had directed the Defense Communications Agency to prepare plans for the prototype network's design and development. DCA promptly set to work, establishing a PWIN element within its joint technical support activity (JTSA—previously called the Joint Technical Support Group). With more than one hundred assigned personnel, the JTSA already had taken on several important tasks in WWMCCS automatic data processing. It was responsible for ensuring that the hardware and software acquired under the WWMCCS ADP update were compatible with the assets of the Defense Communications System. JTSA was also a major focal point for WWMCCS ADP activities, particularly in software, hardware installation, and planning. It served as an ADP technical information clearinghouse for members of the WWMCCS community.[44] These functions made JTSA an obvious choice to coordinate the development of PWIN, and the positions of PWIN project director and operational testing director were created therein.

Consistent with DCA's coordinating function, the actual development of PWIN was not done by the JTSA but by a joint Honeywell Information Systems/Computer Sciences Corporation team. The PWIN design the contractors produced appeared simple enough on the surface. It consisted of three interconnected WWMCCS computer sites, or nodes: the Atlantic Command in Norfolk, Virginia; the Command and Control Technical Center, Reston, Virginia; and the National Military Command Center at the Pentagon.[45] These PWIN sites contained a Honeywell H6000 host computer and its associated front-end processor, a Datanet 355 computer. A number of user terminals were connected to each of the host computers. Some of these terminals were local, meaning that they were physically collocated with the host computer at the PWIN node. Others were remote terminals, located tens, hundreds, even thousands of miles from the host computer. They were connected to the host computer by one of a variety of communications media: microwave, cable, satellite, or landlines. For example, a host computer in Virginia might support a remote terminal in Germany. By simply "dialing up" the host computer, the European user would gain access to its databases, even though he was thousands of miles away.[46]

To provide this sort of interface between remote users and host computers, the PWIN design involved a complex software application called the Network Control Program that was resident in the host computers. This program provided the necessary protocols for establishing the interfaces and for performing security checking, statistics gathering, network flow controlling, and other functions.[47] Another PWIN feature, called TELNET, allowed users to connect to any other network site, access databases, and perform data processing on a timesharing basis. This interconnection capability increased the network's endurance, an important effectiveness criterion, since if one's own host computer was down, damaged, or destroyed, any other site with the same databases and applications programs could substitute for it. In PWIN it was necessary for a user to know the actual location of the databases he wished to dial into, but in the operational network to follow the relevant databases would be accessed automatically.[48]

Another key element of the PWIN design was what was called its communications subnet, as direct a copy of the packet switching design used in the ARPANET as was possible. In this arrangement, a terminal user at one PWIN location would instruct his host computer to send a message to a user at another location, and his host computer would forward the message to a Honeywell H716 minicomputer known as an interface message processor (IMP). The IMP performed the packet switching function, breaking the message down into packets of one thousand bits, which it then passed on to a cryptographic device. The encrypted message packets then would go to a modem, and, finally, to the network's transmission lines, a series of secure, high-speed communications circuits. At first these circuits were dedicated to PWIN use, but as the network evolved, the plan was to make them compatible with DCA's envisioned AUTODIN II. At the other end, the reverse process would take place. The received packets would be demodulated and decoded, collected by the IMP, reassembled into a complete message, forwarded to the host computer, and passed along to the recipient's computer terminal.[49]

Although it sounded reasonably straightforward, it wasn't; and it was in PWIN's complex design that many of the network's subsequent problems had their genesis. Much of the complexity arose because the Honeywell 6000-series computers in use throughout WWMCCS could not perform the packet-switching function. To get the job done required additional hardware—the IMPS. On top of this, the need for communications security dictated that messages had to be coded prior to transmission and decoded once they were received, which caused the addition of even more hardware.[50] To make all of this equipment play together, at least five different software applications had to be employed. Since a greater number of components in a system invariably increases its complexity, and since complexity increases the likelihood that some components will experience failure (a condition that appears doubly true regarding computer software), the prospects for PWIN reliability that resulted from this network design were not at all propitious.[51]

Communications Security

Support for PWIN represented one response by the WWMCCS Council to an abiding concern with communications security, driven in substantial measure by the experience in Vietnam. By one account, the Vietcong had assigned nearly five thousand men to their radio interception units in South Vietnam, and the information they intercepted had made possible practically all of their attacks and ambushes. The cost in American lives had been substantial, but this cost obviously paled in comparison to the price that would be paid if sensitive communications were compromised during a nuclear conflict. In determining the command and control system of the future, then, encryption and such new technologies as narrow beam broadcasting, the use of satellite cross-links, and the utilization of the higher portions of the electromagnetic spectrum that would provide enhanced communications security would be given central importance by the WWMCCS Council. But the security of communications from external penetration represented only one aspect of a broader set of WWMCCS Council security concerns. Equally important in a defense world increasingly dominated by computers was the possibility that unauthorized personnel gaining access to sensitive, classified databases could compromise the nation's security internally. The hope of the council in this area rested on a concept known as multilevel computer security. Under this arrangement computer users with different types of security clearances could use the same equipment in a time-sharing mode, gain access to the information for which they were cleared, and be denied access to that information for which they lacked the appropriate clearance.

The council was keenly aware, however, that the hardware circuitry of the Honeywell 6000-series computers purchased as part of the WWMCCS ADP Upgrade Program was not designed to support a multilevel computer security requirement. The council's attention would thus be directed toward two types of action. The first of these involved finding short-term solutions and, more specifically, determining alternative approaches to the multilevel security problem. Many of these approaches, including the use of dedicated computers and separate databases, would prove exceedingly costly and cumbersome, or unworkable in time-sensitive situations.[52] For the longer term, the council's weight was thrown behind programs that promised to provide a truly workable multilevel computer security capability, something that would prove frustrating since the capability being sought was well beyond the current state-of-the-art software.[53]

Most of the WWMCCS Council's initiatives and concerns involved programs already in the works or already well established. With the possible exception of conceptualizing a new satellite system with nuclear war-fighting capabilities, the council felt the modernization of WWMCCS consisted of building on the work of its predecessors rather than offering revolutionary new ideas. Like the doctrine of strategic sufficiency, these improvements in command and control were in many respects simply putting a new spin on an old problem.[54] The overwhelming tendency was for goals to follow from actions rather than to prescribe them. Extant technologies represented solutions in search of problems, and new programs and new doctrine would be offered to make sense of an already well-established reality.

How would the modern-day WWMCCS be evaluated? The council identified two principal ways. First, exercises specifically designed by the joint chiefs would test the worldwide operations of the system. These tests, which began in the latter half of the decade, pointed up many WWMCCS shortcomings, thereby provoking a chorus of criticism and providing impetus for major system reform. The other means of testing would come in the form of actual crises and emergencies, and they were not long in coming. One "success story" in which WWMCCS met the needs of top decision makers took place during the October 1973 Yom Kippur War.[55] As hostilities between the Arabs and Israelis escalated, US intelligence became aware that the Russians had alerted their airborne forces for possible unilateral movement into the region. A series of emergency meetings were initiated at the White House, and President Nixon ordered Defense Secretary James R. Schlesinger to place US military forces on global alert, the first such alert since the Cuban missile crisis. Schlesinger passed the alert order on to the chairman of the JCS, who issued it through the facilities of the National Military Command Center at the Pentagon. In less than three minutes, all of the unified and specified commanders had received and acknowledged the order.[56] But other accounts suggested that things might have gone more smoothly during the war. Within the first two days of hostilities, the Israelis needed several new canopies for their F-4 Phantom jet fighters to replace ones that had been damaged in the fighting. They contacted the Pentagon, where officials, in turn, contacted the headquarters of the Air Force's Logistics Command, located at Wright-Patterson AFB in Ohio. For an entire day, personnel there searched frantically and unsuccessfully through the command's enormous computerized inventory for the replacement canopies. A warehouse-by-warehouse search was subsequently conducted by hundreds of personnel at a dozen Air Force facilities around the globe. The canopies were eventually located, but by that time, the war had ended.[57]

PWIN Expansion

The hope was that the Prototype WWMCCS Intercomputer Network, then under development by the Defense Communications Agency, would solve these problems. But things were not well with PWIN. DCA personnel were by no means blind to the possibility that it might prove unreliable, and PWIN's test director informed DCA officials that the network's complexity might make it failure prone. When on 29 October 1973 DCA approved the first comprehensive test plan for PWIN, the plan explicitly emphasized that reliability was a major concern. In a 19 November briefing for top DCA management, JTSA officials underscored the reliability problems and potential for network failure. In response, DCA officials began distancing themselves from the potentially flawed network by saying that PWIN was not intended to become an operational network. This message was not one other influential parties wished to hear, however, and the joint chiefs quickly called top DCA officials on the carpet for their lack of enthusiasm and programmatic commitment. Was PWIN not intended to be an operational network? To the contrary, DCA officials were instructed, the whole point was to develop precisely such an operational capability using

PWIN as the foundation. Having received their marching orders, DCA personnel were expected to fall in line, and most of them quickly did so.

Most but not all, and one of the more vocal officials who persisted in making the case for PWIN deficiencies was John H. Bradley, a DCA civilian computer expert. He was so concerned that PWIN's WWMCCS Honeywell computers might never perform as promised that by the end of 1973 he was ready to go over his superiors' heads because they did not appear interested in pressing the issue. But having received specific instructions to proceed with work on the prototype network—WWMCCS Honeywells and all—what were Bradley's superiors supposed to do? When serious concerns over PWIN reliability were also raised in an April 1974 MITRE Corporation report, these too were downplayed by DCA. Bradley kept up the heat, however, causing such irritation that one PWIN project director wrote several letters recommending Bradley's removal from the project. Later that year Bradley was in fact transferred to nonrelated clerical duties. Although removed from direct work on WWMCCS, he continued to be a thorn in DCA's side, forwarding a series of memoranda on PWIN reliability to DCA higher-ups.[58] But as yet nobody was listening.

The next major PWIN milestone came on 4 September 1974, when the joint chiefs recommended to the secretary of defense that the prototype network be expanded from its current three nodes to six. The reasoning behind the request appeared solid enough: The current three-node PWIN configuration was quite limited, a series of experiments and tests was being planned, and including additional WWMCCS sites as part of the network would make the tests more realistic, meaningful, and informative. The joint chiefs' recommendation was accepted, and on 4 December a memorandum was issued approving the PWIN expansion. The new nodes included the Alternate National Military Command Center at Fort Ritchie, Maryland; the Military Airlift Command at Scott AFB, Illinois; and US Readiness Command headquarters at MacDill AFB near Tampa, Florida. The memorandum also pointed out that requirements for the expanding network had not been well defined, and it instructed DCA to prepare both a PWIN development plan and a "concept of failure" plan, specifying in advance what should be done if reliability problems arose during the network testing phase that would begin shortly.[59]

Notes

1. Peter Pringle and William M. Arkin, *SIOP: The Secret U.S. Plan for Nuclear War* (New York: Norton, 1983), 140–41.

2. "Operations," *TIG* [The Inspector General's] *Brief*, no. 20 (1977): 21.

3. House, Committee on Armed Services, *Review of Department of Defense Command, Control and Communications Systems and Facilities*, 94th Cong., 2d sess. (Washington, D.C.: Government Printing Office [GPO], 18 February 1977), 5.

4. *Congressional Record*, vol. 117 (29 October 1971): 38381.

5. I. Cassandra, "C^3 as a Force Multiplier—Rhetoric or Reality?" *Armed Forces Journal International* 115 (January 1978): 18.

6. Edgar Ulsamer, "The Growing, Changing Role of C^3I," *Air Force Magazine* 62 (July 1979): 36.

7. Irving Luckom, "Overview of WWMCCS Architecture," *Signal* 30 (August 1976): 62.

8. General Accounting Office (GAO), *The World Wide Military Command and Control System—Major Changes Needed in its Automated Data Processing Management and Direction, Report to the Congress*, LCD-80-22 (Washington, D.C.: GAO, 1979), 2.

9. Cassandra, 18.

10. "Remember the *Pueblo*," *Government Executive* 3 (December 1971): 15.

11. House, Armed Services Investigating Subcommittee, Committee on Armed Services, *Review of Department of Defense Worldwide Communications, Phase I*, 92d Cong., 1st sess. (Washington, D.C.: GPO, 1971), 18.

12. Edgar Ulsamer, "Command and Control Is of Fundamental Importance," *Air Force Magazine* 55 (July 1972): 43.

13. Donald T. Poe, "Command and Control: Changeless—Yet Changing," US Naval Institute *Proceedings* 100 (October 1974): 27.

14. John P. Riceman, "National Defense Policy and Command and Control Communications," *Signal* 29 (January 1975): 32.

15. Poe, 27.

16. Ibid., 29.

17. Lee M. Paschall, "The Command and Control Revolution," *Air Force Magazine* 56 (July 1973): 40.

18. Lee M. Paschall, "USAF Command Control and Communications Priorities," *Signal* 28 (November 1973): 13.

19. Review of Department of Defense Worldwide Communications, 5–6.

20. *The World Wide Military Command and Control System*, 8–9.

21. *Review of Department of Defense Worldwide Communications*, 29.

22. Hubert S. Cunningham and William E. Kenealy, "The Joint Chiefs of Staff and Command and Control," *Signal* 29 (March 1975): 16.

23. Cassandra, 18.

24. Thomas C. Reed, "Evolving Strategy—Impact on C^3," *Signal* 24 (March 1975): 10.

25. Cunningham and Kenealy, 16.

26. Janne E. Nolan, *Guardians of the Arsenal: The Politics of Nuclear Strategy* (New York: Basic Books, 1989), 100–101.

27. Paschall, "The Command and Control Revolution," 39.

28. "Pentagon Procurement Trends: For What and Why," *Government Executive* 7 (March 1975): 42.

29. James H. Babcock, "Defense Communications System Policy and Fiscal Guidelines and Areas of Future Emphasis," *Signal* 31 (August 1977): 79.

30. Samuel L. Gravely Jr., "The DCA—A Rock and a Hard Place," *Signal* 34, (April 1980): 8.

31. Senate, Committee on Armed Services, Hearings before the Committee on Armed Services, *Fiscal Year 1977 Authorization for Military Procurement, Research and Development, and Active Duty, Selected Reserve, and Civilian Personnel Strengths, Part 1,* 94th Cong., 2d sess., (Washington, D.C.: GPO, 1976), 243.

32. "Man and Machine Must Learn to Talk Together," *Armed Forces Management,* July 1968, 111.

33. Karl E. Weick, *The Social Psychology of Organizing* (Reading, Mass.: Addison-Wesley, 1969), 27.

34. Gravely, 8.

35. Lee M. Paschall, "The Role of the Present and Future DCS in Support of the WWMCCS," *Signal* 29 (March 1975): 39.

36. Ulsamer, "Command and Control," 43.

37. *Fiscal Year 1977 Authorization for Military Procurement,* 245.

38. Lee M. Paschall, "Command and Control: Why the Air Force's New Systems are Revolutionary," *Air Force Magazine* 57 (July 1974): 62.

39. Thomas C. Steinhauser, "Command and Control Improvements . . . on the Front Burner," *Armed Forces Journal* 109 (May 1972): 47.

40. Paschall, "Command and Control," 62.

41. Steinhauser, 47.

42. J. P. McConnell, "Command and Control," *Sperryscope* 17, no. 2 (1965): 5.

43. "*Signal* Interviews Lt Gen Richard P. Klocko, USAF," *Signal* 26 (September 1971): 7.

44. Herbert B. Goertzel and James R. Miller, "WWMCCS ADP: A Promise Fulfilled," *Signal* 30 (May/June 1976): 60.

45. *The World Wide Military Command and Control System,* 43, 81.

46. Frank R. Dirnbauer, Gary Bowles, and Warren Mahaly, "A Milestone in the Development of Data Management Systems for WWMCCS," *Signal* 30 (July 1976): 36.

47. Paschall, "Command, Control and Technology," *Countermeasures*, July 1976, 42–43.

48. Goertzel and Miller, 60.

49. Paschall, "Command, Control and Technology," 42–43.

50. *The World Wide Military Command and Control System*, 42–43.

51. Ibid., 86.

52. Ibid., 28.

53. *Review of Department of Defense Command, Control and Communications Systems and Facilities*, 13.

54. Nolan, 101.

55. Cunningham and Kenealy, 20.

56. Lawrence E. Adams, "The Evolving Role of C^3 in Crisis Management," *Signal* 30 (August 1976): 61.

57. Frank Greve, "Pentagon Calls Super-Computer A 'Disaster,'" *Parameters* 10, no. 1 (March 1980): 96.

58. Pringle and Arkin, 147.

59. *The World Wide Military Command and Control System*, 82–83.

Chapter 9

The WWMCCS Architect and Architecture

During 1973 Deputy Defense Secretary William P. Clements Jr. and his fellow WWMCCS Council members had perceived the need for some sort of framework within which the elements of the worldwide military command and control system could be melded into a coherent system, one cognizant of budgetary constraints and the existing technological state of the art. No such plan then existed, so in December 1973 the council decided the time had come to create one. As Clements phrased it, this framework would "put all of our worldwide military command and control systems into proper perspective."[1] What this meant in practice was that the council agreed, informally at first but more formally later on, to undertake a comprehensive, soup-to-nuts review of WWMCCS which would serve as a first step in producing a master plan for its development in the years to come.[2] This review made eminent sense given the 1971 reorientation of WWMCCS's priority mission as support of the National Command Authorities. The issues it would address necessarily went far beyond any requirement for standardization provided by the then-in-progress WWMCCS Automatic Data-Processing Update Program. What was now required was a systems approach in which the various elements of the WWMCCS "confederation" would be considered in an integrated way, with an eye to future system development. This was precisely where the master plan, called a system architecture, came in.

The ambitiousness of this plan was considerable. Since any effort to engineer major changes obviously required an understanding of what was inside of WWMCCS and what was not, the architecture would begin by specifying precisely where the system's boundaries lay, what systems and people it included, and what their responsibilities were. Since WWMCCS users had a diversity of information requirements, the architecture would then specify their needs. In that any system had to have a purpose, goal, or set of goals, the architecture would clearly define what these were. On the basis of the goals specified, the architecture would define system interfaces and information flow—normative prescriptions for how the various elements of WWMCCS should play together.[3] All of this would permit subsequent acquisitions and growth to be orderly and coherent. Architecture in hand, the ad hoc incrementalism that had characterized the growth of WWMCCS would be brought to a halt and the system's effectiveness would be enhanced. The WWMCCS Council's decision to undertake this sort of basic reevaluation was formalized in January 1974. But architectures do not materialize merely from the collective expression of sentiments, however deeply felt. Something more was required: specifically, a WWMCCS architect.[4]

The architect that Clements and the WWMCCS Council selected in February 1974 was IBM's Federal Systems Division (FSD), and the choice seemed eminently appropriate. FSD had developed as a separate division within IBM because of the company's early work on several Air Force projects, including SAGE, and its work included the development of militarized computer hardware, software, and peripherals. Among its many defense- and aerospace-related projects, FSD had developed the bombing navigation system for the B-52, pioneered the application of airborne digital computers for the B-70, played a key role in the Saturn space program, and developed the FAA's enroute air traffic control system. It also had participated in the development of the Airborne Warning and Control System (AWACS) planes, and had responsibility for integrating the command and control systems on the *Trident* submarine.[5] This experience would surely come in handy when dealing with WWMCCS, perhaps the most complex information processing system in existence.

By agreement, FSD as WWMCCS architect was to provide the WWMCCS Council with a comprehensive document, the WWMCCS Architecture Planning Studies, within two years' time, by the end of 1975. That document was to contain recommended architectural alternatives and plans for evolving from the present system to each of these in the 1980, 1985, and 1995 time frames. The plans were also to include recommended methodologies for reviewing and, if necessary, for changing the long-range architectural plans in light of budgetary, doctrinal, or other programmatic changes.[6] From IBM's multiple alternatives, the council would then select its one preferred option, after which implementation would begin. "Just as in building a house," Clements colloquially quipped, "once we have got the architecture, we'll go into the engineering phase."[7]

Director of Telecommunications and Command and Control Systems

The efficacy with which the WWMCCS "house" would be constructed was soon thrown into serious jeopardy. In previous years, David Packard and other proponents of centralized command and control had fought hard, and apparently held sway, against an entrenched bureaucratic status quo that liked the current decentralized system just fine and resisted any substantive changes. One of their major accomplishments in this effort had been the elevation of the telecommunications post to a full assistant secretaryship. The establishment of the WWMCCS Council and subsequently the council's expressed intention to create a system architecture also seemed to portend the increased centralization to come. But these efforts appeared to suffer a major setback when, in January 1974—the same month that the WWMCCS Council made public its plans to select an architect for the system—Defense Secretary Schlesinger initiated an organizational shake-up with far-reaching implications for WWMCCS.

Among Schlesinger's moves, the Office of the Assistant Secretary of Defense for Telecommunications was redesignated the Telecommunications Office, and the assistant secretary position that Packard had considered so important was downgraded to the position of director of telecommunications and command and control systems (DTACCS). The formal mission of the new office remained ambitious—to manage DOD telecommunications resources to support both the NMCS and the individual services in carrying out their specified missions. But as if to make those already daunting tasks more difficult still, the new director was reassigned to the Office of Legislative Affairs, a completely political division. "Admittedly, legislative affairs are of great concern to the Department of Defense," the House Armed Services Committee would later dryly note.

But the committee found it difficult to imagine any instance in which legislative affairs should be accorded a higher defense priority than command and control, "which is the very reason for the Department's existence."[8] Although less than two years had passed since the assistant secretary position had been created, the importance of the communications function apparently had changed drastically. With Schlesinger's action, the bureaucratic muscle that Packard had worked so hard to institutionalize was noticeably relaxed.

Consistent with a venerable bureaucratic tradition, the best possible public face was put on the situation. Defense journals pointed out how the new office underscored the Pentagon's recognition that the command and control function needed greater consolidation. According to Thomas C. Reed, a former engineer with the Lawrence Radiation Laboratory who assumed the duties of DTACCS on 19 February 1974, his directorate was the result of a fundamental change in national defense policy, a reference to the ascendancy of the doctrine of flexible response.[9] As Reed explained later, his office would concern itself with system efficiency and effectiveness. On the efficiency front, his office would work to increase system capacity even as the per-bit cost of transmitting information was reduced. Effectiveness would be enhanced through improvements in the survivability, flexibility, security, and interoperability of its constituent parts. Reed acknowledged that while such efficiency criteria as channel capacity and cost were relatively easy to measure, effectiveness criteria were not. But whatever the obstacles, he said, his office would pursue its tasks with vigor.[10]

How well the new directorate and its director would be able to do this was an open question. While Reed told members of a House subcommittee that he did not feel the downgrading had any negative impact on the authority of his office, he went on to note that his authority was not structural but rather derivative of his personal relationships with higher-ranking officials, such as the secretary and the deputy secretary of defense. Reed then agreed that the management of defense communications was too important a function to rest upon the vagaries of personal relationships, acknowledging that to work effectively with the military services, other Pentagon offices, and Congress, the head of the telecommunications office required the statutory authority of an assistant secretary. But stepping deftly through the rhetorical minefield he had just created, Reed quickly excluded himself from this requirement. Authority should be commensurate with responsibility, he opined, but this was something that could be achieved over the long term.[11]

The problem was that Reed's weakened directorate had responsibilities requiring substantially more than philosophical promulgation: it needed real authority and needed it immediately. Since the directorate was responsible for supervising the work of IBM and its subcontractors in developing the WWMCCS architecture, a WWMCCS Architecture Management Office was established within DTACCS to aid the new director and the WWMCCS architect in their work.[12] But real authority was indeed lacking, and the tone of urgency that had prevailed under Packard's regime was discernibly muted. Legislative Affairs was hardly the best bureaucratic location from which to launch a major set of programmatic initiatives after all, and the upbeat rhetoric accompanying the establishment of the new office did not accord with the obvious reality of the situation.

The military departments, ever bureaucratically adept, were quick to get the message that the star of command and control was no longer in the ascendancy. Never comfortable with the centralizing trend in command and control anyway, they followed Schlesinger's move enthusiastically by downgrading their own military communications functions and subordinating them to organizations with little communications experience. The bureaucratic axe was wielded handily over the months to come, cutting a broad swath across DOD's communications landscape, including the Joint Chiefs of Staff organization.

Schlesinger's downgrading of the assistant secretary position and the associated moves by the military departments had other predictable consequences. For the individual communicator, the fact that these actions eliminated a number of two-star billets previously filled by communications specialists seriously affected opportunities for promotion to the flag or general ranks. Not unreasonably, this move was interpreted as a clear signal that there was no future in the military for communications specialists.[13] A decline in morale rapidly ensued, and the best officers began departing from military service in droves, seeking the more hospitable career climate of the civilian sector. Their accumulated technical experience went with them, of course, and was lost to the DOD. Promoted in their stead and placed in positions of communications management responsibility were officers who lacked the technical background and experience necessary in the dynamic telecommunications field. In many ways this can be read as an effort—ultimately a futile one—to turn back the clock to an earlier, simpler military time. It was a brief revival of the archaic notion of the "well-rounded officer," an attempt to return to the practices of the past when only such officers had a reasonably assured chance of reaching flag rank.[14] With a system now in place that virtually guaranteed a lack of high-tech expertise at the top, inefficient and ineffective communications programs and procurements were also virtually guaranteed.[15]

With the creation of the new Office of the Director of Telecommunications and Command and Control Systems, the decentralized needs of the services and the centralized needs of the NCA again came into open conflict. To achieve the goals for WWMCCS as outlined by Reed, the cooperation of the services was crucial. Yet, with the devaluing of the command and control function implicit in the creation of DTACCS, cooperation was increasingly less likely to be forthcoming, at least in the short term. The main areas of contention, the related issues of service autonomy and budgetary control, were painfully familiar. The major concern of the services has always been the performance of their military missions, which in turn depended on the weapons systems they possessed. Therefore, the services had always placed considerable emphasis on planning, procuring, and protecting their weapons. Command and control of those weapons was by no means neglected, however, and each of the services had established independent communications commands and technical systems they considered adequate for this purpose. The problem had always been that the human and technological assets the services deployed were not designed to satisfy requirements generated at the national level, and any move that would alter or reduce those systems, by the Telecommunications Office or anyone else, was regarded as a direct interference with their ability to perform their military missions.[16]

Then there was the issue of defense dollars. Since the services continued to control WWMCCS's budgetary purse strings, they had to identify in their budgets and resources for any initiative they planned to pursue, which would then be defended throughout the budgetary process. The way it worked, the process typically began with a list of validated requirements far in excess of the fixed-budget ceiling. Some of these would be for WWMCCS upgrades, but a validated requirement by no means assured that money would be available to meet it. Throughout the year, a number of boards and panels would evaluate the competing validated programs, a process that eventually would result in decisions to fund some programs, defer others, and eliminate still others. In this survival-of-the-fittest programmatic approach, WWMCCS programs were granted no particular preference. So as funding decisions were made, WWMCCS requirements were forced to compete head to head with non-WWMCCS requirements, including major weapons systems, for the same budgetary dollars.[17] While clearly an approach with a measure of merit, it also guaranteed difficulties for major joint-service command and control expenditures; after all, greater prestige and thus better prospects for funding tend to adhere to high-value, high-visibility weapons systems that "belong" to a single service. Understandably enough, WWMCCS did not fare particularly well in this process.

To make things more difficult still, even in the absence of such prejudice, decision makers in the services often found it difficult to see the need for WWMCCS-related programs. "Most people will understand the requirement for a new bomber more readily than the requirement for a new modulation scheme for an existing low-frequency communications system," one Air Force official remarked, and for improvements in WWMCCS to be pursued, their service-specific payoffs had to be made much more explicit. This was unlikely to occur by chance, and so each of the services had been compelled to establish WWMCCS program offices to perform as intraservice advocates for the joint-service system, making clear their value to the services.[18] But with the downgrading of the assistant secretary position and the creation of the Office of the Director of Telecommunications and Command and Control Systems the payoffs all seemed to run in the opposite direction. Now that key defense officials from the secretary on down were de-emphasizing the command and control function, thus implicitly emphasizing the ability of the status quo to meet the nation's military requirements, why divert resources from the more desirable and visible service programs? It was a context that naturally lent impetus to the services' natural tendency to resist centralization. The problem was that the decentralized status quo did not work well when it came to WWMCCS, as would be seen during the *Mayaguez* incident in May 1975.

The *Mayaguez* Incident

The *Mayaguez* incident occurred when military forces of Cambodia's Khmer Rouge government seized the *Mayaguez*, an American merchant ship operating off the southern coast of Cambodia. Diplomatic efforts to secure the release of the ship rapidly ensued, but they failed just as quickly because the Cambodians insisted they had the right to seize the ship because it was inside their territorial waters. Frustrated, the Ford administration decided to mount a joint Navy, Marine, and Air Force rescue operation to free the vessel and its crew.

How successfully WWMCCS performed during the incident depends upon who was asked. On the up side, contemporary accounts described how, as the crisis developed, command centers in Southeast Asia were linked together by secure communications capabilities, providing the national leadership with a timely, almost blow-by-blow assessment of what was taking place. The result, it was said, was positive and precise, real-time Pentagon control, even as the incident developed.[19] And control was something the Cambodians, by their own admission, did not have. Ieng Sary, Cambodia's foreign minister, later acknowledged how his government had been unable to keep up with events, and that American technology had enabled US forces to operate more effectively than Cambodian forces. US officials later credited this command and control advantage for the successful rescue of the ship and her crew.[20]

There was also a down side, for things surely did not work as well as they might have. Indeed, some critics have described the incident as a major WWMCCS fiasco. They noted how, early in the crisis, President Ford queried WWMCCS to learn how long it would take the nearest aircraft carrier, the USS *Coral Sea*, to reach Cambodia. The information Ford wanted was apparently quickly forthcoming, but WWMCCS failed to determine whether the carrier could depart immediately or would have to remain on station to recover its aircraft before steaming to Cambodia. In fact, it was the latter. The *Coral Sea* had to wait around to recover its planes, and it arrived off Cambodia several hours later than the president and his puzzled military advisors had anticipated.[21] And during the joint Navy, Marine Corps, and Air Force operation to free the crew, the WWMCCS computers, apparently unable to keep up with the situation, crashed. Pentagon officials later disputed this conclusion, claiming that the computers had been little used and performed "adequately" during the rescue operation. (A Pentagon spokesman did point out, however, that in another, unidentified crisis that occurred about the same time, the WWMCCS computers had been broken for several hours because regularly scheduled maintenance had not been performed.)[22] Despite assurances from the Pentagon, it is clear that not all went well with the operation to rescue the *Mayaguez*. For as retired Navy Vice Adm Jon Boyes later remarked, "The Marines were getting their butts shot off, and the Navy couldn't talk to the Marines."[23] Such was the ambitiousness and the uncertainty in which the new WWMCCS architecture would be constructed.

The WWMCCS Architecture

Irving Luckom, IBM federal systems division's manager for WWMCCS architecture, recognized that two major uncertainties existed when his company's effort to define the WWMCCS architecture began. The first involved the question that had preoccupied the WWMCCS Council: "What is WWMCCS?" A variety of definitions were available, some little more than general platitudes. Others were technological, emphasizing the system's various assets, its hardware and software. Still others were organizational in their focus, concerned with procedures and with rules and lines of authority. "Depending on whose definition you use," Luckom observed, "WWMCCS could vary from a relatively limited system involving the NMCS and the

CINCs to a system including almost everything but the forces."[24] For the architect to proceed, decisions obviously had to be made regarding the precise nature of the system itself; the location of its boundaries, and the human and technological elements those boundaries encompassed. These decisions by necessity would often be subjective.

If arriving at a definition for WWMCCS was problematical, an equally vexing second question facing the new system architect was "What is an architecture?"[25] To answer this question, it was not sufficient to know what elements constituted WWMCCS. It was also necessary to determine their functions. This process began by asking what the information requirements of system users were to determine the most appropriate systems for meeting them.[26] What equipment did users need? What sort of connectivity to other system users did they require? In practical terms, what all of this meant was determining whether a specific commander would be better served by a cathode ray tube display, a facsimile machine, or a telephone sitting on his desk. It meant determining the types of automatic data-processing support he would need. It meant ascertaining with whom he might need to communicate, and it meant determining these things for a variety of situations and circumstances. And so on, for every commander.[27] The answers to these questions would provide FSD, as architect, with a basis for answering the central question for any architecture: What are the goals of the system? What is WWMCCS actually supposed to do? Once expected outcomes were described, once goals had been identified and documented, a systems engineering effort could commence. It would be directed toward acquiring the technologies and the organizational structures most appropriate for goal attainment. In this way the architecture, a thoroughly normative formulation, would allow the specification of a concrete set of performance criteria and the acquisition of real-world assets. Architecture in hand, at long last it would be possible to determine how effective the World Wide Military Command and Control System actually was.

Or so it seemed. Lost on almost everyone involved in this major effort to rationalize WWMCCS was its backwards, essentially irrational basis. By the time IBM's Federal Systems Division commenced work as system architect, WWMCCS already had been in existence for more than a decade. Billions of dollars had been spent on it, thousands of military and civilian personnel had been involved with it, and numerous programs to improve it were in progress. Plans for still other programs, including a WWMCCS Intercomputer Network, were in the works. In classic solution-looking-for-a-problem fashion, this vast system sprawled across the Department of Defense landscape. Now it was time for IBM to answer a range of questions the indisputable fact of WWMCCS's existence raised, time to define some goals and objectives for it. The notion of organizational theorist Karl Weick that goals tend to follow rather than precede actions, and to a large extent represent rationalizations for actions already taken, has seldom received better press.[28]

IBM's Scenario-Based Approach

The first step taken by the WWMCCS architect was to examine the nature of national defense policies, or doctrine. As it concerned the strategic nuclear forces, that doctrine was flexible response; and flexible response was a tough taskmaster, involving substantial expectations regarding survivability, capability, and connectivity. The demands of doctrine firmly in mind, the WWMCCS architect next examined the existing and projected military force structures and weapons capabilities of the United States and a number of potential enemies. After assessing military resources, the architect examined the national-level decision-making process itself. To get a feel for actual as well as formally stated lines of authority and communication, the architect engaged in extensive consultations with officials throughout the government. Who really makes the judgments and the decisions? Who really talks to whom? Finally, and only when all of this had been done, the architecture addressed the issue of WWMCCS structure and boundaries: What was in and what was out? What resources were available to be tapped when needed? What were the interfaces with other systems?[29]

The parameters of WWMCCS established, the next step was to determine how all of these resources would play together. IBM chose a scenario-based architectural approach that involved identifying a representative set of likely military states, including peacetime, low-level crises such as evacuations, then military buildup, conventional war, theater nuclear war, general nuclear war, and conflict termination. For each of these crises, the architect evaluated exactly what types of information the National Command Authorities would require to meet the demands of the situation. The resulting lists were long and varied. Information requirements might include warning times, availability of communications channels, location of forces, accuracy of impact predictions, damage sustained and inflicted, and a wide range of other information.[30] While admittedly a highly judgmental process, the scenarios were grounded in realistic expectations and devised to cover major geographical areas in which trouble was expected during the upcoming decade.[31]

For each scenario, the architect's approach was to consider the entire range of military response options available to the NCA. Not surprisingly, this list was also long and varied. At that point, and only at that point, a series of WWMCCS requirements was identified to support the national leadership for each situation/response combination. These requirements were considered in terms of five major WWMCCS elements: facilities, warning and intelligence, automatic data processing, executive aids, and communications. For theater nuclear war, for example, the national leadership might require its command centers to be able to withstand a certain level of blast overpressure. Intelligence assets should be able to collect strike results, identify new targets, and monitor the execution of launch orders. Communications between key command facilities and the operating forces should remain intact throughout the conflict, and so on. These necessary capabilities were referred to as functional requirements.

When the functional requirements for each system element had been established, the architect then directed attention toward specific quantitative requirements—the actual capabilities that had to be acquired or developed for each of the major states of crisis and conflict.[32] This was determined by comparing the functional requirements for WWMCCS's five elements to so-called WWMCCS baselines. A baseline was simply a statement for what a given WWMCCS element—automatic data processing, say—would look like in 1985, given existing capabilities and those improvements already funded or in an advanced stage of research and development. If these baseline systems would satisfy the functional requirements, well and good, for no new capabilities would need to be acquired. If, as in most cases, the comparisons pointed up shortfalls in capabilities, then specific architectural solutions aimed at correcting them would be developed. By intent, these were constrained to be both technically feasible and to permit assessment of approximate costs.[33] Since it was not infrequently the case that several remedies were possible for any single deficiency, IBM would develop a series of architectural alternatives, representing a range of WWMCCS capabilities and costs.[34]

Developing the architecture was conceived as an iterative process, involving close contact and a great deal of give-and-take between the architect and the WWMCCS Council. The architect would view each scenario state independently and come up with a set of architectural solutions for each of the five WWMCCS elements. The WWMCCS Council then would review each of these with regard to its capabilities, feasibility, and probable cost. As the review process proceeded, objectives, priorities, and cost estimates were progressively clarified, and the council would suggest modifications. The architect would then go back to work, ultimately proposing new alternatives that would, in their turn, be reviewed. Because the various states were considered separately and their architectural solutions presented sequentially, inconsistencies between them were not infrequent, driving the architect to come up with solutions that were consistent with the requirements of both.[35]

Developing the architecture was a highly judgmental process. It was necessary to determine such general issues such as priorities and likely threats. It was necessary for the architect to specify values for many of the variables used in the calculations, including survivability, capability levels, and accuracies. It was necessary to estimate future needs and costs. It was not a process that could guarantee that actual future situations would be perfectly addressed by WWMCCS, but rather one that upped the probability that WWMCCS would be responsive.[36] WWMCCS effectiveness, like most human endeavor, had been moved out of the realm of mathematical certainty and into that of statistical probability. And yet however subjective this process to be, it is worth underscoring the point that its specification was accomplished by the WWMCCS Council, an organization whose primary concern was the needs of central decision makers, not organizational subunits. Throughout, the WWMCCS architect simply presented the alternatives without recommending which solutions should be implemented. Because of this developmental approach, the material and social technologies that flowed from the architectural effort would, at least in theory, reflect those needs more adequately than in the past, when subunit needs had dominated the definitional process.

In August 1975, after many iterations between the architect, the Joint Chiefs of Staff, the OSD, the commanders in chief, and the military services, FSD presented the WWMCCS Council with a preliminary document, the *WWMCCS Architectural Planning Studies*.[37] Further modifications ensued, and FSD's architectural alternatives for the first state, theater warfare, were presented to the council that November. The alternatives for the other states followed shortly, and the highly classified final WWMCCS architectural plan was submitted to the WWMCCS Council for formal review in June 1976.

The document began by describing in considerable detail WWMCCS's current shortcomings in crisis management, the majority of which were identical to concerns voiced by the WWMCCS Council at its formation. It then outlined a series of specific improvements, to begin in fiscal year 1977, that were considered essential for making the system more responsive to the needs of the national leadership.[38] Not surprisingly, many WWMCCS shortcomings appeared in automatic data processing, which did not meet the full range of needs of its individual users and was only marginally effective during times of crisis.[39] Other needs involved the conversion of the Defense Communications System to an all-digital system and the development of new networks that could make full use of digital capability such as Phase III of the Defense Satellite Communications System. There was the need to pursue programs such as the Navy's extremely low frequency system for communicating with the ballistic missile submarine force and a follow-on to the Navy's problem-plagued Fleet Satellite Communications System that would be capable of using both the ultra high frequency and super high frequency portions of the frequency spectrum. Air Force programs described as essential by the WWMCCS architectural plan included the Emergency Rocket Communications System, the Post Attack Command and Control System, SAC's Automated Total Information System, and AFSATCOM, the Air Force Satellite Communications program.[40] Preeminent was the Advanced Airborne National Command Post, a WWMCCS Council priority program. (Given the large number of its programs that were affected by the architectural effort, the Air Force established a WWMCCS program office at the Electronics Systems Division's headquarters near Boston. This office worked closely with IBM's Federal Systems Division throughout the development process.) [41]

Architectures must be realistic and take into consideration not only the desirability of certain programs but also their technical and fiscal feasibility. Some programs viewed as both feasible and desirable were designated as WWMCCS priorities for the 1977–85 time frame. Other programs were perhaps equally desirable, but because of their cost or the existing technological state of the art they were deferred until some later time or relegated to the status of research and development programs. One such scheme, desirable but far too costly, was the development of superhardened command posts that could survive direct hits by nuclear weapons. Another area of considerable interest was the development of executive or decision-aid technologies. The Defense Advanced Research Projects Agency (DARPA) was heavily involved in this area, specifically in those advanced computer techniques known as artificial intelligence and expert systems.[42] But because the application of these technologies to the command and control area was beyond the current state of the art, the WWMCCS Council considered them of lesser urgency and deferred their pursuit.

In a world of finite resources, a realistic architecture would also have to allow for a reasonable period of transition from the current state to the new one. The natural desire of some interested critics, including Congress, was for a "turnkey" system, where improvements could be implemented all at once, and the new system would begin to function, completely, on a given day. The problem with this approach is its cost. Given the vastness of the WWMCCS undertaking and the reality of budgetary constraints (since the Vietnam spending peak in 1968, the overall defense budget had declined, in real terms, by almost 40 percent), the architectural plan prudently called for the improvements to be introduced sequentially, such that the most serious deficiencies were remedied first. The transition to the mystical WWMCCS city of the future, to an integrated and interoperable national-level system, would thus occur as a process of evolution, not revolution. This approach would guide the present efforts of those throughout DOD with WWMCCS responsibilities and would continue to guide them until the architectural plan was modified.

The architect recognized from the outset that modifications of the plan were likely, indeed inevitable, for any of a number of reasons, including advances in technology, changes in the nature of the threat, doctrinal changes, or an altered budgetary context. Therefore, the WWMCCS architecture was intended from the outset to be a flexible instrument. What this meant in practical terms, in the words of Secretary of Defense Donald H. Rumsfeld, was that a "modest continuing design effort" to promote the development of WWMCCS over the long haul was built into the plan as an integral element.[43] In other words, FSD's job was by no means finished with the delivery of its architectural plan. As a natural follow-on, the company would assist the WWMCCS Systems Engineering Organization in defining specifications to implement that architecture. It would aid the Office of the Director of Telecommunications and Command and Control Systems in monitoring the architecture's overall implementation. It would monitor environmental shifts and changes to determine their likely impact on the plan so that appropriate revisions could be recommended to the council.[44]

All in all, IBM's effort was a vast one, its WWMCCS architecture plan representing the most complex and comprehensive systems engineering effort the Federal Systems Division had ever undertaken. Yet, everything was still on paper only. Whether this architectural framework would actually produce a worldwide military command and control system that responded to the needs of the national leadership remained to be seen.

The WWMCCS System Engineer

Once the WWMCCS Council had made its selections from the architectural menu offered by IBM, the approved architecture was to be implemented immediately. This obviously required that a formal mechanism of some sort be in place to take the architecture and translate it into appropriate system designs. The council had recognized the need for such a mechanism quite early in the architectural process and decided that the best approach was to establish a general system engineering entity for this purpose. Initially, consideration was given to placing the engineering entity in a civilian corporation such as IBM, hardly unreasonable given the Federal Systems Division's role as WWMCCS architect. But it was soon decided that the nature of the task required the system engineering activity to be located within the DOD.[45] Simply put, this was nuts and bolts stuff, and the council judged that only those wearing uniforms would be able to make the bolts turn properly. Consideration was then given to locating the activity within one of the service organizations specializing in command and control, such as the Air Force's Electronic Systems Division. This idea was also rejected because of the not unreasonable fear that it would lead to the ascendancy of service interests over the needs of a truly joint-service, national-level WWMCCS.

To keep the engineering function within the DOD while at the same time minimizing the possibility of WWMCCS being held hostage to service parochialism, the council ultimately decided that the best home for the new organization was the Defense Communications Agency. DCA was tasked to draft a charter and organizational chart for what was called a WWMCCS System Engineering Organization, and the agency promptly began to work. A 15-member WWMCCS system engineering task force was activated in August 1975 to plan the work of the new engineering entity and to do whatever organizational work might be necessary for its activation.[46] Proposals for a charter and organizational chart for the new office were drawn up, along with appropriate modifications by the WWMCCS Council. All in all, the process reflected perfectly the council's view of how the system engineering effort should be conducted: while much of the work would be done by DCA, final authority would always reside with the WWMCCS Council.[47]

On 21 November 1975 the Pentagon formalized the entire arrangement by issuing DOD Directive 5100.79, which initiated a series of important changes for WWMCCS. First, it formally established the WWMCCS System Engineering Organization and located it within the Defense Communications Agency. The original 15-member DCA task force was made permanent, and additional personnel were authorized. To ensure that WWMCCS would be accorded an appropriate measure of importance within DCA, the agency's director was formally designated to wear the dual hat of director of WWMCCS system engineering.[48] Finally, to provide for appropriate technical expertise at the top levels of the new organization, the WWMCCS system engineer, a new position, was created. The directive specified that the holder of this position, who ranked directly below the director and to whom he would report, would be a highly qualified civilian government employee, not a military officer, though he would be assisted by a general or flag-rank deputy. Through this arrangement, the WWMCCS Council hoped that service parochialism would be minimized, appropriate technical expertise brought to bear on WWMCCS-related problems, and a continuing emphasis placed on issues affecting national-level command and control.

Despite his formal position as number two in the hierarchy, the system engineer would function as the new organization's chief operating officer, the person who would actually run the show. The responsibility of this "technical traffic cop" was considerable—to take the architecture approved by the WWMCCS Council and translate it into specific plans, designs, technical procedures, and standards. Once that was done, the system engineer would be responsible for acquiring the capabilities and assets necessary to realize the objectives and meet the demands of the architecture.[49] It was also the system engineer's task to ensure that the evolving WWMCCS was compatible with other command and control systems then in operation throughout the DOD. To this end, DOD Directive 5100.79 authorized the system engineer to specify where the boundaries lay between WWMCCS and related tactical command and control systems (no small task in a system that expands or contracts depending upon the level of crisis) and to pursue a technical issue to as low a level as necessary.[50] This meant that he would have the authority to define performance criteria and specify standards and interface criteria so that service- and agency-unique command and control capabilities were consistent with the WWMCCS architecture, an obvious necessity if they were to be interoperable with, and utilized by, the national-level system. To keep his finger on the tactical as well as the strategic pulse, the system engineer would continuously monitor all WWMMCS-related programs of the services and defense agencies to make certain they were consistent with the WWMCCS architecture.[51]

In addition, there was the issue of the future. As things were outlined in the architectural plan, once the engineering and implementation stages of the process had been accomplished, the activities of the system engineer would by no means cease; a recognition that changes in policy, threats, and technology were an inevitable and ongoing process. The WWMCCS Council would determine the necessary architectural changes in response but once these decisions were made, it would be necessary to translate them into appropriate system designs, and that is where the system engineer came in. In sum, DOD Directive 5100.79 represented a major effort at centralization, with DCA at the forefront of the action.[52]

The WWMCCS system engineer and organization began work at the Defense Communications Agency in early 1976. Their work began immediately, amidst considerable optimism, and a number of WWMCCS-related programs were quickly moved into the spotlight. Debate, however, soon arose over how well the new WWMCCS engineering organization was meeting, or could be expected to meet, its objectives in critical areas, one of the most important of which was ADP. Here, the fault was laid directly at the doorstep of DOD Directive 5100.79. Since the director of DCA was also the director of WWMCCS engineering, divided and often conflicting responsibilities accrued to two separate organizational masters. With respect to organizational and technical matters, the director reported to the director of Telecommunications and Command and Control Systems. But in matters pertaining to doctrine, operational policies and procedures, development and validation of requirements, and warning and intelligence, the director reported directly to the chairman of the JCS. This division of management responsibilities held the potential to seriously impede the coordination of ADP development efforts.[53] Even worse, the dual-hatted director of WWMCCS engineering/DCA had no funding, budgeting, or management authority for the WWMCCS ADP program. Funding authority remained overwhelmingly where it had always been—with the military services. This arrangement represented a bureaucratic impediment to change of almost insurmountable proportions. The services continued to develop hardware and software systems individually under their own budget, and, not surprisingly, they tended to emphasize their own needs and requirements. Therefore, despite the considerable movement toward a coherent, centralized management structure for command and control, the reality was yet to be achieved.

Notes

1. William P. Clements Jr., "Command and Control of Our Forces," *Air Force Policy Letter for Commanders*, sup. no. 6 (June 1975): 15.
2. "Pentagon Procurement Trends: For What and Why," *Government Executive* 7 (March 1975): 42.
3. General Accounting Office (GAO), *Computer Systems: Navy Needs to Assess Less Costly Ways to Implement Its Stock Point System—A Report to the Chairman, Subcommittee on Defense, Committee on Appropriations* (Washington, D.C.: GAO, December 1988), 7n.
4. "Pentagon Procurement Trends," 42.
5. John B. Jackson, "IBM's Federal Systems Division—A Technical Overview," *Signal* 30 (October 1975): 55.
6. Thomas C. Reed, "Evolving Strategy—Impact on C^3," *Signal* 29 (March 1975): 12.
7. "Pentagon Procurement Trends," 42.
8. House, Committee on Armed Services, *Review of Department of Defense Command, Control and Communications Systems and Facilities*, 94th Cong., 2d sess. (Washington, D.C.: Government Printing Office [GPO], 18 February 1977), 29–30.
9. Edgar Ulsamer, "C^3: Key to Flexible Deterrence," *Air Force Magazine* 57 (July 1974): 45.
10. Thomas C. Reed, "Command & Control & Communications RDT&E," *Signal* 30 (October 1975): 6.
11. House, Committee on Armed Services, *Review of Department of Defense Worldwide Communications, Phase III: Report of the Special Subcommittee on Defense Communications of the Committee on Armed*

Services, 93d Cong., 2d sess. (Washington, D.C.: GPO, 1977), 5. (Hereafter, *Phase III Report*.)

12. Reed, "Evolving Strategy—Impact on C^3," 10.

13. *Phase III Report*, 30.

14. "To Moorer, the 'Dom Rep Action' Proved CINCLANT has Command and Control," *Armed Forces Management* 11 (July 1965): 70.

15. *Phase III Report*, 31.

16. Congress, Hearings before the House Committee on Armed Services, *Review of Department of Defense Worldwide Communications, Phase II* (Washington, D.C.: GPO, 1972), 5.

17. Van C. Doubleday, "WWMCCS in Transition: An Air Force View," *Signal* 30 (August 1976): 70–71.

18. Ibid.

19. Lawrence E. Adams, "The Evolving Role of C^3 in Crisis Management," *Signal* 30 (August 1976): 60.

20. Thomas C. Reed, "A Better Horse for Paul Revere: National Command and Control Communications," *Strategic Review* 4 (Summer 1976): 25–32.

21. Frank Greve, "Pentagon Calls Super-Computer A 'Disaster,'" *Parameters* 10, no. 1 (March 1980): 96.

22. Michael Putzel, "Pentagon Warning System Defective, Experts Claim," *Washington Post*, 10 March 1980, A10.

23. Richard C. Gross, "C^3: Fewer Mixed Signals," *Military Logistics Forum* 3 (June 1987): 20.

24. Irving Luckom, "Overview of WWMCCS Architecture," *Signal* 24 (August 1976): 62.

25. Ibid.

26. Lee M. Paschall, "WWMCCS: Nerve Center of U.S. C^3," *Air Force Magazine* 58 (July 1975): 56.

27. Ulsamer, 50.

28. Karl E. Weick, *The Social Psychology of Organizing* (Reading, Mass.: Addison-Wesley, 1969), 63–71.

29. Irving Luckom, "Worldwide Military Command and Control System: An Approach to Architecture Development," *Technical Directions* (Autumn 1976): 11.

30. William P. Clements Jr., "DOD Command and Control Activities," *Signal* 29 (May/June 1975): 19–20.

31. Luckom, "Worldwide Military Command," 11.

32. Ibid., 11–12.

33. Ibid., 12–13.

34. Luckom, "Overview of WWMCCS Architecture," 63.

35. Ibid.

36. Luckom, "Worldwide Military Command," 13.

37. Doubleday, 70.

38. Luckom, "Worldwide Military Command," 10.

39. GAO, *The World Wide Military Command and Control System—Major Changes Needed in its Automated Data Processing Management and Direction: Report to Congress*, LCD-80-22 (Washington, D.C.: GAO, 1979), 62–63.

40. Doubleday, 71.

41. Edgar Ulsamer, "Machine Intelligence Shapes Global C^3 Nets," *Air Force Magazine* 60 (July 1977): 68–71.

42. Ibid., 66.

43. Donald H. Rumsfeld, "A Command, Control and Communications Overview," *Signal* 30 (May/June 1976): 38.

44. Luckom, "Worldwide Military Command," 13.

45. Lee M. Paschall, "WWMCCS in Transition: A WWMCCS System Engineer View," *Signal* 30 (August 1976): 64.

46. Ibid.

47. Clements, "Command and Control of Our Forces," 15–16.

48. Samuel L. Gravely Jr., "DCA's Route to Readiness," *Air Force Magazine* 62 (July 1979): 87.

49. Lawrence E. Adams, "The Evolving Role of C^3 in Crisis Management," *Signal* 30 (August 1976): 59.

50. Paschall, "WWMCCS in Transition," 64, 66.

51. Reed, "Evolving Strategy," 12.

52. Paschall, "WWMCCS in Transition," 65.

53. GAO, *The World Wide Military Command and Control System*, 9–10.

Chapter 10

WWMCCS Intercomputer Network

Throughout this time, results from a series of tests and evaluations were creating doubts in the minds of defense officials regarding the reliability of the prototype WWMCCS Intercomputer Network. In the network's first system integration test, conducted in early 1975, communication failures had been about 50 percent. The results of a study conducted for DCA by the University of Illinois's Center for Advanced Computation in May 1975 had concluded that PWIN's ability to operate in an on-line, real-time environment was seriously limited. The WWMCCS ADP community had a strong "batch orientation," the researchers noted, whereas an intercomputer network was an inherently interactive technology.[1] Two months later, the General Accounting Office raised similar concerns about PWIN's response times, its ability to provide fully interactive operations, and its ability to provide multilevel computer security features. By September 1975 the concerns had been such that Thomas C. Reed, director of telecommunications and command and control systems, was forced to delay final approval of DCA's PWIN development plan.

Serious PWIN reliability problems continued to be reported as the new year began. A RAND Corporation report issued in March 1976, "WWMCCS ADP Communications Interface Requirements," highlighted the reliability problem and pointed out how the WWMCCS standard computers were severely limited in their data-processing capacity. A key problem was that the General Comprehensive Operating System software installed in the Honeywell 6000s was designed for batch processing of data and was unable to handle the increased communications loads accompanying interactive network operations. As a consequence, RAND concluded, the network's interrupt rate would be higher than otherwise would be the case. Additionally, the WWMCCS architectural plan submitted to the WWMCCS Council for formal review in June 1976 detailed a series of system shortcomings relevant to network operations. Included among these were the fact that several of the

WWMCCS software applications were so large that only one could be loaded into memory at a time, that information was too voluminous and difficult to extract under time-sensitive conditions, and that users were not guaranteed availability when they required it. Numerous indications were also being received throughout this period that multilevel computer security, a key requirement for intercomputer networking, was beyond the current software state of the art. Things were looking bleak indeed as the summer of 1976 approached, the time scheduled for the PWIN operational experiments.[2]

The PWIN Operational Experiments

As part of the buildup for the experiments, a system demonstration (a sort of dry run) was scheduled for 24 June 1976. PWIN personnel held practice sessions for several weeks before that date, but as the director of PWIN operational testing pointed out, throughout that time not a single full run of the planned demonstration could be completed because of system hardware and software problems. Believing that a network that functioned so poorly would seriously compromise the experiments, the director demanded that the joint chiefs' Command and Control Technical Center demonstrate the network's reliability on 12 and 13 July, a week before the PWIN operational experiments were scheduled to begin. If an acceptable level of reliability could not be shown, he said, the experiments would have to be delayed. As anticipated, PWIN lived up—or, perhaps better said, down—to its expectations, experiencing several hardware and software failures and a serious instability in the communications links between network nodes. Originally scheduled for 19–30 July, the PWIN operational experiments were postponed for several months.[3]

During that time, the scramble began among PWIN participants to find what was wrong with the fledgling network and to fix it. Emergency "patches" were applied to WWMCCS's General Comprehensive Operating System software as engineers desperately tried to find a way to work around the limitations of the Honeywell 6000 computers so that the network's nodes could be effectively internetted. But the problem was so vast, meaning that the patches proliferated to such an extent that DCA officials were soon expressing almost as much concern with the fixes as with the problems they were intended to fix. To make things more complex still, during this time a number of additional WWMCCS facilities were added to the network: Army Forces Command at Fort McPherson, Georgia; Tactical Air Command at Langley AFB, Virginia; and the Pentagon headquarters of the military services.[4]

The rescheduled PWIN operational experiments took place in September and October of 1976. In each of those months, a major set of tests was conducted involving a different application of PWIN. In the September tests, known as Experiment 1, PWIN was employed in a crisis scenario that already had been used and carefully evaluated in a previous JCS exercise. This experiment, it was believed, would provide a highly controlled assessment of the networking concept. In the October tests, called Experiment 2, PWIN was used during a military exercise called Elegant Eagle 76.[5] The point in this far less controlled use of the network was to see how well it supported users' demands, as operational plans had to be modified to meet the requirements of an unfolding crisis.

How well PWIN performed during the operational experiments depended once again upon whom you asked. First, the good news: According to some officials, use of the network substantially increased the level of interaction among the network nodes. Because it allowed planning information to be shared simultaneously, networking allowed for greater coordination among participating units. Computer internetting was also said to have accelerated the transfer of databases between sites and to have increased the accuracy of the transfers. Therefore, PWIN was said to have provided its users with an enhanced ability to identify and resolve problems.[6] The bad news was that not everyone had these positive experiences; or, if they did, it was apparently only when the network happened to be working. For as subsequent evaluations showed, PWIN reliability during the operational experiments had been poor; indeed, several commands had considered it a "critical problem." As to the network's ability to transfer databases successfully among the computers at the various nodal sites, some accounts maintain that this had simply not happened.[7] Top Pentagon officials were briefed on the progress of the prototype

WWMCCS Intercomputer Network during the first two months of 1977. By all accounts it was a candid series of briefings, pointing up the problems that had been encountered before and during the PWIN operational experiments. Despite these problems, the desire and hopes for the network were such that the strong recommendation was to proceed apace with an operational network.

Prime Target 77

During the period of 1–16 March 1977, the Joint Chiefs of Staff conducted a military exercise called Prime Target 77. This exercise provided PWIN's next opportunity to prove its networking mettle in an uncontrolled operational environment. Six PWIN sites participated in the exercise, including the Atlantic Command, Readiness Command, Tactical Air Command, and National Military Command Center at the Pentagon. Also participating were two recent additions to the expanding prototype network: the headquarters of US European Command in Vaihingen, West Germany; and DCA's Command and Control Technical Center in Reston, Virginia.

The participating sites used PWIN primarily for teleconferencing and transferring data during Prime Target, and in both cases the network's reliability proved considerably less than had been hoped. For example, the European Command attempted to use the network 124 times, but experienced 54 "abnormal terminations" due to hardware or software failures—a failure rate of some 44 percent. Matters went similarly for the Atlantic Command, which logged on to PWIN 295 times and experienced 132 failures, a failure rate of 45 percent. The Tactical Air Command, for its part, tried to use the network 63 times and failed 44, a rate of 70 percent. The worst record by far, however, was that of the Readiness Command, whose 247 failures in 290 attempts to use PWIN represented an 85 percent failure rate. Collectively for these four sites, the only ones for which statistics were kept, 772 efforts to use PWIN had resulted in 477 hardware- or software-related communications failures. In other words, PWIN worked only about 38 percent of the time. To make matters even worse, the duration of the outages tended to be longer and more widespread than those experienced the previous fall during the exercise Elegant Eagle. Now, it seemed, if one PWIN site went down, the entire network went down with it.[8]

Both the operational experiments and exercises served to underscore the earlier unfavorable comments of in-house critic John Bradley, then still working at the DCA. Disturbed that all the indicators pointed to the conclusion that PWIN simply did not work and unable to get DCA officials to act, Bradley decided to go over his bosses' heads directly to the White House.[9] He arranged a meeting with a National Security Council official, Col Robert A. Rosenberg, and laid out what he considered WWMCCS's many flaws and shortcomings. Rosenberg expressed concern. He asked Bradley to put it all in writing, which Bradley did in a two-page letter, dated 29 April 1977.[10]

As might be anticipated, DCA officials were far from pleased to learn about this. For a number of reasons beyond the appearance of reports critical of PWIN, sensibilities were particularly raw at this juncture. After all, a new administration with a suspected antimilitary bias had just arrived in Washington. An evaluative group called the President's Reorganization Project was just gearing up, and nobody expected praise of WWMCCS to be one of its principal findings. The results of the PWIN operational experiments and PWIN's poor performance during the recently concluded Prime Target 77 exercise, while known, had yet to be made public, and DCA had hoped to limit the scope of their impact. Things were tough enough without DCA insiders going around bad-mouthing PWIN to officials in other departments and agencies. Bradley was fired two months later. The reasons given were "inefficiency, resistance to competent authority, and making false and misleading statements" about the Prototype WWMCCS Intercomputer Network.[11]

Despite the efforts of critics such as Bradley, the thoroughly counterintuitive consequence of Prime Target 77 was an immediate demand to keep PWIN alive and move it toward operational status.[12] How can one explain this remarkable result? On the one hand, a simple lack of information concerning PWIN's problems might be responsible. While major network problems had indeed been experienced during Prime Target

77, the full scope of difficulties would go unreported for several months. In addition, the final consolidated report on the earlier PWIN operational experiments had not been released. In this view, the network's champions were preempting, hoping to muster support for PWIN before the bad news struck.

An alternative explanation for this support in the face of adversity that should also be considered concerns the management philosophy of PWIN's advocates. Many of these advocates, the JCS apparently among them, embraced an evolutionary approach to system development that was just then becoming the rage in the Pentagon. This logic sought to discount problems with PWIN, even serious ones. The network was experimental, after all—a prototype. No one denied that there were problems with developing distributed database technologies, with automating security, and with making network operations easier. Problems are to be expected with new technologies. They can be fixed incrementally, as matters evolve. For PWIN supporters, then, whether they were true evolutionary believers or, as with many in DCA, bureaucratic opportunists whose career stars had been attached to the network, embracing the evolutionary approach made a great deal of sense. No need to worry about the problems; they can be solved, will be solved, or can only be solved by letting the system evolve.

The Joint Chiefs of Staff approved PWIN's operational requirements on 18 July 1977. Following the DCA's development plan, PWIN would be expanded to include a number of other WWMCCS sites, becoming an operational WWMCCS Intercomputer Network, or WIN, in the process. And so despite a number of documented procedural problems, and with no changes having been made in its hardware or software, the problem-plagued prototype network had been given approval to move forward to full operational status.[13]

A lengthy shadow of doubt was promptly cast over the wisdom of the joint chiefs' decision when the report on PWIN's reliability during Prime Target 77 became available. Thirty-eight percent reliability was hardly salutary, of course, and to head off criticism as well as to advance their cause of moving the network toward operational status, the joint chiefs directed the Defense Communications Agency to conduct a series of studies. Among other things, DCA was tasked to determine the precise nature of the hardware, software, or procedural problems that had produced such poor reliability. The agency was also to identify alternative ways for users to gain access to the network in case of continuing hardware- or software-induced abnormal terminations. In addition, DCA was to identify which network elements required monitoring and then to determine which monitoring methodologies were most appropriate to use.[14]

The problems that the DCA studies identified fell into several general categories. First, there were problems with the network design itself. For reasons of cost containment, each PWIN site had been designed so that alternative access to the network was not possible if the site's computers failed. In other words, if a WWMCCS host computer or IMP went down, the site was isolated from the network. DCA next identified a number of specific problems with hardware, especially the IMPs, which proved to be especially troublesome and prone to failure. Detecting the cause of the failures was often exceedingly difficult, DCA noted, meaning that once the IMPs went down, they were difficult to get back on line. Making matters even worse, IMPs were quite fragile, going down during electrical storms and when voltage fluctuations occurred. An absence of adequately trained operators exacerbated this problem by requiring that outside personnel be brought in when problems arose. Another problem was a lack of spare parts at PWIN sites; but even when parts were available, they were often defective. These problems led to excessive computer downtime.

Then there was the software, a network area identified as especially pernicious. DCA engineers found that PWIN's host software contained errors that frequently resulted in aborts and loss of data when personnel tried to use the network's teleconferencing features. Functional software in the Honeywell interface message processors was also found to contain errors, making it difficult for IMPs to perform their promised interface function. In addition, the emergency patches that had been applied to correct these and other software problems were themselves filled with errors, resulting in repeated, and often inexplicable, network failures.[15] If these problems with network structure, hardware, and software could just be fixed, DCA engineers noted, the network's reliability problems would be reduced substantially. But this was no small task, and would do nothing to address a final problem, a conspicuous lack of computer security within PWIN.

Multilevel Computer Security

Networking was essential if WWMCCS was to perform its mission adequately. DOD Directive 5100.30 had specified that the system would provide a range of necessary information to the National Command Authorities so that timely and appropriate decisions could be made. Consequently, there was a need to collect a vast quantity of data, process it, and disseminate the resulting information to commanders at all levels. An intercomputer network was the obvious way to accomplish this objective, but a major concern was one of access. Many of the WWMCCS databases contained highly sensitive information. The WWMCCS ADP community was large, diverse, and distributed. Since by definition networks exist to facilitate the movement of data among sites, the question was how to provide personnel with access to the information they required while denying them access to information for which they were not cleared.[16] Answering that question was clearly necessary if networking was to become a reality, and networking, in turn, had already been judged essential for the successful performance of the WWMCCS mission.

Protecting classified information in a multi-access computer environment was a concern of long standing. As early as June 1967, the ARPA had assembled a task force to study what hardware and software improvements would be necessary to achieve such a capability.[17] When the WWMCCS Council was created in late 1971, computer security promptly became a key council concern. The council's efforts were channeled thereafter into two streams of action. The first involved coming up with interim solutions to the security problem that would serve until a permanent solution could be found. For the longer term, the council's weight was thrown behind a concept known as multilevel computer security (MLS), an arrangement by which numerous users could access a computer simultaneously and run programs at several classification levels. Throughout, the computer would provide them with access to those types of information for which they had the appropriate security clearances while denying access to other information. The benefits of such an approach—indeed, its necessity for network operations—were obvious.

But it soon became apparent to the WWMCCS Council that the WWMCCS Honeywell 6000-series computers simply could not support a multilevel security requirement. Two reports issued in June and October 1974 by the System Development Corporation (SDC) helped to raise everyone's consciousness in this regard. Everyone was thinking "network" at this time, and the Joint Technical Support Activity, DCA's in-house software specialists, had contracted with SDC to evaluate WWMCCS ADP security as part of that effort. SDC's dour conclusion was that major security deficiencies at all WWMCCS sites seriously affected future considerations for system internetting, including resource sharing and remote interactive processing—precisely the capabilities deemed essential for the prototype WWMCCS Intercomputer Network.[18]

Based in large measure on SDC's findings, the General Accounting Office forwarded two letters, dated 21 July 1975 and 20 April 1976, to the secretary of defense expressing its concern regarding security deficiencies in the prototype WWMCCS Intercomputer Network. Judging existing hardware and software inadequate for meeting security requirements, the GAO recommended that major changes be implemented before final approval was granted for an operational WWMCCS Intercomputer Network. A specific suggestion in the hardware area involved upgrading the WWMCCS Honeywell computers. As to software, GAO suggested that an operating system software application then being developed by the Air Force's Electronic Systems Division, called the Multiplexed Information and Computing Service, be considered for use in the network. Reed agreed in principle with the GAO recommendations, promising to examine alternative means of achieving multilevel computer security before the prototype network was declared operational.[19]

The quest for a secure, truly workable multilevel computer security capability for PWIN would be continually frustrated and ultimately prove chimerical since the features being sought for the network were simply not available for WWMCCS or anyone else because they were beyond the existing state of the art in computer software technology.[20] With no breakthrough in sight, many DOD programs requiring this capability would flounder on these same shoals, among them DCA's AUTODIN II. The goal of multilevel security was by no means wholly abandoned, however, and a number of new programs were soon initiated by MITRE Corporation, DARPA, UCLA, SRI International, and others. But they found no workable solution to the multilevel computer security problem.[21]

Many defense officials were confident a solution would be found in the future. Their concern was, however, what to do about security in the interim. They were aware that if the computers themselves could not restrict access to sensitive information, environmental and procedural security controls would have to serve. After all, a combination of physical and environmental security protected the WWMCCS computers, which were located in restricted areas in shielded rooms to block transmission of signals through the walls. Uniformed guards hovered nearby to prevent unauthorized use.[22] Such controls could obviously be continued. Additionally, at least three procedural techniques were used at WWMCCS sites to protect sensitive information that could be continued or extended. Dedicated computers could be used for data at each security level. This practice was used at the Alternate National Military Command Center, which used two wholly separate computer systems for its data processing.[23] This was obviously an expensive practice and precluded the efficient sharing of databases; and it was cumbersome, since it required manual updating of files. At other locations, a technique known as periods processing was used. As the name suggests, this practice involved processing data at different security levels at different times. The major problems with this approach was the need to sanitize the entire area each time the security level was changed and the fact that it was fundamentally inconsistent with the simultaneous utilization of computing resources that is the essence of time-sharing. Another technique was system high operations, the technique of (necessary) choice for AUTODIN II, where everyone and everything involving the computer was simply cleared to the highest security level used on the system. Taken together, these controls were indeed adequate to prevent unauthorized users from obtaining classified information from the WWMCCS computers, but they also effectively precluded the sort of computer internetting that was considered essential.

In PWIN the problem of multilevel security was similarly addressed by using the system high approach. During the PWIN operational experiments, for example, machines, terminals, and personnel were cleared to the highest security level being used, Top Secret, to allow for data processing and for teleconferences to be set up. Of course, this meant that even personnel performing routine functions had to be cleared to that security level.[24] Although this approach was inefficient and costly, it did not prove to be a major problem when things were limited to a relatively small prototype network.

But the lack of adequate provision for the security of sensitive information certainly would create substantially greater concerns in the context of a fully developed WWMCCS Intercomputer Network. One major problem was that WIN would be extended to Europe, and this would necessarily include headquarters of the Supreme Allied Commander in Europe (SACEUR). The problems arose because SACEUR's headquarters was not an exclusively American-run show. All other NATO countries had personnel there, and they used WWMCCS to access command and control information relevant to the European theater. This had not proved an especially serious security problem when dealing with an individual WWMCCS site, for sensitive databases could simply be withheld from foreign personnel by using periods processing, that is, scheduling separate computer operating sessions for each of the various security levels. But things became dramatically different when access to sensitive information had to be limited on a network whose whole purpose was to permit ready exchange of data between sites. The dominant concern focused on how to protect sensitive information while permitting our NATO allies to use WWMCCS. That was where multilevel computer security was supposed to step in, but since it remained beyond the state of the art at the time WIN was declared operational, users had to find alternative techniques to protect sensitive information.[25] The solution was to develop various types of security filters, but these were costly and cumbersome stopgap measures designed to serve only until true multilevel computer security could be brought on line.

Many Pentagon officials considered none of these shortcomings particularly serious. Problems always occur with any new technology, and they can be fixed as they occur. So problems and all, the prototype network was transitioned to operational status. But as we shall see, two events at the end of the 1970s underscored the many shortcomings of WIN: a full-scale mobilization exercise and an actual crisis, both of which produced a cacophony of WWMCCS criticisms and subsequent calls for reform.

Notes

1. Comptroller General, *The World Wide Military Command and Control System—Major Changes Needed in Its Automated Data Processing Management and Direction, Report to Congress,* LCD-80-22 (Washington, D.C.: General Accounting Office, 1979), 64.
2. Ibid., 62–63, 83–84.
3. Ibid., 83–84.
4. Ibid., 43, 85–86.
5. C. J. LeVan, "WWMCCS Automation—A Team Effort," *Signal* 32 (November/December 1977): 12.
6. Ibid., 13.
7. Comptroller General, *Report to Congress,* 51, 861.
8. Ibid., 51–52.
9. Michael Putzel, "Pentagon Warning System Defective, Experts Claim," *Washington Post,* 10 March 1980, A10.
10. James North, "'Hello Central, Get Me NATO': The Computer That Can't," *Washington Monthly,* July–August 1979, 52.
11. Peter Pringle and William M. Arkin, *SIOP: The Secret U.S. Plan for Nuclear War* (New York: Norton, 1983), 148.
12. LeVan, 12.
13. Carol Hamilton, "Worldwide C^2 System Networks Strategic Data for Joint Chiefs," *Defense Electronics* 20 (June 1988): 57.
14. Comptroller General, *Report to Congress,* 87.
15. Ibid., 87–89.
16. Fred J. Shafer, director, Logistics and Communications Division, US General Accounting Office, letter to the secretary of defense, 21 July 1975, 1–2.
17. Ibid., 5 April 1978, 1.
18. Ibid., 21 July 1975, 2.
19. Ibid., 5 April 1978, 2–3.

20. House, Committee on Armed Services, *Review of Department of Defense Command, Control and Communications Systems and Facilities*, 94th Cong., 2d sess. (Washington, D.C.: Government Printing Office, 1977), 13.

21. Comptroller General, *Report to Congress*, 28.

22. House, *Review of Department of Defense Command, Control and Communications*, 13.

23. Shafer, 21 July 1975, 2.

24. *Modernization of the WWMCCS Information System (WIS)*, (Washington, D.C.: Department of Defense, 1981), 7.

25. Hamilton, 55.

Chapter 11

The Carter Administration and the Evolutionary Approach

Defense Secretary James R. Schlesinger's January 1974 downgrading of the assistant secretary of telecommunications to the position of director, telecommunications and command and control systems, and the associated moves on the part of the military departments, had relegated communications to essentially a support function. It seemed obvious that something had to be done to arrest this deterioration in authority, and in 1977 a Command, Control, and Communications Panel of the House Armed Services Committee outlined a series of moves designed to effect an organizational about-face, putting into place a more centralized management structure.

Assistant Secretary of Defense for C^3I

The most important of the Armed Services Committee panel's recommendations was that the telecommunications position be immediately restored to its previous assistant secretary level. Once that was done, the panel said, it was necessary that the occupant of the new position be given supervisory authority over all tactical and strategic communications systems, as well as for all command and control programs, including responsibility for related programs involving automatic data processing. Since in the hardball world of Pentagon politics real managerial authority had to come with budgetary teeth, the panel also recommended that the newly restored assistant secretary be given budgetary authority for these programs. Indeed, things had deteriorated so badly that almost no amount of authority, up to and including "absolute authority," was adjudged excessive for the new assistant secretary if that was what was required to compel the services to participate in joint-service programs. For until the existing fragmented authority structure was replaced with a radically revamped management structure of this sort, the panel concluded, duplicative efforts were inevitable, and dollars and efforts would continue to be wasted.[1] It was difficult to imagine a stronger call for centralized control.

It was far easier to imagine that not all members of Congress would share the sentiments of their colleagues on the Command, Control, and Communications Panel. After all, if some congressmen were calling for dramatically enhanced authority at the assistant secretary level, there were others who believed that things already had devolved too far in the opposite direction of centralized control. Their complaints centered around a management process that they felt gave WWMCCS-related programs a sort of "special advocate" within the Office of the Secretary of Defense, a brand of bureaucratic preferential treatment that allowed these programs to avoid the stringent, highly competitive budgetary reviews that were the norm for most non-WWMCCS programs.[2]

It was, in fact, the long-familiar face-off between the proponents of centralization and decentralization, just played out on another stage at another time. Yet with the passage of each successive year, the very terms of the command and control debate were being altered. No longer were arguments cast in the stark, dichotomous, either/or, yes/no terms of years gone by. The overall trend was unmistakable by the mid-1970s: the importance of effective national-level command and control had been firmly established as part of the conventional wisdom, and it was difficult to overstate its importance.[3] That is, command and control had moved rhetorically to center stage. What remained was a sort of mopping-up exercise in which the debate increasingly involved procedural issues: the best way to go about doing what almost everyone now acknowledged had to be done, rather than the more basic issue of whether it should be done in the first place. Despite their short-term successes, then, those who deprecated the entire WWMCCS concept found themselves increasingly on the defensive. As the decade advanced, they would find themselves in full retreat.

The movement back toward centralization began, most visibly, in early 1977, with the arrival of a new administration in Washington. Throughout his presidential campaign, former Georgia governor Jimmy Carter had portrayed himself as an antigovernment outsider, untainted by the Watergate scandal and uncompromised by any history of dealings inside the Beltway. His victory over sitting president Gerald Ford was perceived by many members of the new administration, as well as by Carter himself, as a public mandate to bring organizational change and other basic reforms to a number of areas of government.[4] Given Carter's background as a Navy officer, his personal abhorrence of nuclear weapons, and his campaign pledge to "banish" those weapons, it was hardly surprising that one of the key organizations he targeted for reform was the Department of Defense. And given the criticisms then issuing from the House's Command, Control, and Communications Panel, the General Accounting Office, and a number of other influential fora—most of which emphasized the need to establish some sort of organizational center of gravity for the Pentagon's disparate command and control programs—a great deal of the specific pressure for reform would soon be directed toward WWMCCS.

The first major step in this effort came in the middle of that same year, 1977, when, in a dramatic reversal of his predecessor's actions, Carter's new defense secretary, Harold Brown, began an organizational shake-up at the highest levels of the OSD. Brown ordered that two existing offices, the director of telecommunications and command and control systems and the assistant secretary of defense for intelligence, be consolidated. The resulting single new office, designated the assistant secretary of defense for communications, command, control, and intelligence (C^3I), appeared to be both a significant symbol and a portent of things to come. It suggested, respectively, the importance that the Pentagon's civilian leadership attached to the area of command and control, and it underscored the secretary's intention to direct developments within it.[5] The person Brown appointed to fill the new assistant secretary position was an electrical engineer, Gerald P. Dinneen, and by all criteria he seemed an excellent choice. A professor at the Massachusetts Institute of Technology (MIT), Dinneen had worked on a number of major command and control projects at MIT's Lincoln Laboratory, including the Lincoln Experimental Satellite (LES) program. To complement his scientific and technical credentials, Dinneen brought management experience to his new position, having served as the Lincoln Laboratory's director.

Perhaps equally important, Dinneen was well acquainted with the labyrinthine politics of the Pentagon bureaucracy, having served on a number of defense advisory committees in the past, including the Defense Intelligence Agency's Scientific Advisory Committee.[6]

The new command and control assistant secretaryship was pointedly intended to play a major role in the administration's effort to bring organizational coherence to the command and control community, but it by no means completed the bureaucratic overturn. Simultaneously with the creation of the new assistant secretary position, the undersecretary of defense for research and engineering, William J. Perry, established a new, parallel office within his organization. Its day-to-day management was to be given to a new deputy undersecretary of defense for research and engineering (C^3I), who would also serve as Perry's principal deputy. While on the surface this new position might seem to represent an even further fragmentation in command and control authority within the Pentagon, it, in fact, was intended as quite the reverse. To consolidate that authority, Dinneen was given the dual hat as Perry's deputy, thereby setting the stage for a more active involvement in command and control issues by the Pentagon's top civilian officials.[7] The hope was that the consolidation would produce "a more effective and more efficient operation," and Dinneen lost no time in making known what he meant by effective.[8] For the strategic forces, it meant the unequivocal ability to deliver emergency action messages during the pre-, trans-, and postattack phases of a nuclear conflict. It meant maintaining communications between the various nuclear commanders in chief and the ability to direct the strategic forces under all conditions. It also meant having a report-back capability so that the status of one's own forces could be continuously monitored. In other words, Dinneen's first set of criteria for command and control effectiveness included survivability both to physical and electronic attack and endurance.[9]

Equally important were concerns for compatibility and interoperability. To achieve these goals, Dinneen announced that his office would advocate and pursue the use of such relevant technologies as digital operations throughout the command and control environment. He would also pay careful attention to interoperability among the command and control systems of the United States and its allies, especially among the member nations of NATO. Responsiveness was another important criterion given the time-sensitive nature of intelligence information. Reliability was also important, since critical information must not suffer degradation during transmission. Flexibility was also a must, and so Dinneen emphasized the need for improvements in satellite communications systems. Naturally the security of communications, satellite or otherwise, also had to be guaranteed. Finally, Dinneen forcefully underscored the point that the whole purpose of the system was to serve the needs of central decision makers, whether in managing crises, conducting war, or handling conflict termination.[10] The command and control centralizing impetus that had been reversed during the Nixon and short-lived Ford administrations was now to be reversed again.

So that these ambitious goals could be pursued, a central concern for Dinneen was to effect a tight coupling between system requirements and acquisitions. To this end, he advocated and aggressively pursued what he described as a "general systems approach" for command and control development. To a very considerable extent, this approach paralleled the logic of the WWMCCS architect in its recognition that planning for command and control systems must begin by identifying military capabilities, policy objectives, and the nature of the existing threat. Once these had been established, Dinneen noted, general command and control requirements for the military forces could then be generated for a series of situations ranging from peacetime to general nuclear war. These requirements, in their turn, would serve as the basis for specifying technical requirements in automatic data processing, communications security, survivability, and other areas. Only at that point could specific technologies capable of meeting the requirements be identified. Since more than a single technology would presumably be capable of doing the job, a range of alternatives would be developed, varying in capability and cost, which would then be presented to Pentagon decision makers for their consideration and selection.[11] How well then was this systems approach currently being adhered to in the

DOD? Dinneen and his boss, William J. Perry, proposed to find out.

The Defense Science Board Report

In September 1977 Perry asked the Defense Science Board (DSB) chairman Eugene G. Fubini to establish a task force to review the ways in which various defense elements developed and deployed their command and control systems. The task force, chaired by Bell Laboratories vice president Solomon J. Buchsbaum, was commissioned the following December. A number of command and control luminaries were appointed to it, including Clay T. Whitehead, who had served as director of the Office of Telecommunications Policy during the Nixon administration; Richard D. DeLauer, an executive vice president at the TRW Corporation, who himself later became assistant secretary of defense for C^3I; Charles A. Zraket, executive vice president of the MITRE Corporation; and, a variety of other academic and defense intellectuals and retired military officers. Buchsbaum's task force forwarded its final report to Fubini in July 1978.

The DSB task force identified a series of basic problems with the Pentagon's ability to develop and deploy command and control capabilities, the most basic of which was that practically no commonly understood conceptual framework that then existed for designing, analyzing, and evaluating command and control systems. Lacking any agreed-upon definition of what the system should do and how resources should be organized—indeed, lacking any agreed-upon vocabulary for articulating the issues specific measures of effectiveness were obviously impossible to formulate.[12] The results of this were the system's inability to provide appropriate information to commanders, a lack of responsiveness to the national leadership during crises, and the inherent inability to control the crisis situations this implies. The task force report's dour conclusion was that the United States had failed to deploy command and control systems "commensurate with the nature of likely future warfare, with modern weapons systems, or with our available technological and industrial base."[13] It was language hauntingly reminiscent of earlier critical system evaluations, suggesting that little material progress had taken place in WWMCCS at all.

Several broad recommendations, described as a "useful conceptual framework" for action, were then presented to address these problems.[14] First, to ensure the compatibility and effectiveness of joint-service programs, it was recommended that the DOD charter a new centralized command and control agency to manage their design and acquisition in a more coordinated way. This agency, working in conjunction with DARPA, would also be responsible for undertaking a major research effort to identify and develop the command and control technologies of the future.[15] The centralization that this recommendation bespoke was, however, quickly qualified by the next one, which was that the new agency establish general programmatic guidelines only. Program specifics, as well as the resources necessary to adapt and modernize their own command and control system to meet specific mission requirements, were to remain with the military services. Another recommendation similarly intended to enhance user participation in system definition and development was to strengthen the power of the services and the unified and specified commands to operate and evaluate their command and control systems. The task force's final recommendation, one also intended to provide greater user input into system development, was that the Pentagon issue new directives explicitly recognizing the unique character of command and control systems, and thus develop them in an evolutionary fashion.[16]

The services and the joint chiefs were asked to formally review the task force's recommendations, and they were unanimous in their resistance to an expanded, more powerful command and control agency, despite the conscious effort to ensure input and influence by the operating forces.[17] The services, not surprisingly, believed that their traditional dominance in the command and control area would be diminished if control were given to a central agency. The irony, of course, was that it was precisely because of long-standing service aggrandizement at the expense of joint-service functions that there was a need for a review of the practices used in planning and procuring command and control systems in the first place. Responding to the pressure, Defense Science Board chairman Fubini endorsed all of his task force's recommendations except the call for the creation of a new defense agency. He suggested instead, as a compromise, that the functions the new agency was to have performed could be accomplished equally well by expanding the charter of the Defense Communications Agency.

The revised report was forwarded to Secretary of Defense Brown in August 1978. The following month, Deputy Secretary of Defense Charles W. Duncan Jr. moved to implement its recommendations by instructing Undersecretary William J. Perry to take immediate action. Perry, in his turn, began with the task he considered most important: the revision of two DOD acquisition directives to account for the special evolutionary nature of command and control systems. These directives, numbers 5000.1 and 5000.2, were titled *Major System Acquisitions* and *Major System Acquisitions Procedures,* respectively. The need to begin here was manifest, Perry felt, for all of the other recommendations for enhanced user input into the planning, development, deployment, and operation of command and control systems was underpinned and justified by evolutionary logic. Because of its internal contradictions and the fact that it would put the brakes on the movement toward greater centralization, that logic merits a more detailed examination.

The Logic of Evolution

"I think it is time we faced up to the fact that command and control systems are, by their very nature, impossible to completely specify at the time development is begun," Albert Babbitt, the WWMCCS system engineer, remarked during the late 1970s. And even if by some alchemy it were possible to fully specify them, he went on, the incessant pace of technological advance and environmental change would preclude fielding command and control systems that were completely adequate. "We should accept this," Babbitt concluded, "and develop our systems in a way to accommodate growth and change."[18]

As the Defense Science Board Task Force saw things, command and control systems possess characteristics not found in other complex systems. They are "information rich," the task force said, meaning that they are highly dependent upon the information they contain and the demands placed upon them. They require integrating a wide range of users with diverse needs and perspectives and demand interoperability with a number of other systems. These purportedly unique features, it was argued, necessitated a special type of management structure, one in which command and control systems could evolve naturally over time. Under such an evolutionary approach to system development, users could identify their requirements, then develop the systems to meet them in incremental fashion. They could test off-the-shelf and prototype equipment in an operational environment and then select what worked best. If changes in the technology, doctrine, or threat occurred (and this was certain), commanders could redefine their needs and adjust their systems accordingly without having to begin again from scratch.[19] Given the fluidity of environmental conditions, why bother with exhaustive, and necessarily imperfect, a priori considerations of what one needs and what a system should do? When considered in this fashion, the evolutionary approach appears to have its advantages and its logic.

Indeed, the task force argued that these unique characteristics were sufficiently compelling that command and control systems should be exempted from the usual principles of life-cycle management, long recognized as a fundamental tenet of effective program management. The point was to have in place a management structure that could identify the roles and responsibilities of key individuals throughout the system and over time, emphasize management accountability for the success or failure of system development, promote interoperability of system assets through such strategies as standardization, and establish some sort of cost-control mechanism. To many, command and control systems should bend to this logic, but here was a distinguished task force saying that such systems were somehow fundamentally different and suggesting that the logic should not apply.

If the recommendation to exempt command and control systems from the usual type of management oversight was unusual, it also appears suspect, and for several reasons. First, the assertion that these systems are somehow unique, requiring a special type of management structure, is dubious on its face. Many systems are complex and information rich, have extensive demands made upon them, require interoperability among a variety of users, and still follow the principles of life-cycle management. Examples from the commercial sector include banking, aviation, and telecommunications. Many other systems have to deal with high levels of environmental uncertainty, and yet still engage in rigorous planning. The sense of this effort has been captured nicely in discussions of news organizations and their "emergency routines," carefully planned patterns of action explicitly designed to allow them to deal with unpredictable occurrences in a routine way.[20] Law enforcement, hospitals, and fire and rescue organizations appear to operate using similar routines. Private industry had similarly developed comparable structures to facilitate management's decision making in a world of volatile consumer preferences and technological surprises.[21] In this sense, there appears to be nothing unique about military command and control systems.

A second reason to question the appropriateness of the evolutionary approach was that it was unlikely to produce the "best" or "most optimal" system. This is because evolution as applied to defense systems operated in a different way from evolution in the natural world. The defense version of evolution involved a series of choices made incrementally over time, each decision at each stage of the process involving conscious deliberation and trade-offs. If the level of analysis is the subunit level, it is simply an iterative version of the standard engineering approach, in which human foresight and knowledge of constraints are ever at work, directing the selections that are made. It is a top-down approach, with direction proceeding from on high, imposing local rules and setting in motion a series of projects. That is simply how things human function, and, as philosopher Daniel Dennett has noted, this top-down approach is so common to large-scale human projects that alternatives are difficult to imagine. This approach stands in contrast to evolution in the natural world, which is not top-down and purposive but bottom-up and lacking in insight altogether. Whereas the evolutionary approach always has involved making a series of choices between alternatives, actual evolution is profligate and costly, throwing out and testing alternatives in countless myriads. "Mother Nature has no reason to avoid high-risk gambits," Dennett notes, "she takes them all, and shrugs when most of them lose."[22] Of the few that succeed, most do so in ways that were impossible to predict a priori.

The evolutionary approach thus misses the whole point of actual evolution. Where attempts are made to minimize unforeseen consequences in the former, in the latter unforeseen consequences represent precisely the stuff of evolutionary advance. In a world of limited resources, it is of course impossible for defense systems to be constructed in a similarly wasteful fashion. But if not all available alternatives are played out, it is not possible to find out which one might work best, thus severely restricting the possibilities for the serendipitous appearance of a better evolutionary form.

The evolutionary approach also misses the point of the standard engineering (weapons system) approach, which involves the rational pursuit of a capability based on a clear specification of user requirements and programmatic goals. Neither fish nor fowl, the evolutionary approach thus reaps neither the benefits of actual evolution nor those of rational planning.

As a consequence of using the evolutionary approach, inappropriate technologies were acquired and ill-considered organizational changes made, resulting in inadequate system performance. So unless funding were somehow to become limitless, this state of affairs appears to argue for a more rational management approach of the sort called for in the WWMCCS architecture. But the recommendation of the Defense Science Board Task Force was precisely the reverse: to avoid adequate goal specification and to permit the system to evolve "naturally," formalizing the de facto strategy of ad hoc incrementalism that had proved deficient in the past.

The notion of command and control system evolution approach was by no means original to the Defense Science Board Task Force. Robert S. McNamara had championed the evolutionary approach more than a decade before, describing how "changes in the command and control systems will be, of necessity, evolutionary, and the systems must be flexible enough to adapt to changes in the world situation and U.S. strategy."[23]

He believed that unlike systems whose designs had been frozen earlier, those which emerged from an evolutionary process of development would be more effective, possessing greater capabilities, better reflecting users' needs, and more closely aligned with the requirements of national military policy. Indeed, many of McNamara's ideas had been most forcefully articulated by none other than DSB Chairman Fubini himself, who, while wearing the dual hats of assistant secretary of defense and deputy director of defense research and engineering (DDR&E), had emphasized flexible development over inflexible "standardization" of assets.[24] His advocacy of an evolutionary approach to command and control system development proceeded directly from this logic.

Fubini's advocacy also derived from his disenchantment with such command and control products of the weapons system approach as Project 465L, the Strategic Air Command's Automated Command Control System. In Fubini's view, in the real world of national defense, a world characterized by incessant change, the whole notion of a fixed-system concept, the "old idea that a command and control system could be developed, built, tested, installed, and finally turned over to the users on some magic date," had been rendered anachronistic.[25] A different approach was called for, he believed, one that not only would recognize the fact of change but also would incorporate it as an integral element of the system's management structure. The evolutionary approach surely did, and Fubini had been one of its earliest, most forceful proponents.

The problem was that the evolutionary approach was at odds with the sort of centralized control of forces that was increasingly called for by recent changes in US strategic doctrine. Fubini himself had recognized this back in 1965, pointing out how the approach would likely result in a system that was "at best, a harmonious conglomerate of elements of different size, loosely but effectively federated."[26] Many critics of the approach would base their opposition on precisely this point, contending that "loosely but effectively federated" was an oxymoron. But despite this limitation, or perhaps precisely because of it, the idea of evolutionary development became fashionable with the release of the Defense Science Board's report. All the rage, it became "one of the most widely used 'buzz words' of recent years" in the words of DCA director Samuel L. Gravely Jr.[27] But unless one believes that illogic had all of a sudden become epidemic within the halls of the Pentagon, it follows that support for the evolutionary approach must reflect some other agenda.

In determining what that agenda might be, consider that the evolutionary approach had its strongest appeal to those who liked the notion of a "loose federation" just fine, those who were interested in maintaining their own autonomy, those whose interest lay in restraining the growth of centralized control. Whereas a firm requirement to name specific criteria and goals up front would inevitably have led to greater centralized control, the evolutionary approach worked in the opposite direction—decentralizing authority for command and control development to system subunits. It was a recipe for suboptimization, and as such, it held considerable appeal to self-interested subunits such as the military services, who, pursuing their own programmatic interests, adopted and defended it with gusto; often at the expense of the larger organizational entity that was WWMCCS.

The approach was also appealing because it increased the chances that favored programs would receive funding. After all, if program costs were fully elaborated at the outset, they might well scare cost-conscious members of Congress. But with the evolutionary approach, it was necessary to identify costs only for the relatively near-term, for a core capability, obviating the need to identify other expenses until some later time, when the system had "evolved." It was in many respects the services' analogue of the contractor's buy in, wherein an artificially low price could be initially offered to get a contract, with additional costs being added on later, after a major financial commitment had already been made to the program. The evolutionary approach also set the stage for *actual* contractor buy ins, with firms understanding full well that under the evolutionary logic prices could be raised, and raised again as the project evolved.[28]

In other words, the evolutionary approach appears to have been attractive because it allowed its proponents to have it both ways: they could cast themselves as forward-looking innovators even as they advocated an approach which maintained a status quo that worked to their benefit.

This subunit mind-set, part self-interested pragmatism and part Machiavellianism, was articulated well by one Air Force general who complained that Congress always seemed to demand a complete system architecture in hand before work on a program could proceed. The search for an optimal system was obviously desirable, he noted, "but few are willing to address the resources necessary to transition to the 'Mystical City,' the ultimate architecture, if the slums must first be cleared." So if the search for a perfect system resulted in elevated costs and program delays such that some programs were never initiated at all, the preferable approach, obviously, was to engage in step-by-step evolutionary progress.[29] If users could not identify their requirements with precision, nor contractors the types of technologies they could produce, no problem: "This will not prevent progress down the development path," remarked one Army general.[30]

The enthusiasm with which the Defense Science Board Task Force's report was received, then, and the pervasiveness of its subsequent acceptance, appears to have had less to do with its appropriateness in the larger programmatic scheme of things than with its usefulness to powerful WWMCCS subunits. Perry moved expeditiously to implement the evolutionary approach as *the* approach for command and control development, and it quickly became the newest front in the decades-old battle between the forces of command and control centralization and decentralization.

Nifty Nugget

Nifty Nugget 78 was a secret governmentwide mobilization and deployment exercise conducted by the JCS during the fall of 1978. Essentially a massive computerized war game, the exercise was, in many respects, a series of firsts: It was the first exercise of its kind ever to be conducted. It was the first time any mobilization effort had been mounted since the real mobilization that had taken place during World War II. Nifty Nugget was, finally, the first military exercise to test the operational capabilities of the new WWMCCS Intercomputer

Network, and, as such, it represented another opportunity for the Pentagon to demonstrate the value of the networking concept in support of command and control during a major (albeit fabricated) crisis.[31]

The scenario that Nifty Nugget employed was ambitious, to say the least: an all-out attack against Western Europe by the Warsaw Pact nations, igniting a major conventional war with NATO forces. The United States, as the senior member of NATO, was drawn into the conflict from the outset. As the scenario unfolded, US forces stationed in West Germany immediately found themselves engaged in pitched combat with vastly superior attacking Pact forces. Reinforcements were urgently required, and so some four hundred thousand American combat troops were to be mobilized and deployed to the plains of Central Europe rapidly. That was the general framework provided to the more than one thousand military and civilian Nifty Nugget players, located both in the United States and abroad. But it was a context only. In the war game, just as in a real war, things were not immutable or fixed. By design, many of the situations that arose were unanticipated, issuing from actions taken and decisions made earlier in the game. These were the conditions under which the one thousand players waged fictitious war for a full month that fall. (Throughout the exercise, by some scenario-mandated magic, the nuclear threshold was never crossed by either side.)

As it turned out, the war did not go well for the United States. In fact, what happened was that in remarkably short order, the mobilization plans of the United States simply fell apart. Many of the fictitious troops and much of the equipment that were to be deployed to the battlefront could not get there because of logistical snarls. Of the soldiers who did make it, most died, but not for want of proper training or weaponry. To the contrary, under the terms of the scenario, the troops were assumed to be highly trained and their weapons top notch. The problem was a lack of proper supplies and support. One exercise planner described how the Army "was simply attrited to death": artillery pieces had no shells, tanks had no fuel or spare parts, soldiers had no bullets or food.[32]

By no means was WWMCCS the source of all these difficulties; serendipity also had a hand to play. In one particularly bizarre incident, a civilian game participant working for the Department of Health, Education, and Welfare (HEW) retired just before Nifty Nugget began. As in reality, during the exercise HEW was responsible for handling civilian evacuees from the war zone after they arrived in the United States. But the problem for the exercise was that this responsibility had been invested in a single individual, now retired. He never had been replaced, and the not inconsequential result of this for other Nifty Nugget players was that right in the middle of the deployment operations they had to contend with almost a million civilian evacuees flooding into Army bases.[33]

Other problems were more directly attributable to the WWMCCS Intercomputer Network. According to participants, the response time of WIN was often unsatisfactory, in part because of the semi-independent nature of many of the software applications in use throughout the system. They also noted how the WWMCCS standard computers could not, in effect, walk and chew gum at the same time; they were simply incapable of keeping up with the simultaneous demands of the exercise and routine operations and maintenance. Many players were not provided with the level of automatic data-processing support they required. There also was a problem of securing alternative computing capability if the computers at one's own site went down, because other sites were generally operating near their limit, with little in the way of excess capacity to offer to anyone else.[34]

A final problem revolved around the issue of planning. One of the major lessons that was quickly drawn from Nifty Nugget concerned the lack of flexibility in the Pentagon's computerized mobilization plans. WWMCCS computers were programmed so that once a decision was made, a whole series of orders would be issued automatically and simultaneously to combat, transportation, and support units. It made a great deal of sense and speeded things up, provided that all eventualities could be anticipated. The only problem was that Pentagon planners, like everyone else, were far from omniscient. So when the unexpected occurred, when the war game diverged from the scenario as originally scripted, the computers were frequently caught flat-footed. In one example, an unexpected decision to redeploy a Marine unit to Iceland resulted in the loss of six full days of airlift capability. The original plans no longer valid, they had to be removed from the computer and recalculated by hand.[35] Problems such as these resulted in what Army vice chief of staff, Gen Walter T. Kerwin, the game's official overseer, described as "great gaps" in players' understanding. "You wonder," Kerwin mused ruefully, "whether they were playing the same exercise."[36]

Nifty Nugget was thus far from an unmitigated success, either in general terms or for WIN specifically. It disclosed a series of fundamental command and control shortfalls, including an inadequate ability to coordinate the use of transportation resources and difficulties in collecting the information necessary for crisis decision making. Also noteworthy were the insufficient automatic data-processing capabilities demonstrated by the World Wide Military Command and Control System and its new WWMCCS Information Network. And in much the same way that the three WWMCCS failures a decade earlier had initiated a series of critical assessments of the command and control structure and produced subsequent demands for reform, Nifty Nugget signaled the beginning of a similar process a decade later. What to do? One move in direct response to the deployment problems identified during the exercise was the creation of a new joint deployment agency within the JCSO to coordinate and manage the actual deployment of forces.[37] Other exercise-inspired activities included a series of DOD workshops and the creation of several new defense journals.[38] There was certainly reason enough for all this: Nifty Nugget raised the sobering possibility that the United States might just lose the next conventional war.[39] (Things proved little better two years later in Proud Spirit 80, an updated and much less ambitious version of Nifty Nugget. Despite the fact that it almost seemed that Proud Spirit had been designed to prove a point about the capabilities of WWMCCS, the performance of the WWMCCS computers continued to be considerably less than optimal. Indeed, in the words of one exercise participant, WWMCCS "just fell flat on its ass.")[40]

Notes

1. House, Committee on Armed Services, *Review of Department of Defense Command, Control and Communications Systems and Facilities*, 94th

Cong., 2d sess. (Washington, D.C.: Government Printing Office, 18 February 1977), 31–32.

2. I. Cassandra, "C^3 as a Force Multiplier—Rhetoric or Reality?" *Armed Forces Journal International* 115 (January 1978): 16.

3. Irving Luckom, "Worldwide Military Command and Control System: An Approach to Architecture Development," *Technical Directions*, IBM Federal Systems Division, Autumn 1976, 9.

4. C. Kenneth Allard, *Command, Control, and the Common Defense* (New Haven: Yale University Press, 1990), 194.

5. Edgar Ulsamer, "Machine Intelligence Shapes Global C^3 Nets," *Air Force Magazine*, July 1977, 71.

6. Gerald P. Dinneen, "Industrial Luncheon Address," *Signal* 31 (August 1977): 48.

7. ———"C^3I: An Overview," *Signal* 33 (November/December 1978): 11.

8. Dinneen, "Industrial Luncheon Address," 48.

9. Dinneen, "C^3I: An Overview," 13.

10. Dinneen, "Industrial Luncheon Address," 48.

11. Dinneen, "C^3I: An Overview," 12.

12. Kenneth L. Moll, "The C^3 Functions," in *Selected Analytical Concepts in Command and Control*, eds. John Hwang, Daniel Schutzer, Kenneth Shere, and Peter Vena (New York: Gordon and Breach, 1982), 23.

13. Gerald P. Dinneen, "C^2 Systems Management," *Signal* 34 (September 1979): 16.

14. Victor J. Monteleon and James R. Miller, "Another Look at C^3 Architecture," *Signal* 42 (May 1988): 82.

15. Comptroller General, *The World Wide Military Command and Control System—Major Changes Needed in its Automated Data Processing Management and Direction*, LCD-80-22 (Washington, D.C.: General Accounting Office [GAO], 14 December 1979), 62.

16. Dinneen, "C^2 Systems Management," 16.

17. Ibid., 18.

18. Albert E. Babbitt, "New Communications and Technology in WWMCCS," *Signal* 31 (August 1977): 84.

19. Comptroller General, 14–18.

20. Gaye Tuchman, *Making News: A Study in the Construction of Reality* (New York: Free Press, 1978).

21. Thomas C. Reed, "Evolving Strategy—Impact on C^3," *Signal*, March 1975, 9.

22. Daniel C. Dennett, *Darwin's Dangerous Idea: Evolution and the Meanings of Life* (New York: Simon & Schuster, 1995), 224–25, 259.

23. John B. Bestic, "No More Confused Situations," *Signal* 21 (March 1967): 56.

24. "Dr. Fubini Stresses DDR&E's Desire for System Compatibility," *DATA*, February 1965, 8–9.

25. Ibid., 9.

26. Ibid.

27. Samuel L. Gravely Jr., "The DCA—A Rock and a Hard Place," *Signal* 34 (April 1980): 8.

28. Bernard L. Weiss, "Keys to Success in C^3I," *Signal* 36 (August 1982): 53.

29. Richard C. Henry, "Space Systems Communications," *Signal* 34 (February 1980): 29.

30. Robert W. Zawilski, "Evolutionary Acquisition of U.S. Army Tactical C^3I Systems," *Program Manager* 15 (May/June 1986): 8.

31. John J. Fialka, "The Pentagon's Exercise 'Proud Spirit': Little Cause for Pride," *Military Review* 11, no. 1 (1981): 38.

32. Ibid., 38–39.

33. Ibid., 41.

34. Comptroller General, 54–55.

35. Fialka, 39.

36. William J. Broad, "Philosophers at the Pentagon," *Science* 24 (October 1980): 409.

37. GAO, *Deployment: Authority Issues Affect Joint System Development*, NSIAD-86-155 (Washington, D.C.: GAO, July 1986), 10.

38. Roger A. Beaumont, "Perspectives on Command and Control," in *Principles of Command and Control*, eds. Jon L. Boyes and Stephen J. Andriole (Washington, D.C.: AFCEA International Press, 1987), 15.

39. Broad, 412.

40. Fialka, 39.

Chapter 12

Crises and Criticisms

The fall of 1978 was a busy one for the World Wide Military Command and Control System. Following the Defense Science Board Task Force's report and the Nifty Nugget affair, the ADP battle was again joined, this time by the National Security Team of the President's Reorganization Project. At the outset of the Carter administration several years before, a series of efforts collectively known as the President's Reorganization Project had been initiated to improve the management and operation of a wide range of government programs. One of these, the Federal Data Processing Reorganization Study, was an effort to improve the acquisition, management, and use of computer systems. Responsibility for the study was given to the Office of Management and Budget (OMB), and OMB identified 10 general areas of interest with respect to government ADP. Ten separate teams of experts were then established to examine each of these areas, and one of these, the National Security Team, focused on the computer systems used by the DOD, particularly those used in WWMCCS and related systems. Its final report was released on 25 October 1978, and its numerous criticisms fell into three general areas: the adequacy of deployed ADP assets, ADP management within the DOD, and the quality of the personnel employed in ADP-related functions.

The National Security Team first found quality and adequacy of deployed computing assets to be severely unsatisfactory. Many of the WWMCCS ADP sites were not hardened, the team reported, and they offered little or no protection against nuclear attack or sabotage. The system's continued reliance upon batch-processing technology meant that users were restricted in their ability to operate in an on-line, real-time environment. Many of system's data-processing installations were operating near their limit, meaning that there was little surge capacity that could be drawn upon in the event of crises, when an increased volume of communications traffic and greater ADP load were inevitable. The situation was especially critical in the logistics and supply areas, where the WWMCCS computers already were operating at their limit.[1]

Arguably, many of these limitations could have been overcome by using backup computers, either on-site or at remote locations. The problem here was that existing arrangements for this sort of fallback computer support were themselves woefully inadequate. For the Air Force, except for a few critical commands such as NORAD and SAC, the availability of on-site backups was almost nonexistent. As for the use of backups at remote locations, some Air Force organizations had in fact entered into agreements to have this sort of service provided on an emergency basis. But the organizations that were to provide this service were themselves operating near their systems' capacities, meaning that they were unlikely to be able to satisfy substantially increased demands. In any case, the varying computer configurations between ADP sites virtually guaranteed that a backup capability, even if available, would likely not be fully appropriate. And, if things were bad for the Air Force, they were even worse for the Army, which had no surge or backup capacity whatsoever. The National Security Team noted that in discussing the backup issue with the military commands, a not uncommon response was "a shrug and a comment that manual procedures will have to do."[2] Such assessments were promptly characterized as "alarmist" by such insiders as John M. Carabello, the Pentagon's director of data automation, who claimed that existing backup systems were more than sufficient to take care of any shortcomings that might show up. When asked what systems he was referring to, however, Carabello responded by saying they were classified and could not be discussed with the reorganization project members.[3]

Still another serious problem was that few of the major WWMCCS facilities had uninterruptable power supplies or a provision for auxiliary power. The National Military Command Center in the Pentagon, the most vital WWMCCS installation, was found to be totally dependent on commercial sources of power.[4] Other critical installations also relied on commercial power, and often with serious consequences. During the course of the review, team members learned that such key commands as the Military Airlift Command and NORAD had their computers go down every time the commercial power lines on which they relied were struck by lightning, something that did not bode well for their ability to provide continuing service during a nuclear war.[5] While military officials were quick to point out that both NMCC and NORAD had diesel generators that allowed them to rapidly restore power in the event of outages, concern was hardly allayed; after all, most WWMCCS facilities did not have that capability.

The second general area addressed by the National Security Team concerned the management of defense computer resources, which it found to be fragmented and not managed as a coherent whole. Within the Office of the Secretary of Defense, for example, a large number of officials had been given responsibility for various aspects of the automatic data-processing function. These included the DOD comptroller, the assistant secretary for communication, command, control, and intelligence (C^3I); the undersecretary for research and engineering; and two of his deputies (research and advanced technology and acquisition policy). And if the civilians were bad off, things were even worse among the services. Lacking any central authority capable of setting guidelines and establishing standards, the forms ADP management took were all over the landscape. The Navy ADP program, for example, followed the traditional Navy management concept of centralized policy direction and decentralized execution. What this meant in practice was rampant suboptimization, with users developing and operating their systems essentially on their own without regard for the larger needs of the service. For the Air Force, ADP management was essentially a part-time affair that resulted in poor coordination, long approval cycles, and poor communications between the various agencies and individuals responsible for automatic data processing. Overall, it was concluded that the ability of ADP to meet a variety of critical defense functions was severely deficient.[6]

Another consequence of fragmented management was that the hardware and software technologies that were acquired were frequently obsolete. The National Security Team found the average age of DOD computing equipment to be some six years older than comparable equipment used in private industry—that is, a full generation behind. Many of the defense ADP assets used were no longer in production, meaning that premium prices had to be paid to repair and support outdated technologies. Poor management also meant that it was often impossible for defense organizations and agencies to determine what they were actually spending on ADP. The Army, for example, had always treated automatic data processing as a support service; costs were carried in other lines of its budget, and it was virtually impossible to break them out to provide a comprehensive look at Army ADP expenditures. Obviously, there was also no way to ascertain the total extent of the Defense Department's investment in ADP technologies and systems, and only broad estimates were possible.[7] This was precisely the situation that David Packard had faced at the beginning of the decade, which he had hoped to resolve by creating the position of assistant to the secretary of defense for telecommunications, and instituting related changes in management structure. But things appeared to have improved little with the passage of the years.

It was a condition that had not gone unnoticed by Congress. If the Pentagon was unable or unwilling to manage its automatic data-processing programs itself, many congressmen reasoned, Congress had no choice but to fill the vacuum. And fill it they did, especially Rep. Jack Brooks of Texas, who used the investigatory powers of the General Accounting Office like a weapon in his war against what he viewed as widespread Pentagon ADP mismanagement. The results were civilian micromanagement of military ADP and considerable military resentment. The only way to put an end to this acrimonious situation, the National Security Team concluded, was for the Pentagon to restore confidence by putting its managerial house in order.[8]

The final area of attention was the personnel who worked in defense ADP, and here the major concerns were career advancement opportunities for ADP professionals and the quality of personnel that those prospects engendered. As to the first, from the beginning, one of the most serious impediments to the utilization of ADP technologies had been widespread institutional resistance. In those early years, computers and those who operated them were considered a sort of necessary evil. Computer experience did nothing to enhance career opportunity, and it was only the "well-rounded officer" who would eventually get a shot at flag or general rank.[9] The National

Security Team discovered that things had not changed much. There was severe and widespread dissatisfaction with training and the prospects for career advancement, arising most notably from a conspicuous lack of flag or general rank billets to be filled by officers with an ADP background and a considerable resistance to making any more slots available. "There are three ways to make a career in the Navy: under the water, on the water, and in the air," one admiral remarked, exemplifying the prevailing view, "I'd really wonder about an officer who wanted to make a career in computers."[10]

The results of such an atmosphere of professional disdain were entirely predictable: Many of the brightest and most ambitious officers avoided the ADP area like the plague, and others left military service during their peak years of productivity, while the quality of those who remained diminished. To make matters worse, these conditions had their analogue among civilian computer specialists. Why, after all, would the best and brightest want to work with obsolete technologies when private industry offered them the chance to do cutting-edge work? In addition, industry provided its ADP specialists both with better career opportunities and considerably higher salaries. In short, obsolescence in ADP hardware and software translated directly into personnel whose ADP skills were themselves obsolescent.[11]

Coming up with ways to improve this dreary situation was the National Security Team's final task, and recommendations were quick to follow. The Pentagon should forcefully implement life-cycle management policies for developing computer systems. The services and defense agencies should be required to make all information technology costs explicit in their budgetary requests. The problem of obsolescent ADP equipment should be given prompt attention. Adequate career paths for ADP specialists should be established. A complete reorientation of the Pentagon's relations with Congress should be vigorously pursued. The military had been rendered operationally vulnerable because of its ADP shortfalls, the team concluded, and only such actions, vigorously pursued, could begin to remedy the situation.[12]

A reorganization of ADP management was clearly called for. Despite the warning at the outset, saying, "We do *not* recommend that some form of radical reorganization take place," the report's central recommendation was fairly substantial nonetheless. Noting that previous recommendations for change, including the 1970 blue ribbon panel and a 1975 RAND study, had been largely ignored, the team urged a vigorous restructuring of ADP management both on the Pentagon's civilian and military sides. Along the lines of the Defense Science Board proposal submitted just a few months earlier, what the team called for was a substantial centralization in ADP resource management—specifically, the creation of the Office of Information Technology within the OSD with responsibility for all defense ADP programs, including those implicated in WWMCCS.[13]

Even before the National Security Team's report was issued, the services, anticipating its criticisms, began taking preemptive action. The Air Force formally increased the frequency of the meetings of its ADP review board, established new equipment acquisition programs to replace obsolescent equipment, and implemented a new ADP career path. The Army, which created an assistant chief of staff for automation and communications, was actively considering changes in its ADP acquisition process, and was working on establishing new ADP career fields for officers. As for the Navy, the service secretary issued an instruction reorienting the Navy Automatic Data-Processing Management Steering Committee to a more strategic oversight function. A special project to develop a new Navy long-range ADP plan was undertaken, and a number of efforts to replace obsolescent ADP equipment were initiated.[14] None of these "good citizen" reforms addressed the team's central recommendation for a new centralized ADP authority, of course. Indeed, they could reasonably be interpreted as moves designed precisely to prevent the establishment of such an office and to maintain ADP control at the subunit level. But as had happened so often in the past, a crisis would cast doubt upon the wisdom of decentralization.

Jonestown, Guyana

The prototype WWMCCS Information Network and its operational successor WIN had experienced their fair share of difficulties during their initial period of controlled testing and uncontrolled use in JCS training exercises. How well WIN would perform under actual crisis conditions remained a question mark as Nifty Nugget ground to its unhappy conclusion in the fall of 1978. The WWMCCS community would not have long to wait in having that question answered, however. WIN would be put to the test that November because of the actions of a religious zealot by the name of the Reverend Jim Jones.

Several years before, Jones and more than one thousand members of the People's Temple, as his religious following was called, had moved from San Francisco to Jonestown, Guyana, reportedly because of the negative publicity Jones had been receiving about his unorthodox preaching and his claims to heal the sick and raise the dead. *Moved* is probably not the most apt term to employ, however, for what took place was more like a disappearing act on a vast scale. As one reporter described it, "Spouses who were in the Temple left husbands and wives who were not. Children dropped out of school. Homes for the elderly, which were run by the Temple, were suddenly emptied of patients and staff. Wealthy members sold their homes and other possessions, or simply left them behind."[15] Even when worried relatives and friends discovered where their loved ones had gone and attempted to contact them by radio, all they got amidst the static were recriminations and demands to be left alone. For those left behind and bewildered by this mass exodus, there was substantial cause for concern.

The concern was particularly strong among a group of family members who, immediately following the Guyana relocation, organized themselves into a group called the Concerned Relatives. Most of these people disliked Jones intensely, and many of them publicly dedicated themselves to doing everything necessary to rescue their loved ones from his influence. One element of their concern involved appealing to Rep. Leo J. Ryan of California for assistance, with group members describing to him how their relatives were being held in Guyana against their will. Ryan responded by organizing a fact-finding tour of Jonestown on their behalf.

That tour began on Wednesday, 15 November 1978, the day Ryan arrived in Georgetown, Guyana's capital. Among the congressman's entourage were two staffers, eight reporters, and a 14-person delegation from the Concerned Relatives group. Ryan and a smaller group left the capital for Jonestown the following day. Things apparently started well enough, but had clearly deteriorated by Saturday, the day Ryan departed Jonestown, taking with him a small group of cult defectors. Shortly after the congressman left, Jones broadcast a radio message saying that Ryan had persecuted him, and that "avenging angels" had been sent after him. The angels caught up with Ryan and his party at the Port Kaituma airstrip, some six miles from Jonestown. Ryan was killed by a shotgun blast in the face, becoming only the second US congressman ever to be assassinated. Four of those with him were also killed, three of them members of the press, and 10 others were wounded.

The carnage had only just begun, however. Once word was received that Ryan had been killed, everyone in the Jonestown colony was ordered to gather at the colony's pavilion. After they had assembled, Jones ordered them to drink cyanide-laced grape Fla-Vor-Aid. Those who refused to drink it received injections. All except Jones, that is, who either shot himself or had someone else do it for him. More than nine hundred people died, 260 of them children.[16]

Sorting out the details of what actually happened that day would take place only later. At the time, it was a real-life crisis—a small one, to be sure—but the type of crisis that WWMCCS had been designed to manage. As soon as the first report of the attack on Ryan's group reached Washington, the JCS immediately assembled a crisis action team. Sending a military force to Guyana was under serious consideration, meaning that the chiefs needed information concerning the availability of planes, troops, and medical aid. To help them in their search, the chiefs turned to the Readiness Command, a WWMCCS node located at MacDill AFB near Tampa, Florida.[17] A Guyana crisis team was promptly created there, and a teleconference between the Washington and Florida teams commenced by way of the WWMCCS Information Network. And it was with WIN's teleconferencing software that the problems began.

The lengthiest and most embarrassing WIN failure during the Guyana crisis started with a typical Florida thunderstorm.

At the height of the crisis, the storm created a power outage that interrupted the teleconference. Power was quickly restored but not the teleconference. When members of the joint chiefs' crisis action team tried to rejoin the conference, the WWMCCS computer in Florida, which was the host computer for the conference, would not accept their request to sign on. The reason, as a subsequent inquiry would determine, was that despite the disconnect, the joint chiefs' remained "signed on" as far as the host computer was concerned, and the WWMCCS software would not allow them to sign on again. A condition bordering on panic ensued as automatic data-processing specialists at the Pentagon frantically tried to figure out what had gone wrong. The solution they arrived at was to create a new code name for themselves, one the Florida WWMCCS computer would accept, thus allowing the teleconference to be re-established. The fix worked, but by the time all of this had taken place, the joint chiefs' crisis action team had been out of touch with the Readiness Command for more than an hour.[18]

In a related incident, the National Military Command Center's automatic data-processing liaison officer attempted to use WIN to enter the teleconference. But when he tried to access the Readiness Command's computer by way of WIN, the message "Remote Host Dead" appeared on his terminal. The computer was clearly down, so he picked up the telephone and called the Readiness Command to find out what was wrong. He was told that there was no problem, the computer was up and fully operational. Confused, the ADP liaison officer spent the next 20 minutes making repeated, unsuccessful attempts to access the Readiness Command's WWMCCS computer by way of WIN. The real problem was not the computer, it turned out, but rather the dedicated communications lines that were being used in WIN. When this possibility occurred to the ADP officer, he switched his computer terminal to another, non-WIN communications line and established contact with the Readiness Command's computer. But this was in spite of, not because of, the WWMCCS Intercomputer Network. To make matters worse, bugs in WIN's software caused headaches for teleconference participants who attempted to log off to perform data-processing functions; when they did so, unexpected computer failures occurred. Things got so bad that the

Readiness Command finally directed its personnel not to use WIN's teleconferencing capabilities until the problems could be fixed.[19]

This was obviously not how things were supposed to work. The problems experienced during the Guyana crisis and Nifty Nugget seemed to represent a late-1970s analogue to the *Liberty*, *Pueblo*, and EC-121 crises of the late 1960s. But a key difference was that many members of the press had heard of WWMCCS by this time and knew at least in general terms what the system was supposed to do, so when fresh word of its failures began leaking out, they were ready. One critical press account began by noting how the Guyana crisis had simply overwhelmed the WWMCCS computers and went on to say how their malfunctions had been so frequent "that the joint chiefs were stymied." Although the system was supposed to have permitted the chiefs to manage the Guyana crisis, what happened instead was a situation "just like an airline ticket counter when all the computers go down." WWMCCS was publicly denounced as a disaster, a system with apparently intractable technical and managerial problems.[20] Not surprisingly, Pentagon officials were quick to disagree, dismissing the press reports as "horror stories" and the situations they portrayed as of dubious relevance for actual military operations. They pointed out how throughout the Guyana crisis, the average availability for WIN's 12 network nodes had been on the order of 95.5 percent.[21] However that may be, the 5 percent or so of non-availability clearly seemed to have major ramifications for crisis management and would soon lead to some of the sharpest criticism ever of WWMCCS and its automatic data processing.

DCA and Centralization

It was in a dour mood in January 1979 that Assistant Secretary of Defense for C^{3}I Gerald Dinneen instructed Defense Communications Agency director Samuel L. Gravely Jr. to prepare a plan for expanding his agency's charter along the lines suggested by the Defense Science Board's report. Gravely was enthusiastic, and he consulted numerous experts both inside the DOD and in the private sector. His report was submitted to Dinneen and the JCS that February.

Symbolic of the proposed expansion of his agency and its functions, management responsibilities, and emphases, Gravely's plan began by recommending that the name of his agency be changed to the Defense Communications, Command, and Control Agency (DC^{3}A). A series of specific recommendations followed: First, the expanded agency would be given complete control over the Defense Communications System, including funding and actual operations. Second, the agency would assume budgetary and operational control for a variety of joint-service command and control programs, including TRI-TAC, the Joint Tactical Communications program, and general program guidance over other tactical programs such as the E-3A Airborne Warning and Command System program.[22] DCA's previous responsibilities for program monitoring and technical support for a number of national-level command and control programs would be expanded similarly. The agency would assume sole responsibility for WWMCCS system engineering. It would exercise management control and have funding responsibility for such priority elements of WWMCCS as the National Military Command System and the Minimum Essential Emergency Communications System. It would assume control over WWMCCS automatic data processing, secure voice communications, electronic counter-countermeasures, and a range of other vital system functions. In a number of other areas in which DCA had no previous involvement, including the National Emergency Airborne Command Post program, the Precision Acquisition of Vehicle Entry and Phased Array Warning System (PAVE PAWS) and Ballistic Missile Early Warning System (BMEWS) radar systems, and the Navy's Take Charge and Move Out (TACAMO) planes, it was recommended that the agency be given responsibility for general program guidance.[23]

Gravely was clearly suggesting an approach to the development and operation of command and control systems in which the activities of the military services would be more tightly coupled to, and directed by, national-level requirements. While to some extent the Defense Communications System already provided the national leadership with this sort of top-down view, Gravely noted that as far as WWMCCS as a whole was concerned, there was still a long way to go.[24] Truly a clarion call for centralization, DCA's plan was intended to help

WWMCCS advance at least some distance in that direction; and this, after all, was what Gravely thought Dinneen had asked him to do. The plan also could not have been further removed from the ad hoc incrementalism of the evolutionary approach advocated in the Defense Science Board's report.

Dinneen promptly asked the Joint Chiefs of Staff and the military services to review the DCA plan. Their formal and informal responses, coming in over the ensuing weeks, were so blistering in tone, so emphatic in their recommendation that an expanded DCA not be created, that Gravely found himself questioning "our mutual understanding of what the C^3 community is really trying to do." The services, specifically, interpreted the DCA report as a power play, an illicit attempt by the agency to strengthen its bureaucratic hand and feather its nest at their expense. One officer typified this view when he remarked sardonically how DCA would, no doubt, "happily volunteer to take over the reins" of command and control, and then "require some 300-plus additional billets and control of service funds." Although Gravely made it a point to say that DCA was not on its own initiative seeking additional responsibilities for WWMCCS (he had, after all, been asked to prepare his report), service suspicions were hardly allayed when he went on to say that his agency would not shrink from any new tasks that might be assigned to it.[25]

Much of the resistance stemmed from the fact that many service officials continued to consider the Defense Communications System to be strictly a general-user, peacetime system. By giving DCA a more central role in WWMCCS, by giving it responsibility for key WWMCCS assets such as Minimum Essential Emergency Communications Network, by emphasizing such issues as survivability and a concern with command and control counter-countermeasures, Gravely's report clearly suggested that DCA's war-fighting responsibilities were to be substantially expanded. But the services considered the conduct of warfare to be their exclusive province, and they jealously guarded that mission and the funds, personnel, and hardware that followed directly from it.

A closely related concern from the services' perspective was that the changes Gravely advocated would result in a further blurring of the distinction between the strategic and tactical worlds. Formulation of the next iteration of US strategic doctrine, countervailing strategy, was already well in the works in the higher levels of the Carter administration (it would find formal expression in July 1980, with the issuing of Presidential Directive 59). The doctrine's central tenet, that the United States should be capable of waging and prevailing in a protracted nuclear war, carried with it an obvious demand that the NCA be able to communicate directly with commanders on the scene for a protracted time under all sorts of conditions. It implied that the intelligence traffic on which command and control relies would flow unimpeded across both strategic and tactical assets and whatever boundaries might formally separate them. Under this new set of rules, everything would in effect become "strategic."

There were major problems with this from the perspective of many service officials. First, they believed countervailing strategy to be little more than the most recent in a series of increasingly elaborate civilian doctrinal fantasies. Many of the forces and technologies necessary to make it work either were not in place or simply did not exist.[26] Of those assets that did exist and that could be marshaled in support of the new doctrine, many were currently considered "tactical" assets and the property of a single service. By redefining everything as strategic, such control would presumably be lost, meaning that the costs of pursuing the doctrinal fantasy would come directly out of the services' hide. So when Gravely called for a more powerful Defense Communications Agency, one in which the "needs of the national level take precedence over the parochial needs of the individual services," one in which the traditionally distinct boundary separating service and agency would be blurred (to the services' detriment), it is hardly surprising that everyone balked.[27] Indeed, so strong was the resistance, so energetic the efforts to protect the bureaucratic status quo, that Gravely would soon exclaim in dismay, "I feel, at times, that we let management boundaries become the primary concern rather than national capabilities."[28]

Thrown onto the defensive, Gravely would cause additional irritation by publicly airing his view that many of the system's shortcomings could be directly traced to its subunit-dominated character. The substantive issues addressed by the Defense

Science Board Task Force, as well as the recommendations for reform that it had proposed, were being lost in the shuffle, he said, as the services hastened to protect their roles and their missions. As for himself and the DCA, Gravely noted that by calling a spade a spade he risked suffering the fate of the Young Lady of Kent:

> There was a young lady of Kent
> Who always said just what she meant.
> People said, 'She's a dear,
> So unique—so sincere—'
> But they shunned her by common consent.[29]

Gerald Dinneen closed the matter, presumably sparing Gravely that ignominious fate, by creating in March 1979 a new command, control, and communications systems directorate within the Joint Chiefs of Staff organization (JCSO). Lt Gen Hillman Dickinson, its new director, would serve as principal advisor to the undersecretary of defense for policy and to the chairman of the JCS for command and control matters. His directorate would serve a master planning function, establishing priorities for research and development initiatives, as well as for operational programs. In this, the directorate would represent the military analogue to the Office of the Assistant Secretary for C^3I, with which its activities would be coordinated. In other words, if there was going to be increased centralization of the command and control function, the joint chiefs wanted to make certain it would occur on their terms. As a complement to the changes in the JCSO, Dinneen began plans to reorganize his own office, and these were approved in July of 1979.

As for the Defense Communications Agency, although Gravely would not get his new, enhanced DC^3A, his efforts to enhance the authority of his agency would not be wholly in vain. As bureaucratic recompense to the irate DCA director, Dinneen and his bosses subsequently proposed that the World Wide Military Command and Control System Engineering Organization be integrated into the DCA. When this occurred, the WWMCCS system engineer would be given the additional title of deputy director for command and control systems and be given responsibility for providing architectural and systems support for interservice command and control systems. This included responsibilities for establishing defensewide architectures for computer-to-computer communications, secure voice communications, and the move to an all-digital world.[30] Additionally, DCA's current deputy director for command and control would be retitled deputy director for command and control technical support, emphasizing his responsibility for directing the Command and Control Technical Center that supports the JCS organization. The center was given a series of WWMCCS responsibilities, including developing and maintaining plans for WWMCCS's further development, managing WWMCCS standard software applications, and providing an organizational point of contact for database administration.[31] Dinneen remarked that all of these changes were the best way to pursue the DSB task force's recommendations to create a central focus for command and control systems management, while at the same time involving the unified and specified commands in the process.[32] All in all, the amount of organizational shuffling and reshuffling had been considerable. Whether any of these new arrangements would ultimately make for a more responsive WWMCCS was, of course, an entirely different matter.

The 1979 GAO Report

Of all the criticisms of WWMCCS that were mounted during the 1970s, none was harsher or more unremitting in its attack than a General Accounting Office report released just two weeks before decade's end. The report's genesis was in John H. Bradley's firing from the Defense Communications Agency for his persistent criticisms of PWIN reliability. Believing that the Pentagon's only interest was in sweeping WWMCCS's problems under the rug and outraged that his concern for the nation's security was earning him nothing but bureaucratic disapprobation and vindictiveness, Bradley decided to go public. Speaking to reporters on the record, he had outlined PWIN's many problems and shortcomings. His interviews resulted in a series of articles in the popular press, all of them highly critical of WWMCCS, which had the effect of focusing public attention on the system for the first time since the command and control failures of the late 1960s. Perhaps more significantly, congressional interest was aroused, and the watchdog GAO was directed to conduct a full-scale investigation into WWMCCS's automatic data processing.[33]

The GAO report, titled *The World Wide Military Command and Control System—Major Changes Needed in its Automated Data Processing Management and Direction*, was released in December 1979, and on almost every one of its more than one hundred pages were calls for precisely such major changes. GAO's overarching criticism, echoing in many respects the findings of FSD's architectural studies, was that the Department of Defense had never adequately defined command and control systems users' information requirements. Various reasons for this were cited, one of the more important being that WWMCCS's ADP management structure was so complex and fragmented that no single individual or organization had responsibility for such matters as budgeting, funding, and management. Responsibilities were all over the landscape, it was said, and accountability hard to pin down. As a consequence, nobody was responsible for ensuring the coordination and operation of the system. Nobody was accountable for system deficits or for making necessary changes when things went wrong. Indeed, things were so diffuse and uncoordinated that no one even had a thorough general understanding of the program. It was this lack of centralized management, the GAO concluded, that impeded defense efforts to design, develop, implement, and operate a command and control system responsive to users' needs at the local level, or a national-level system capable of meeting the declared policies of the United States.[34]

In other words, nobody was really in charge of WWMCCS automatic data processing, and it showed. The GAO report then proceeded to describe in graphic detail the shortcomings of the WWMCCS standard computers and their associated system software, noting how these failed to support the command and control function. Most of the computers were found to lack independent and uniform sources of electrical power, making them vulnerable to power outages. Nor were the computers survivable, and there was little provision for backup in the event of accidents, acts of nature, or aggression. The computers' main memory was limited and the machines relied on batch-processing technology, both of which resulted in serious difficulties when the system was called upon to operate in a real-time environment. It was these conditions, the GAO said, that had produced the problems encountered in exercises such as Nifty Nugget, and in real crises like Guyana. The overall conclusion contained in the report could hardly have been clearer: the objectives of the WWMCCS ADP program had not been achieved, and there had been little, if any, improvement in the program since its inception back in 1966.[35]

What should be done? To create a more effective system, the GAO underscored the recommendation of virtually all past studies: a single central organization should be given project management authority for all WWMCCS and WWMCCS-related computer-based information systems. This WWMCCS project manager, as the GAO called it, would assist in the identification of users' information requirements, prepare plans for the development of ADP systems responsive to those needs, develop systemwide standards for WWMCCS, and implement cost-accounting mechanisms throughout the WWMCCS community. Since much of this was frankly inconsistent with the evolutionary approach to system development in vogue in the Pentagon, the GAO urged that such an approach be immediately abandoned. For it was the evolutionary approach, perhaps more than anything else, that limited proper system development and impeded appropriate management practices for WWMCCS ADP. In its stead should be substituted the principles of life-cycle management, emphasizing rational planning and clear accountability. Finally, it was recommended that Congress withhold funding both for WIN and for upgrading the WWMCCS standard computers until these necessary reforms were made.[36]

The GAO concluded with a discussion of how Pentagon officials had tried to restrict the scope of its investigation: "We were unable to fully discharge our statutory responsibilities because the Joint Chiefs of Staff denied us complete access to documents we considered to be pertinent to the evaluation." For example, no information at all was provided about the alleged WWMCCS fiasco that occurred during the seizure of the American merchant ship *Mayaguez* in May 1975, when the WWMCCS computers reportedly crashed.[37] Only partial information was provided about the Guyana crisis, it was said, and funding figures for a number of WWMCCS-related agencies and programs were not provided. All told, only about two-thirds of the information requested from the Pentagon was actually made available.

For the GAO investigators, it added insult to injury that much of the withholding apparently took place on the flimsiest of bureaucratic pretexts. Navy documents concerning Nifty Nugget, for instance, were withheld because they did not represent the "official position" of the Department of the Navy. *The Preliminary National Military Command System Master Plan* was not provided because it was not an "official" document—this despite the fact that it had been provided to FSD three years earlier to aid in its work as WWMCCS architect. Another set of documents, the *Technical Support Requirements for the Command and Control Technical Center*, were withheld because they were "internal working documents." The clear position of the Pentagon was that drafts, working papers, and so on, would simply not be made available. To make matters more inconvenient still, Pentagon officials stood firmly on bureaucratic formality and required that all requests for documents and interviews be made in writing. The result was lengthy delays, many of which were doubly unnecessary because Pentagon officials apparently decided to sit on information for as long as possible, for days or weeks after it was ready for release. The GAO concluded that these sorts of actions, many "without legal justification," had adversely impacted its ability to provide a thorough, timely evaluation of WWMCCS automatic data processing.[38] But given the extensiveness of the WWMCCS shortcomings described in the report, it is difficult to imagine what a more thorough evaluation might have revealed.

But criticisms there were aplenty that had to be addressed by the Pentagon, and in so doing, the most common response was the one used in the wake of the Guyana crisis: acknowledge that some problems exist, while shifting the ground of argument from what is wrong with the system to what is right with it. (Yes, there were some problems with WIN during Guyana, but the network had an average component availability of over 95 percent.) General Dickinson, the newly appointed director of C^3 systems in the Joint Chiefs of Staff organization, exemplified this approach by acknowledging up front that a number of improvements to the system were still necessary. But he quickly followed this up by noting how significant improvements in performance had been achieved in the recent past, most of them as a result of better management. For example, evaluations had shown that WWMCCS automatic data processing was generally quite satisfactory for routine operations. The computers were "doing more and more [of] what they were designed to do," Dickinson said. They performed "very useful functions," and "no using command would give up its WWMCCS ADP service."[39] A Pentagon document leaked to the *Washington Post* defending WWMCCS similarly claimed that the computers provided generally effective support to their users. If there were any problems, it was only when extraordinary demands were placed upon the system.[40] Throughout, the flap over WWMCCS's deficiencies was attributed more to critics' characteristics—their naivete, bias, opportunism, even maliciousness—than to any actual problem with the system. The official view was that the system had gotten a bum rap, when in fact "WWMCCS works."[41]

But the glare of the media spotlight was relentless. Dismayed by what was interpreted as unfair press treatment of WWMCCS, the Pentagon's public stance soon became one of defensiveness or silence. Irritated by the congressional meddling in defense automatic data-processing matters that the criticism had provoked, suspicious of congressional motives, the Pentagon began to experience increasingly strained working relations with Congress. It was a mistrust that cut both ways, of course. The treatment of the GAO during its inquiry had led not a few members of Congress to question the credibility of Pentagon officials' motives and objectives in their dealings with Congress. Indeed, things had deteriorated so badly, relations with some congressional groups had become so strained, that if another major incident involving WWMCCS ADP occurred, it did not appear unlikely that the whole enterprise, budget and all, might become vulnerable to congressional reprisal.[42]

Such was the context when a team of 30 academics was assembled at the Pentagon in 1979. These outside consultants—anthropologists, systems theorists, mathematicians, and all the rest—were not being turned to in a search for new technologies, perhaps not surprising given the frequent inability of new technologies to solve WWMCCS's problems. Rather, what the Pentagon was looking for from these "academics with a philosophical bent" was an organizational solution; some new, nontraditional command and control management technique that would be consistent with all other defense resources and functions.[43] At bottom, the difficulty was the complexity of command and control, its essential ambiguity—problems made more difficult by the phenomenal growth that had taken place since the time WWMCCS was established. The whole enterprise was now incomparably larger in quantitative terms: more people, organizations, facilities, and assets. It had changed qualitatively as well. Driven by changes in doctrine and the strategic threat, there had grown up an apparently insatiable demand for information, of new and different sorts, that was needed ever more quickly. Both driving this demand and a consequence of it was the explosive growth of automatic data processing at every level of the system.[44] The need was to find a way to fuse the disparate elements of WWMCCS into a coherent whole, and it had at last become clear to many officials that a technological fix was probably not the answer.

True reform would likely come hard, however, because of the subunit-dominated nature of WWMCCS and the fact that change would affect powerful constituencies in ways they found inimical to the performance of their missions. The services tended ever to go their separate ways, their efforts generally lacked positive synergistic effects, and the Pentagon hoped the academics would "find the means of orchestrating the cacophony now rampant in the US command and control domain," as one of the consultants who helped organize the meetings phrased it. But these hopes would be quickly dashed; no easy answers were out there just waiting to be discovered and implemented. To be certain, all sorts of recommendations were offered up, some of them drawing upon elaborate biological models or theories from the hard sciences. But in the words of the consultant, what was lacking was "critical examination of the dominant paradigm which condones the expenditure of vast resources without even a semblance of a conceptual rationale for the effort."[45] For without such a rationale, a truly effective WWMCCS would likely remain chimerical, with additional system failures a virtual certainty. These were, in fact, not long in coming. The next failure, or, more appropriately, series of failures, took place at one of WWMCCS's most important nodes, the Cheyenne Mountain headquarters of the North American Aerospace Defense Command.

Notes

1. President's Reorganization Project, *Federal Data Processing Reorganization Study: National Security Team Report*, 25 October 1978, 11.
2. Ibid., 11, 22.
3. Frank Greve, "Pentagon Calls Super-Computer a 'Disaster,'" *Parameters* 10, no. 1 (1980): 95.
4. William J. Broad, "Computers and the U.S. Military Don't Mix," *Science*, 14 March 1980, 1187.
5. President's Reorganization Project, 11.
6. Ibid., 4, 6, 40, 43.
7. Ibid., 6.
8. Ibid., 38.
9. "To Moorer, the 'DomRep Action' Proved—CINCLANT has Command and Control," *Armed Forces Management*, July 1965, 70.
10. President's Reorganization Project, 36.
11. Ibid., 5, 9, 18, 22, 27, 30.
12. Ibid., 45.
13. Ibid., 48.
14. Ibid., 66–69.
15. Lawrence Wright, "Orphans of Jonestown," *The New Yorker*, 22 November 1993, 69.
16. Ibid., 66.
17. Broad, 1183.
18. Comptroller General, *The World Wide Military Command and Control System—Major Changes Needed in its Automated Data Processing Management and Direction*, LCD-80-22 (Washington, D.C.: General Accounting Office, 14 December 1979), 55.
19. Ibid., 55–56.
20. Greve, 94.
21. Comptroller General, 55–56.
22. Ibid., 17–18.
23. Ibid.
24. Samuel L. Gravely Jr., "The DCA—A Rock and a Hard Place," *Signal* 34 (April 1980): 8.
25. Ibid., 7–8.
26. Janne E. Nolan, *Guardians of the Arsenal: The Politics of Nuclear Strategy* (New York: New Republic Books, 1989), 138.

27. Samuel L. Gravely Jr., "DCS at the Crossroads," *Signal* 33 (May/June 1979): 54.
28. Gravely, "The DCA—A Rock and a Hard Place," 11.
29. Ibid., 7.
30. Gravely, "DCS at the Crossroads," 54.
31. Comptroller General, 97.
32. Gerald P. Dinneen, "C^2 Systems Management," *Signal 34* (September 1979): 19.
33. James North, "'Hello Central, Get Me NATO': The Computer That Can't," *Washington Monthly,* July–August 1979, 52.
34. Comptroller General, 12.
35. Ibid., 5.
36. Ibid., 17, 71–73.
37. Richard C. Gross, "C^3: Fewer Mixed Signals," *Military Logistics Forum* 3 (June 1987): 20.
38. Comptroller General, 67–69.
39. "Improving C^3 Systems and Requirements," *Signal,* May/June 1981, 75–76.
40. Michael Putzel, "Pentagon Warning System Defective, Experts Claim," *Washington Post,* 10 March 1980, A10.
41. Perry R. Nuhn, "WWMCCS and the Computer That Can," *Parameters,* September 1980, 20–21.
42. President's Reorganization Project, 8.
43. William J. Broad, "Philosophers at the Pentagon," *Science,* 24 October 1980, 409.
44. Ibid., 410.
45. Ibid., 412.

Chapter 13

Failures at NORAD

Understanding the failures that brought to a close the second decade—and the second phase—of the World Wide Military Command and Control System's operations requires that we take a brief tour backwards in time.

NORAD and the Program 427M

The North American Air Defense Command's Cheyenne Mountain complex, a key WWMCCS node, became operational in 1966. Although clearly a substantial improvement over the systems NORAD had used in the past, it was apparent, even at the time the new facility opened its blast doors for business, that major improvements would soon be necessary if NORAD were to meet its mission responsibilities, which were shifting away from atmospheric threats and toward space. Implicit in this was an increasing flood of sensor data and the need for modern automatic data processing to make sense of it. The ADP capabilities of the second-generation computers used in Cheyenne Mountain's 425L Command and Control System and the 496L Spacetrack System did not provide the sort of computing power that NORAD's changing mission increasingly demanded, and an effort to provide a follow-on ADP capability formally began in December 1968.[1] This effort was designated the 427M computer improvement program, but a major thorn in its side would prove to be the WWMCCS automatic data-processing upgrade program.

Defense Secretary Melvin R. Laird had laid down two WWMCCS-related conditions when he approved Program 427M. The first was that overall management responsibility for the program would be given to the Air Force Systems Command's Electronic Systems Division (ESD). More than any other Air Force entity, ESD had a central stake in the development of WWMCCS and presumably could be counted on to further the goals of the larger national-level system. The second condition was that NORAD use the standard WWMCCS computer hardware and associated software then in the process of being procured.[2]

As designed, Program 427M included three major elements, or system segments. The first of these was the NORAD Computer System (NCS), which involved replacing the existing computers and related equipment of the 425L command and control system with new equipment. The purpose of the NCS was to provide NORAD with missile-warning data, nuclear detonation reports, weapons and sensor systems status, aircraft surveillance and warning reports, and other information. The second system segment, the Space Computational Center (SCC), the intended replacement for NORAD's 496L Spacetrack System, was the focal point for US and Canadian efforts to detect, track, and catalog all man-made objects in space from the moment of launch through the final moments of orbital decay. In addition to generating impact predictions, the SCC would provide an integrated system with enhanced data-processing capabilities. The final segment of Program 427M was the Communications System Segment (CSS). If new computers were to be the brains of NORAD's new program, the CSS would be its central nervous system. Consolidating a variety of currently separated functions into a single integrated system, the CSS would provide complete message processing, monitor relevant data circuits and equipment, supervise the automatic rerouting and restoration of circuits, and provide for message storage and record keeping.[3] This would include the dissemination of aerospace defense warning to other command centers, including the NMCC, ANMCC, and SAC headquarters.[4] All of this would be done by the WWMCCS standard computers, which were installed in Cheyenne Mountain in 1972 as part of the larger WWMCCS ADP upgrade program.

But all manner of computer problems quickly began cropping up. In the area of hardware, it was immediately apparent that the WWMCCS standard computers simply did not have the data-processing capabilities necessary to meet NORAD's mission requirements. But being saddled with the Honeywells, NORAD was forced to improvise. The contractor's written agreement was revised, and a considerable quantity of additional computer hardware and associated equipment for data processing and communications switching was procured. This arrangement naturally added to system cost and complexity and had a negative impact on overall reliability.[5] The primary villain in the program's escalating cost, however, was contractor overruns in the software area, where major difficulties were surfacing. Because the WWMCCS computers' standard software was written to operate in a batch-processing mode, complex software modification and retrofit operations were necessary to make the computers operate in the real-time interactive fashion demanded by NORAD. As with more complex hardware, more complex software both increased costs and raised the likelihood of system failures.[6]

But the additional effort, equipment, and money by no means resolved the problems encountered by each of Program 427M's three system segments. Consider the Space Computational Center, which was originally intended to host a large scientific computer. This plan had been abandoned when NORAD was instructed to use the WWMCCS Honeywell 6080 standard computers—essentially business machines—that were incapable of performing many of the data-processing functions that were described in the baseline requirements for the SCC. A second Honeywell computer was acquired, but by no means was this a solution to all of the problems. With its computers unable to meet specified performance standards, NORAD had no choice but to downgrade those standards. Things appeared equally dark for Program 427M's second segment, the NORAD computer system, where by 1977 NORAD's deputy commander for operations was pointing out how—largely because of the limitations of the WWMCCS standard computers—some 49 NCS program and modification requests were outstanding that were considered absolutely essential if the Combat Operations Center was to perform its missions effectively.[7]

Finally, there was the Communications System Segment, the linkage between NORAD and the outside world. Plans had called for the CSS to use two WWMCCS-standard Data Net 355 computers, but serious reliability and interface problems necessitated the incorporation of two Honeywell 6050 computers as replacements. Still the problems continued. A variety of additional equipment was added, increasing the likelihood of breakdowns and lowering system reliability. In the area of software, things were so bad that they appeared to preclude the program's ever achieving initial operational capability.[8]

The problems themselves being intractable, the solution NORAD eventually arrived at was simply to declare rhetorical victory and move on. A new standard of acceptability was promulgated, a so-called equivalent operational capability, and in September 1979 the 427M system was said to have achieved it. To many, however, equivalent operational capability was a grim reminder that after years of effort, NORAD's new system provided capabilities to its users no better than the outdated, 1960s-vintage systems it had replaced. For if the primary criterion for determining a program's effectiveness was whether the final product met users' needs, the 427M Computer Improvement Program had to be considered wholly ineffective. Something clearly had to be done. The General Accounting Office recommended to Congress that NORAD be exempted from the requirement to use WWMCCS hardware and software, and that all Program 427M's hardware and software be replaced with modern, state-of-the-art ADP equipment.[9] This was certainly in accord with the wishes of NORAD officials, who promptly requested the exemption. But as had been the case almost a decade earlier, the exception was not granted.[10] Such was the context in which the next series of WWMCCS failures occurred at Cheyenne Mountain, many of them directly attributable to the problem-plagued 427M system.

Failures at NORAD

To understand the WWMCCS failures that occurred at NORAD, it is necessary to consider both the way in which the command is linked to other command centers and the nature of the information that passes over some of those linkages. All US early warning radars and satellites feed their data to NORAD headquarters in Cheyenne Mountain, and for many sensors, such as the Ballistic Missile Early Warning System, that is the only place data is sent.

But things operated somewhat differently with respect to the data from those sensors designed to detect missiles launched from submarines, including the defense support program infrared satellites and radars of the Ballistic Missile Early Warning System and the then-under-construction PAVE PAWs radars for submarine-launched balistic missile (SLBM) detection and warning, the first of which came on line in mid-1979. True, NORAD receives data from those systems, which it then processes and transmits in near real time to four key command centers: the National Military Command Center, the Alternate National Military Command Center, the Strategic Air Command headquarters, and the Canadian Federal Warning Center in Ottawa. But because of the short-warning times involved in a SLBM attack, the command centers also receive the raw data directly from the sensors. Should SLBMs be detected, duty officers at NMCC, ANMCC, and SAC would have two separate displays of the same attack data.[11]

The communications circuits linking NORAD to the other command centers for this purpose were dedicated; that is, they were used for no other purpose. One might also imagine, missile attacks being rare, that the circuits would not be much used, but this is not the case for two reasons. First, so that the other command centers could continuously check that the circuits linking them to NORAD were open and properly functioning, NORAD had to continuously transmit a routine message to the three command centers. This message, just filler consisting of a series of zeroes, said in effect, "No missiles for you today." Second, while real missiles might indeed be rare, possible missile threats were considerably more common. The infrared sensors of the defense support program satellites, for example, continually detect a wide range of phenomena—fires, natural gas explosions, and reflected sunlight as well as burning rocket motors—that require identification and evaluation. At such times, the zeroes in NORAD's filler message would be replaced by actual numbers, and the duty officers at the various command centers would go into action. All possible indications of attack, even those involving ambiguous data, were sufficient to initiate a formal evaluative process.[12]

The first step of this process was known as a missile display conference, and these conferences were remarkably common, indeed routine. During the 18-month period from January 1979 through the end of June 1980, some 3,703 such conferences were held, averaging almost seven a day. During a missile display conference, duty officers at the command centers would evaluate the situation, and, in virtually all instances, would quickly determine that no credible threat to the North American continent existed. NORAD's commander would then terminate the conference. But if things remained in doubt and the possibility of a threat persisted, the commander would up the ante, convening a threat assessment conference. Its purpose was twofold, both to determine the nature of the threat and to take preliminary steps to enhance force survivability. With this type of conference more senior officers would be brought in, including SAC's commander and the chairman of the Joint Chiefs of Staff. During the 18-month period just mentioned, four such conferences took place. If the situation could not be resolved during the threat assessment conference, the final step involved convening a missile attack conference. Here, all of the most senior military and government officials would participate, including the president. As far as is known, no missile attack conference has ever been held.[13]

Of the four threat assessment conferences held during the 18-month period in 1979–80, two were attributable to an old FSS-7 radar located at Mount Hebo, Washington. As one of the six original "fuzzy sevens" designed to detect SLBM launches, it was seriously outdated. The Mount Hebo radar was slated to be replaced in the early 1980s by the new PAVE PAWS radar at Beale AFB in California, but at the time, the phased array radar was not yet fully operational.

The first of the threat assessment conferences took place on 3 October 1979, when the Mount Hebo radar picked up a decaying rocket body in low orbit. Just like a live missile, it showed up as a triangle on the "green monster" at NORAD, a washing machine-sized green phosphor radar monitor—surplus from the FAA that had been designed in the 1950s to track airplanes, not missiles.[14] The Mount Hebo radar then generated a false launch and impact report.[15] NORAD's commander was immediately notified, and a threat assessment conference was convened before it was determined that no real threat existed. Five months later, the Mount Hebo radar did it again. On 15 March 1980 the radar detected four Soviet SLBMs that had been launched from the Kuril Islands, north of Japan, as part of a troop-training exercise. In this instance, the radar falsely predicted that one of the missiles had an impact point in the United States.

Another threat assessment conference was convened before the threat was cleared.[16] Arguably, data that is received from a known offender such as the old Mount Hebo fuzzy seven should be viewed with skepticism, but NORAD could not afford to take chances. This was especially so since, beginning in the late 1970s, Soviet Yankee-class ballistic missile submarines had begun approaching to within a few hundred miles of the US coasts, a position from which many command and control facilities and military installations could be destroyed almost without warning.

The other two threat assessment conferences were more directly attributable to problems with NORAD's computers. The first of these, which took place during the morning hours of 9 November 1979, was certainly dramatic enough. The alert-code status board lit up as NORAD's WWMCCS computers indicated that a massive Soviet attack was in progress. A number of SLBMs had apparently been launched off the West Coast of the United States, and a missile display conference was immediately initiated. With the attack continuing, this conference was promptly elevated to a threat-assessment conference; appropriately enough, since, by all indications, military commanders had only about five minutes to act before the first missile detonated on US soil.[17]

Things were hardly unambiguous, however, and many conference participants quickly doubted that an actual attack was in progress. Their reason was a good one: NORAD had the ability to initiate instantaneous conferences with all of the sensor sites; it had promptly done so, and none of them had detected any missile launches.[18] With missiles being reported at NORAD headquarters but not by the sensors, it just had to be an error. But on the off chance that it wasn't, prudence dictated that precautionary actions be taken. SAC B-52 bomber crews were instructed to stand by and await further instructions, and the Minuteman missile force was placed on low-level alert. Jet interceptors were scrambled from US and Canadian air bases. The president's National Emergency Airborne Command Post plane took off from Andrews AFB (by one account, without having heard from President Carter).[19] To clear the way for all the military air activity that might follow, buzzers were sounded at air traffic control centers across the country, and controllers began radioing all commercial aircraft to be prepared to land.[20] But the skeptics had it right: the missiles were not real. What had happened was that an Air Force technician at NORAD's Combat Operations Center had inadvertently loaded a training tape into the on-line WWMCCS computers. The tape's scenario was a dandy, a Soviet SLBM attack that, by one description, "annihilated the world in high fidelity."[21] An interface problem had allowed the training data to enter the operational portion of the missile warning system.

Given all the heat WWMCCS had been taking recently, the incident was precisely the type the Pentagon would have loved to have avoided altogether—or, at any rate, to have kept firmly under wraps. The problem was that a *Washington Star* reporter, working on an unrelated story, happened to be at an air traffic control facility when the alert took place. The story broke—and with a vengeance. However naive or exaggerated the subsequent criticisms may have been, and military officials suggested they were both, the implications of the incident were sobering. One was the suggestion, reiterated endlessly in the press, that accidental nuclear war could actually happen. The strategic nuclear forces had been alerted, after all. The title of a *U.S. News and World Report* article, "Nuclear War by Accident—Is It Impossible?" provides a sense of what the Pentagon had to face in the public relations area.[22] Another suggestion was that the WWMCCS was simply incapable of dealing with an actual crisis. In this "Six Minute War," it was not until a minute after the first nuclear warheads were supposed to have detonated that Air Force technicians determined with certainty that an error had taken place. Throughout those minutes, President Carter and Defense Secretary Harold Brown had never been contacted. What would have happened, it was asked, if the attack had been real?

An official response was obviously necessary, and early Pentagon explanations attributed the incident to a "mechanical malfunction" in the computer's electronic routing of the war game. Subsequent explanations invoked human error. But such explanations did little but further fuel critics' fears of accidental nuclear war. Unable to respond in any meaningful way to such charges—after all, saying that "things don't always work" or that "the system is ninety-five percent reliable" is hardly adequate when nuclear war is at issue—the Pentagon quickly resorted to silence. It was a silence that would be broken only once, three weeks later, when it was announced that the problem had been fixed.[23] There had been no accidental nuclear war, but the damage to WWMCCS's public image inflicted by NORAD's Six Minute War had been considerable.

The major fix introduced by NORAD to guarantee against the recurrence of such an incident was a prudent one. NORAD established an off-site test facility in Colorado Springs so that software development, testing, and training no longer used the on-line computer system in the Cheyenne Mountain complex. A "functional equivalent" of Cheyenne Mountain's 427M system, the $16 million test facility featured leased Honeywell computers and other equipment purchased from Ford Aerospace Communications Corporation on a delegation of procurement authority—meaning that the need for the facility was considered sufficiently urgent by top Pentagon officials that the normal approval and procurement process, which ordinarily takes up to 18 months, was expedited to two months. An operational capability was fielded within a year. With their new facility, NORAD officials argued confidently, the type of false alert that had occurred on 9 November 1979, would never happen again.[24]

NORAD was right: that type of false alert would not recur. But an equally serious failure (indeed, perhaps an even more serious one considering how US-Soviet relations had deteriorated as a result of the Soviet invasion of Afghanistan) took place at NORAD only seven months later. The time was 2:30 A.M., 3 June 1980, when SAC headquarters at Offutt AFB received warning that the United States was under attack. The warning, transmitted to SAC over NORAD's dedicated circuits, showed that two SLBMs had been launched against the United States. The indication was that the missiles had been fired using a depressed ballistic trajectory, meaning that they would strike the United States in as little as three minutes.[25] Eighteen seconds later, the display showed that two additional SLBMs had been launched. SAC personnel did not actually see the missiles; what they saw was that some of the zeros in the standard NORAD filler message had been replaced with the number two. SAC duty personnel immediately turned to their other displays, the ones receiving data directly from the early warning sensors, but no missile launches were indicated. It was ambiguous, but things appeared sufficiently serious that the SAC duty controller ordered B-52 and FB-111 bomber crews and supporting tanker crews to go to their aircraft and start their engines. Alert messages were also flashed to the *Minuteman* and ballistic missile submarine forces. After another few seconds, however, the twos disappeared from SAC's display screen. SAC officers promptly got on the telephone to NORAD to find out what was going on. They were told that NORAD had received no indication of any missile launches from the early warning sensors, and was not transmitting such information to other command posts. The SAC crews were ordered to shut down their engines but to remain in their aircraft.[26]

This watchful peace would not last long. Within minutes, the warning display at SAC again indicated that the United States was under attack, this time by two land-based Soviet missiles. At about the same time, the NMCC received indications that two more SLBMs had been launched. As before, these indications came over the dedicated NORAD circuit, not directly from the sensors themselves. Also as before, the alleged missiles were being launched in pairs, as twos replaced the zeros in the NORAD-generated filler message. The duty officer at NMCC promptly convened a missile display conference, the lowest type, which brought in personnel from SAC, ANMCC, and NORAD. All of the command posts were convinced the attack data were erroneous, since there were no indications of attack from the sensors, the attack did not follow any logical pattern, and different command posts were receiving different indications of attack. Had the system been working properly, all four command posts should have had the same data in front of them.[27]

The attack data appeared to be random, suggesting a computer malfunction of some sort at NORAD. But because its source had yet to be ascertained, the duty officer at NMCC decided to convene a threat assessment conference. His reason for making this apparently paradoxical move—raising the level of seriousness to reduce it—was so that NORAD's commander could personally inform all concerned parties that no real crisis was in progress. NORAD's commander promptly did so, the NMCC duty officer terminated the conference, and SAC's bomber and tanker crews were sent back to their barracks. The threat had been cleared, but the threat assessment conference had one final effect. As part of the programmed responses to the conference, CINCPAC's airborne command post took off from its base in Hawaii, ready to help fight World War III.[28]

NORAD's missile attack warning system had again indicated that the United States was under attack, and again no Soviet attack had taken place. Unlike the insertion of a war game into the computer the previous November, however, it would not be so easy for NORAD to determine the cause of the error this time. It was apparently a random computer error, its possible causes manifold, and tracking it down would be difficult. What NORAD technicians decided to do was to continue to run their equipment in the same configuration used on 3 June in the hope that the error would repeat itself. Sure enough, three days later NMCC and SAC again received indications of attack over the NORAD circuit but not directly from the early warning sensors themselves. As a precaution, SAC bomber and tanker crews were again sent to their planes and their engines were started. Duty officers contacted NORAD, which again confirmed that no threat existed. Shortly thereafter, additional indications of missile attack were received by both SAC and NMCC over the NORAD circuit. There being no question of an error this time, SAC promptly ordered its crews to stand down. Not in the business of sending false reports of attack to the strategic nuclear forces, NORAD switched operations over to its backup WWMCCS computer.[29]

A massive effort ensued by NORAD, government, and industry computer experts to try to locate the source of the error. The conclusion they ultimately arrived at was that it had been caused by a faulty integrated circuit—a computer chip—in one of NORAD's NOVA 840 communications multiplexers. The multiplexer in question, produced by the Data General Corporation, was part of NORAD's Communications System Segment, the system that takes information from NORAD and puts it into message form for transmission to other command posts.[30] The false indications of attack had been caused by a computer chip about the size of a dime, costing forty-six cents.[31]

While in technical terms the failure was small, the political ramifications were substantially larger. As soon as the incidents of 3 and 6 June became known, a flurry of press attention, a cacophony of criticism—international in scope—and subsequent demands for system reform rapidly ensued. The reason was exactly the same as before: in the minds of many, the failures raised the specter of accidental nuclear war. Members of Congress publicly voiced their concern that incidents of this type were precisely the sort of things that could trigger such a war. The Soviet Union, for its part, accused the United States of harboring a "nuclear persecution complex."[32] Throughout, the Pentagon found itself on the defensive, unable to do much more than provide lame-sounding assurances that the United States had in no way come close to accidentally initiating a nuclear war.[33] But even stalwart Pentagon supporters such as Sen. John Tower of Texas found such assurances unpersuasive, and predictions were that a congressional investigation would surely follow.

Feeling the critical heat, NORAD promptly took steps to prevent the recurrence of similar incidents. New software was added that could trace a message through the entire preparation phase, ensuring that what goes in is what comes out. New displays were added inside the Combat Operations Center so that NORAD could monitor its transmissions to other command posts. Had such a capability been in place earlier, NORAD duty personnel would quickly have been able to see that they were transmitting erroneous indications of attack to other command centers. (In a related move, SAC duty officers were instructed to compare NORAD data to those received by the sensor systems themselves before ordering alert crews to their aircraft.) NORAD also changed its protocol, so that now all outgoing warning messages had to be approved by NORAD's commander. Finally, the format that NORAD used in its transmissions to the other command centers was changed. Instead of a message consisting of numbers of missiles detected, with zeros indicating an all-clear status, the filler message would be a standard communications test pattern indicating the status of the system.[34]

Perhaps as well as any example from WWMCCS's troubled history, the responses to the computer failures at NORAD pointed to one of the central ambiguities surrounding the concept of effectiveness—the fact that there were different constituencies with different sets of goals evaluating the system. In this case, strikingly different sets of evaluative criteria were applied to NORAD's performance by those inside and outside of the Department of Defense. For their part, NORAD and Pentagon officials who dealt with the criticisms were frequently bewildered by all of the controversy. After all, a few minor failures in a highly complex system—failures that were easily remedied—were of small consequence. They had no major consequences and hardly warranted characterizing a vast, multi-billion dollar operation as ineffective. In this rational administrative calculus, effectiveness seems to represent a sort of balance sheet upon which the positive and negative aspects of organizational activity are added or subtracted, yielding an overall sense of the adequacy of performance. If the failures are small and the successes large, the logic runs, the overall result will be gauged acceptable. NORAD performs its functions admirably most of the time, and, failures notwithstanding, the June incidents never had come close to unleashing America's strategic nuclear forces.

NORAD's critics in the Congress, press, and public were clearly wielding a quite different evaluative yardstick as they considered these incidents. Implicit in their uproar was the notion that these failures, indeed any failure at all at such a critical facility as NORAD, were intolerable. To critics, even minor failures demonstrated that the system intended to control the nation's nuclear weapons was out of kilter, certainly not effective, and possibly dangerous. Failure raised the specter of accidental nuclear war, and the very existence of that possibility, however remote, was adjudged intolerable. For them, effective performance necessarily connoted perfect performance.

Within two weeks of the June false alerts, Senate Armed Services Committee chairman John C. Stennis asked two of his committee members, Gary Hart of Colorado and Barry Goldwater of Arizona, to investigate incidents at NORAD. Their final report was released the following October, and its overarching conclusion, one that echoed earlier Pentagon statements, was that the United States had never come close to unleashing an accidental nuclear war. Yes, some actions had been taken in response to the erroneous indications of attack, such as SAC bombers and tankers crews being ordered to their aircraft, but these were merely precautionary steps to enhance force survivability. Despite the computer's problems, the human part of the system had detected the problem. "In a real sense," the senators concluded, "the total system worked properly."[35]

What was found to work less well was the overall management of early warning assets by the Air Force, a condition the senators described as "disturbing." Specifically, management responsibility was found to be divided among a number of commands. The Strategic Air Command had responsibility for managing, maintaining, and operating all of the nation's early warning sensors. The Air Force Communications Command (AFCC) had responsibility for those related communications and electronics systems used to transmit the sensor data. The Aerospace Defense Command, the US portion of NORAD, was responsible for operating and managing the Cheyenne Mountain facility and for interpreting the data made available to them by SAC and AFCC. A final organization, the Air Force Systems Command, had responsibility for developing new early warning assets. Because of this fragmentation, the senators found that the missile warning function was not being treated as a "true overall system." The organizational structure that was in place, they said, fragmented management responsibility and led to less effective system performance.[36] Though focused now on a single WWMCCS element, NORAD, it was a criticism that had haunted the larger system since its inception.

Senators Hart and Goldwater found the shortcomings in one other area sufficiently troubling to warrant comment: automatic data processing. In particular, they were concerned with the way ADP systems were acquired and how this acquisition process influenced the systems then in use by NORAD. The problem, they said, was Public Law 89-306, the 1965 Brooks Act, which assigned responsibility for ADP procurements not only to the agency that would actually use the equipment—here the Department of Defense—but also to the Office of Management and Budget and the General Services Administration.

The problem with this arrangement for a critical command such as NORAD was that it was simply too slow. A typical ADP upgrade might take as long as seven years, and, given the extraordinary pace of technological advance in the computer field, this meant that ADP equipment would often be functionally obsolete before it was ever brought into operation. To make matters worse, the need for multiple approvals specified in the Brooks Act had the effect of creating procurement by committee, resulting in a least-common-denominator capability to satisfy all involved constituencies, not the best capability available. What was needed was a mechanism by which critical commands such as NORAD could circumvent the act's cumbersome procedures when necessary, but no such mechanism was then in the cards.[37] Thus, the second decade of WWMCCS's existence ended in much the same way it had begun, with a series of system failures, subsequent criticisms, and calls for reform.

Notes

1. Comptroller General, *NORAD's Information Processing Improvement Program—Will It Enhance Mission Capability?: Report to the Congress,* LCD-78-117 (Washington, D.C.: General Accounting Office [GAO], 21 September 1978), 7.

2. Ibid., 13–14.

3. Comptroller General, *NORAD's Missile Warning System: What Went Wrong?—Report to the Chairman, Committee on Government Operations, House of Representatives,* MASAD-81-30 (Washington, D.C.: GAO, 15 May 1981), 3, 9–10.

4. Frederick W. Eisele, "NORAD Updates C^3," *Signal* 29 (September 1974): 10.

5. Comptroller General, *NORAD's Information,* 14–15, 26.

6. ———, *NORAD's Missile Warning System,* 9.

7. ———, *NORAD's Information,* 16.

8. House, Committee on Appropriations, *Attack Warning: Better Management Required to Resolve NORAD Integration Deficiencies—Report to the Chairman, Subcommittee on Defense,* IMTEC-89-26 (Washington, D.C.: GAO, July 1989), 9–10.

9. Comptroller General, *NORAD's Information,* 29.

10. ———, *NORAD's Missile Warning System,* 17.

11. Senate, *Recent False Alerts From the Nation's Missile Attack Warning System: Report of Gary Hart and Barry Goldwater to the Committee on*

Armed Services (Washington, D.C.: Government Printing Office, 9 October 1980), 3.

12. Ibid.

13. Ibid., 4–5.

14. "U.S. Uses '50s Radar to Track '80s Missiles," *Washington Times*, 16 June 1989, 5.

15. Senate, *Recent False Alerts*, 5.

16. Comptroller General, *NORAD's Missile Warning System*, 3–4.

17. Richard Thaxton, "Nuclear War by Computer Chip," *The Progressive*, August 1980, 29.

18. "Nuclear War by Accident--Is It Impossible?" *U.S. News & World Report*, 19 December 1983, 27.

19. Thaxton, 29.

20. William J. Broad, "Computers and the U.S. Military Don't Mix," *Science*, 14 March 1980, 1183.

21. Bruce Gumble, "Air Force Upgrading Defenses at NORAD," *Defense Electronics* 21 (August 1985): 86.

22. "Nuclear War by Accident," 27.

23. Broad, 1183.

24. Comptroller General, *NORAD's Missile Warning System*, 13.

25. Thaxton, 29.

26. Senate, 5.

27. Ibid., 6.

28. Ibid.

29. Ibid., 7.

30. Ibid.

31. Thaxton, 29.

32. Ibid., 30.

33. Richard Halloran, "Senators Report False Warnings of Soviet Missiles," *New York Times*, 29 October 1980, A16.

34. Senate, 8.

35. Ibid., 8–9.

36. Ibid., 9, 13.

37. Ibid., 13.

PART III

Implementation

Chapter 14

Strategic Modernization

The World Wide Military Command and Control System had really taken a beating during the 1970s. It had failed to provide timely, reliable service during a series of military exercises and had broken down during actual crises. The failures at NORAD had rendered problematical the effectiveness of WWMCCS in what was certainly its most important role: providing the national leadership with accurate and timely information on nuclear attack. Many in the defense establishment began openly expressing their dismay at the WWMCCS's vulnerability, its marginal capabilities, its "gaping holes," and its "scary gaps" in coverage.[1] Others publicly and unfavorably compared WWMCCS to the ostensibly more survivable and enduring Soviet system with its hardened facilities and mobile command centers, the suggestion being that these asymmetries undercut deterrence and threatened America's national security. WWMCCS "verges on a national scandal," others said, forcing commanders to "live in the dark ages."[2] Command and control was portrayed as a glaring military weakness, as America's "Achilles' heel."[3]

Some criticisms came from the most authoritative of sources: people who had seen capabilities suffer severely at the hands of service parochialism and the evolutionary approach. DCA director Samuel L. Gravely Jr. pointed out how the command and control establishment that had evolved under such an orientation was an "ill-defined and unbounded monstrosity . . . an enigma with more people planning, programming, preaching and engineering it than we have understanding it or effectively using it." This had, he said, resulted in serious problems of interoperability and effectiveness when the system was called upon to perform in a joint-service environment. Quoting the comic strip character Pogo, Gravely concluded gravely, "We have met the enemy and they is us."[4] As if to underscore his concerns, and even as the failures at NORAD were in progress, an incident took place that underscored in a dramatic and public way the disconcerting fact that jointness in military operations, and by extension the capability for effective command and control, were far from a reality.

Operation Eagle Claw

That incident took place on 24 April 1980, when members of the military's elite Delta Force undertook an operation to rescue 53 Americans held hostage in Tehran, Iran. The American Embassy there had been taken over by Iranian extremists on 4 November the previous year, and President Carter had promptly ordered the Pentagon to draw up plans for a rescue operation that could be put into effect should diplomatic efforts to free the hostages fail.[5] That mission was given the code name Eagle Claw. The plan called for six planes—three MC-130 transport planes and three EC-130 tankers for refueling—to carry the Delta Force, its equipment, and fuel from Egypt to Oman. There the aircraft would be refueled, and, flying a ground-hugging route to avoid detection by radar, then proceed to a secret landing strip in Iran called Desert One, about three hundred miles from Tehran. There they would rendezvous with eight RH-53D Sea Stallion helicopters, launched from the aircraft carrier USS *Nimitz*. The troops would then transfer to the helicopters, which would fly to Desert Two, a hide site located 50 miles outside of Tehran; there they would rendezvous with two defense intelligence agents who would provide them with vehicles.[6] Plans called for Delta Force to be driven to the embassy compound that night. The hostages would be freed and loaded on to helicopters for a brief flight to a nearby air strip, where they would be transferred to C-141 transports and flown to Egypt.[7]

Anyway, that was the plan. As it soon became known to the world, what in fact happened was that the effort to rescue the hostages came to a dismal and premature end. Only six of the Sea Stallion helicopters ever made it to Desert One; one was forced to land because its instruments showed that one of its main rotors was about to malfunction, while another had to turn back after flying through a severe sandstorm. Of those choppers that made it to Desert One, one had hydraulic problems and was inoperable. The loss of three helicopters meant that there was now insufficient airlift to carry Delta Force, and that 20 men critical to the mission would have to be left behind. But even if the mission proceeded, were five choppers sufficient? Perhaps, but what if some wouldn't start the next day, or were otherwise disabled? Delta Force ground commander Col Charlie Beckwith concluded that the margin of error was now "just too close," and he was forced to abort the mission.[8]

The evacuation from Desert One added serious injury to insult. One of the helicopters banked sharply to the left as it lifted off, slicing into the fuselage of one of the fuel-laden transport planes. Flames soared hundreds of feet into the sky, joined by Redeye missiles ignited by the fierce heat. Five airmen and three Marines died in the inferno. The remainder of the force boarded the transports and flew to safety, leaving behind in the desert five intact helicopters, communications equipment, weapons, secret documents, and the bodies of their comrades.[9]

In the wake of Eagle Claw, the joint chiefs commissioned a special operations review group headed by Adm James L. Holloway (US Navy, Retired), to examine the whole operation. The group's report, afterwards known as the Holloway Report, concluded that compartmentalization and "ad hoc arrangements" had compromised the organization and planning of the operation. Highly placed observers would later acknowledge that a driving force behind the operation had been to ensure that all four of the services were given their piece of the action, and what resulted was Marine pilots flying Navy helicopters carrying Army troops, all the while supported by the Air Force. While it all certainly smacked of "jointness," the participation of all four services meant that parochial interests dictated the nature of the force that was used, to the detriment of mission cohesion and integration. Marine helicopter pilots were selected to join Navy pilots in flying the Sea Stallion helicopters from the *Nimitz* into Iran, whereas Air Force helicopter pilots experienced in long-range flying would have been a more logical choice.[10] The problem was succinctly summed up by Colonel Beckwith: "In Iran we had an *ad hoc* affair. We went out, found bits and pieces, people and equipment, brought them together occasionally and then asked them to perform a highly complex mission. The parts all performed, but they didn't necessarily perform as a team."[11] Until they could do so, many believed, the problems so painfully apparent in Operation Eagle Claw would recur in the future.

Corrective Actions

Despite the characterizations of the previous decade as one of neglect of the command and control function, it had of course been far from that. Despite the problems and the failures that in fact occurred, despite the lack of "jointness" evidenced by incidents such as the hostage rescue operation, numerous changes had taken place in both the technological and organizational areas. That shortcomings remained nobody denied. As President Carter's secretary of defense, Harold Brown, pointed out, while WWMCCS's ability to meet the doctrinal demands of survivability, flexibility, and endurance still fell short, a broad-gauged effort to remedy the situation was under way.[12] Defense journals seconded this, reporting how command and control, while by no means the most heavily funded defense area, had emerged as a key area of attention for the National Command Authorities.[13] Thus, many of the deficiencies the Reagan administration said it found upon its arrival in Washington in January 1981—and which it went to extraordinary lengths to remedy—had already been recognized, with corrective action in progress.

Corrective measures were also being pursued by Congress. During the months following the new administration's arrival, Congress was taking action to improve the "necessary evil" of command and control. In mid-1981 the Senate Armed Services Committee recommended on its own initiative—that is to say, before the administration itself had proposed any improvements—that some $340 million be targeted for WWMCCS enhancements and upgrades; thus anticipating many of the administration's recommendations that would come later that year. The committee specifically recognized an urgent need for hardening command and control assets against the disruptive effects of electromagnetic pulse and for improving their resistance to signal jamming and interception. Relevant programs were endorsed in the Senate/House authorization bill for fiscal year 1982.[14]

At the same time, the Congressional Budget Office (CBO) was considering what to do about WWMCCS modernization. The CBO presented three options to Congress, all of them concerned with enhancing command and control responsiveness in the transattack period. The major emphases of Option One involved improvements in attack warning and surveillance; specifically, upgrading the existing two PAVE PAWS radar sites and deploying another two, fielding more sophisticated airborne command posts, and upgrading communications links. Option Two was directed toward system survivability and endurance, particularly a need for greater mobility for command posts and communications facilities that could be met by installing them on trucks capable of avoiding Soviet missiles by randomly and covertly roaming the countryside during times of crisis. The final and most ambitious option was to pursue the objectives of Options One and Two.[15] All in all, it hardly seemed a case of Congress suffering from chronic command and control neglect, and it was hardly surprising that there was little debate or congressional resistance when the administration presented its own plan to modernize America's strategic forces that October.[16]

This is not to say that anyone was particularly sanguine about conditions as they then stood, and nobody denied that serious problems remained to be solved. But many of these problems had their origins as much in bureaucratic politics—historical differences between the military services, existing force structure, and the desire of the military services to maintain control over existing command and control assets—as in any real or imagined period of neglect. If, as some claimed, military commanders were in fact living in the command and control Dark Ages, it was in large measure because their own services had determined that was where they should live. Not that any of this mattered much to the triumphant members of the Reagan administration; they attributed their election to Reagan's characterization of their predecessors as "soft" on defense. Self-confident and assured of its mandate and its vision, the new administration publicly dedicated itself to making America's military capabilities second to none, with command and control—especially strategic command and control—quickly assuming a new prominence within the overall panoply of defense priorities.

The Strategic Modernization Program

On 1 October 1981 the Reagan administration issued National Security Decision Directive 12, a document addressing the total strategic forces program of the United States. The following day, President Reagan publicly outlined a comprehensive strategic modernization program to improve the capabilities of those forces. The program had five parts, three of which involved modernization of strategic weapons systems. Examples included strengthening and improving the accuracy of the land-based intercontinental ballistic missile (ICBM) force, including the deployment of the MX missile; modernizing and improving the penetrating capability of the strategic bomber force, including the deployment of the B-1 bomber; and pursuing new, more accurate SLBMs that would give a hard-kill capability to US submarines at sea. The fourth item concerned the modernization of US air defenses. The final item involved major improvements in command, control, and communications assets, most of them falling directly under the WWMCCS umbrella. Whereas WWMCCS had previously been an area lacking a strong constituency in the defense establishment, and hence of relatively low priority, the administration planned to change that in a most visible and dramatic way. Of the five concerns, command and control was designated the "highest priority element," equal in importance to the much better known weapons systems it supported. Defense Secretary Caspar Weinberger underscored the new orientation when he told the Congress, "I can't think of any higher priority than improving the [command and control] aspects of this whole program."[17] Thus commenced the most ambitious and far-reaching effort to improve command and control since the dawn of the nuclear age.

Documents distributed by the White House provided specifics about the command and control improvements to be undertaken under the broader rubric of strategic modernization, many of them deriving from a strategic connectivity study conducted by the Pentagon the previous April. These documents fell into four major categories, the first of which was a warning and attack assessment. Here, the performance, survivability, and coverage of the array of surveillance radars and satellites that provided early warning of missile attack and subsequent attack assessment were to be substantially enhanced. The second area was command centers; and the initiatives here involved upgrading the capabilities and survivability of the three national-level command centers, the CINC's 15 command posts, and the airborne command centers that would direct the strategic forces during crises up to and including general nuclear war. Third was the area of strategic communications, where the focus was on enhancing and restructuring a wide range of existing assets, plus the deployment of new assets so that commanders could be linked with the strategic forces under all conditions of peace, crisis, and war. Finally, what was described as a "vigorous and comprehensive" research and development program would be undertaken with the goal of fielding a command and control capability that could survive the first salvo of a nuclear exchange and endure for an extended period thereafter. Specifics here included the hardening of assets against the effects of electromagnetic pulse and improving their resistance to jamming.[18]

Administration officials were quick to point out how the Strategic Modernization Program represented a quantum leap forward in capabilities and how it had been designed to provide the National Command Authorities with a number of courses of action not currently available. With the new capabilities in place, it was said, the president would have a "launch under attack" capability, allowing him to order a retaliatory nuclear strike after detecting Soviet missiles heading toward the United States but before they had detonated. Such an approach had never been considered feasible in the past because the sensor and surveillance systems providing warning of attack were simply too fallible and error prone. Launching the missiles before unambiguous evidence of attack was received (in the form of actual nuclear detonations) ran the risk of accidental nuclear war, a possibility dramatically underscored by the recent NORAD failures. With the improvements in radar and satellite sensor systems called for by the administration's program, the sensors would be made sufficiently reliable that evidence of an attack would be unambiguous and launch on warning a viable military option.

The other goals of the program were equally ambitious. The improvements would permit the United States to engage in a protracted nuclear conflict over a period of days or even weeks, it was said, and this was where the new, highly reliable and survivable communications capabilities came into play. Among others, initiatives in this area included communications satellites mounted on MX missiles that could be launched into orbit to replace satellites that had been destroyed, plus mobile command centers and communications facilities that could escape destruction and permit military action to continue after an initial nuclear exchange. With guaranteed communications capabilities in place, the NCA could elect to fire only a limited number of missiles, for example, while withholding the rest as a negotiating chip or for use in subsequent strikes. Similarly, the improvements would give far greater flexibility to the new B-1 bomber force, allowing bombers returning from an attack on the Soviet Union to be directed to surviving air bases, where they could take on additional fuel and weapons and receive orders for subsequent missions. "Over the past decade, we have not modernized communications and control systems fast enough," White House officials said, "As a result, these systems are not as survivable as we would like, and they could not operate reliably over an extended period after a Soviet attack, if that proved necessary."[19] The major criteria of command and control effectiveness being advanced by the administration in the Strategic Modernization Program were thus precisely those that had concerned the WWMCCS Council, the WWMCCS Architect, and others: reliability, survivability, and endurance.

To these criteria must be added interoperability, one of the areas "most neglected over the past ten years," in the words of James P. Wade Jr., principal deputy undersecretary of defense for research and engineering.[20] Here, the administration's point was to ensure communications across network boundaries—between military and civilian leaders, say, or between US and NATO forces, the latter being an area where interoperability of systems was considered especially deficient.[21] To address this need, Secretary Weinberger appointed Wade to head an executive committee whose purpose was take a systems approach to the issue of strategic connectivity throughout the Department of Defense, an effort to view command and control as a totality spanning traditional service and agency lines.[22] In the past there had been talk of such things, of course, but an administrationwide effort to actually effect it was unprecedented.

Donald C. Latham, the new assistant secretary of defense for C^3I, immediately began stressing the centrality of command and control in the overall defense equation by proposing what he described as a "C^3I Triad." Following the White House lead, the first leg of the triad was a warning and attack assessment. It was concerned with early warning satellites, radars, and other assets used to unambiguously ascertain whether an attack was taking place, and, if so, its precise nature. Information from these sensors would be passed along to various command centers, the second leg of the triad. The final triad leg involved the supporting communications networks that provided connectivity between the NCA and the strategic forces. All three legs were crucial, Latham pointed out, and the nature of the Soviet threat and the requirements imposed by US strategic doctrine required that all three be absolutely credible under all possible military circumstances.[23]

Latham underscored his point by noting the essential duality of WWMCCS. WWMCCS was a "two system concept," he said, one part concerning military needs through the level of conventional war, while the other was oriented specifically toward nuclear conflict. The requirements imposed by each concept were often dramatically different. WWMCCS had been designed and developed primarily with the first concept in mind, Latham argued, and, as far as it went, the United States had a better system than the Soviet Union. But in Latham's view such a system, one in which "you simply don't have to have it all," was necessary but not sufficient.[24] Many of WWMCCS's peacetime functions would not require prompt reconstitution in the event of damage or destruction, for instance, whereas almost all of the wartime functions would have to meet that requirement. Certain capabilities—high frequency communications, for example—were generally appropriate only as a System I capability, given their susceptibility to disruption in a nuclear environment.

The other WWMCCS, System II, had to be capable of supporting military operation during times of theater and strategic nuclear warfare. As such, it had to emphasize hardening, reconstructing assets following a nuclear strike, and maintaining an ongoing capability for force and status assessment. Examples of strictly System II capabilities would be the proposed ground wave emergency network, or the Navy's extremely low frequency and Take Charge and Move Out (TACAMO) systems for communicating with the ballistic missile submarine force. By far the most serious deficiencies with WWMCCS were said to lie in its System II capabilities. The Soviets had emphasized hardened and mobile command and control facilities to a greater extent than had the United States. The Strategic Modernization Program was intended to redress the imbalance, and this was where Latham and his colleagues would focus much of their considerable talents. As they articulated the rationale, America's nuclear deterrent was at last to be made credible, something that in turn required a credible command and control system with nuclear war-fighting capabilities.

Their interests were soon complemented by the concerns of defense analysts outside of government, who began pointing out with almost religious fervor that while the centrality of command and control in the implementation of US strategic policy was everywhere implicit, plans seldom reflected key vulnerabilities and real-world system limitations. Numerous articles and several books appeared during the early 1980s that focused their attention precisely on the lack of survivability, redundancy, flexibility, endurance, and interoperability that characterized the command and control area. Almost without exception, the point of these publications was to sketch out the extraordinary disjunction between rhetoric and reality, the gulf separating the requirements placed on the command and control system and its actual capabilities. A sense of the interest and of the concern that prevailed is captured nicely in the title of a Yale dissertation, "Headless Horsemen of the Apocalypse," by Paul Bracken, which, retitled *The Command and Control of Nuclear Forces*, became one of the better-known command and control books of the period. In this new thinking, command and control systems were absolutely central, "force multipliers" in the buzzword of the time, assets that not only enhanced the effectiveness of other weapons and forces but that made their use possible in the first place.

Such were the interest and the plans, but good intentions had in the past frequently foundered on the shoals of resistance by the services and their supporters in Congress. The way the Reagan administration endeavored to avoid this problem was by explicitly linking the funding for command and control programs to the missiles, bombers, and other high-visibility weapons systems the services coveted. As Latham pointed out, when the budget packages for the weapons systems were sent to Capitol Hill, they were accompanied by a funding request "with a protective fence around it" for upgrade of the relevant command and control systems.[25] But as the administration was quick to learn, the timing was propitious; the tide had changed. Whereas in the past the hardest task was "making everybody believe the problem is real," as Latham described it, now—thanks to the WWMCCS failures and criticisms of the previous few years—there was remarkably little debate over the command and control initiatives in Congress.[26]

For those in the command and control community, it was a time of guarded euphoria and jubilation. Their area was at last receiving the recognition they believed it deserved, and, perhaps more importantly, a number of strong WWMCCS advocates were now in top government posts.[27] But this did not fully allay fears that the traditional antipathy toward command and control might reassert itself and begin to erode the gains. Some feared that, like weapons systems, command and control systems might become bargaining chips in future arms control negotiations. Former DCA director Jon L. Boyes summed up the concerns when he noted how "unlike weapons, forces, and platforms that can be laid on the bargaining table, the central nervous-sensory system that holds them together and gives them their credibility cannot."[28] What to do? True, command and control advocates such as Boyes began pushing for a more substantial voice in policy formulation. Command and control was finally being viewed as more than a technical detail among policy makers, but this was seen as a poor substitute for having people with specific expertise in the area actually making the decisions. The Strategic Modernization

Program had ushered in a new era, and what it demanded were management changes.

A major effort along these lines involved the creation of three new organizational entities to create closer cooperation between the Office of the Secretary of Defense, the Joint Chiefs of Staff organization, and the military services. The first was the C^3 Executive Committee (EXCOM), a top-level group whose purpose was to make recommendations regarding the allocation of resources for command and control programs. Top level was indeed the way to describe the committee, for EXCOM consisted of the JCS chairman and the deputy secretary of defense, with the undersecretary of defense for research and engineering serving as executive secretary. The second group was the C^3 Review Council, a senior-level group chaired by the assistant secretary of defense for C^3I, and including key officials from the OSD, the JCSO, and senior officers from the military services. The purpose of the review council was to provide senior-level resolution of important command and control issues, and to submit its recommendations to EXCOM.[29] Finally, eight specialized panels were established to assist the review council, their foci including military satellite communications, strategic command and control, the Defense Communications System, the WWMCCS Intercomputer Network, and other major areas of concern. Another management initiative involved reorganizing the Command, Control, and Communications Systems Directorate created within the Joint Chiefs of Staff organization in early 1979. Whereas before, the directorate had been broken down into two principal elements, strategic and tactical, it was now expanded to three. The first was the Unified and Specified Command C^3 Support Element, established in substantial measure in response to calls from the commanders in chief and intended as the JCS focal point for command and control matters of relevance to them. Next was the Defense Wide C^3 Support Element, the purpose of which was to pursue improvements in programs intended precisely for broader defense use. Included among these were the National Military Command System, AUTOSEVOCOM, and military satellite communications systems. Equally important, this element was to serve as the JCS focal point for command and control ADP programs, including WIN and its follow-on

WWMCCS Information System (WIS). Finally, there was the C^3 Connectivity and Evaluation Support Element, intended to evaluate existing systems, identify shortcomings, and develop specific programs to address them.[30]

Would these technical improvements and organizational changes prove sufficient? Surely not in the minds of many officials, for whom sufficiency represented a sort of ever shifting rainbow's end. But despite the ever-present need to balance capabilities with costs, even during the munificent early years of the Reagan administration, the times would prove rich indeed for WWMCCS. "We put our money where our mouth is," quipped Donald Latham.[31] And their money would stay there throughout the Reagan years. Despite CIA reports in the early 1980s describing a slowdown in the rate of Soviet military growth, and despite a 1984 NATO report noting that the rate of growth of Soviet military spending had been cut in half, the spending on command and control would continue to increase.[32] When Latham arrived in Washington, he inherited the Carter administration's FY 1982 command and control budget of approximately $9.1 billion, or some 5.9 percent of the total defense budget. By the time Latham departed the Pentagon in 1987 to become a vice president of Computer Sciences Corporation, the total for command and control funding had increased to $21.7 billion, some 7.8 percent of the vastly expanded defense budget; and it would remain fairly constant for the remainder of the decade until the end of the cold war.[33]

Notes

1. James W. Canan, "Steady Steps in Strategic C^3I," *Air Force Magazine* 70 (June 1987): 44.

2. Julian S. Lake, "C^3 Software Myopia," *Countermeasures* 2 (8 December 1976): 62.

3. Tony Velocci, "C^3: Strategy for Survival," *National Defense* 65 (December 1981): 49.

4. Samuel L. Gravely Jr., "The DCA—A Rock and a Hard Place," *Signal* 34 (April 1980): 8.

5. Senate, *Defense Organization: The Need For Change—Staff Report to the Committee on Armed Services* (Washington, D.C.: Government Printing Office [GPO], October 1985), 359.

6. Charlie A. Beckwith and Donald Knox, *Delta Force* (New York: Harcourt Brace Jovanovich, 1983), 253.

7. Ibid., 254–55.

8. Ibid., 276–77, 283.

9. Senate, *Defense Organization*, 361.

10. Ibid., 359, 361–63.

11. Beckwith and Knox, 295.

12. Edgar Ulsamer, "Electronics Takes to the Offensive," *Air Force Magazine* 63 (July 1980): 42.

13. R. J. Raggett, "U.S. Defense Communications: Links to Everywhere," *NATO's Fifteen Nations*, June–July 1979, 101.

14. House, *Strategic Force Modernization Programs: Hearings before the Subcommittee on Strategic and Theater Forces of the Committee on Armed Services*, 97th Cong., 1st sess. (Washington, D.C.: GPO, 1981), 113.

15. Harry V. Martin, "Communications Vulnerability," *Military Science & Technology* 2 (June 1982): 44.

16. Richard Halloran, "Nuclear Plans of Reagan Stir Capital Debates," *New York Times*, 5 October 1981, A1.

17. House, *Strategic Force*, 232–33.

18. "Background Statement from White House on MX Missile and B-1 Bomber," *New York Times*, 3 October 1981, A3.

19. Ibid., A3.

20. House, *Strategic Force*, 211.

21. "Growth in Funding Yields Strategic, Tactical Benefits," *Aviation Week & Space Technology*, 9 December 1985, 46–47.

22. James P. Wade Jr., "The Two Requirements for C^3," *Air Force Magazine* 65 (July 1982): 66.

23. Donald E. Fink, "C-Cube; So Far, So Good . . .," *Aviation Week & Space Technology*, 9 December 1985, 11.

24. House, *Strategic Force*, 236.

25. "Why C^3I Is The Pentagon's Top Priority," *Government Executive* 14 (January 1982): 14.

26. House, *Strategic Force*, 226.

27. Bernard L. Weiss, "Keys to Success in C^3I," *Signal* 36 (August 1982): 52.

28. Jon L. Boyes, "The Maturing of C^3I Policy," *Signal* 38 (April 1984): 13.

29. Robert G. Lynn, "C^3 Management Evolution," *Signal* 39 (November 1984): 21.

30. Ibid., 22.

31. Edgar Ulsamer, "Top Priority for C^3I," *Air Force Magazine* 69 (September 1986): 137.

32. "NATO Cites Soviet Military Funding Slowdown," *Aviation Week & Space Technology*, 13 February 1984, 140.

33. Phillip J. Klass, "C^3 Emerges from Budget Gauntlet with Funds to Meet Inflation," *Aviation Week & Space Technology*, 14 March 1988, 234.

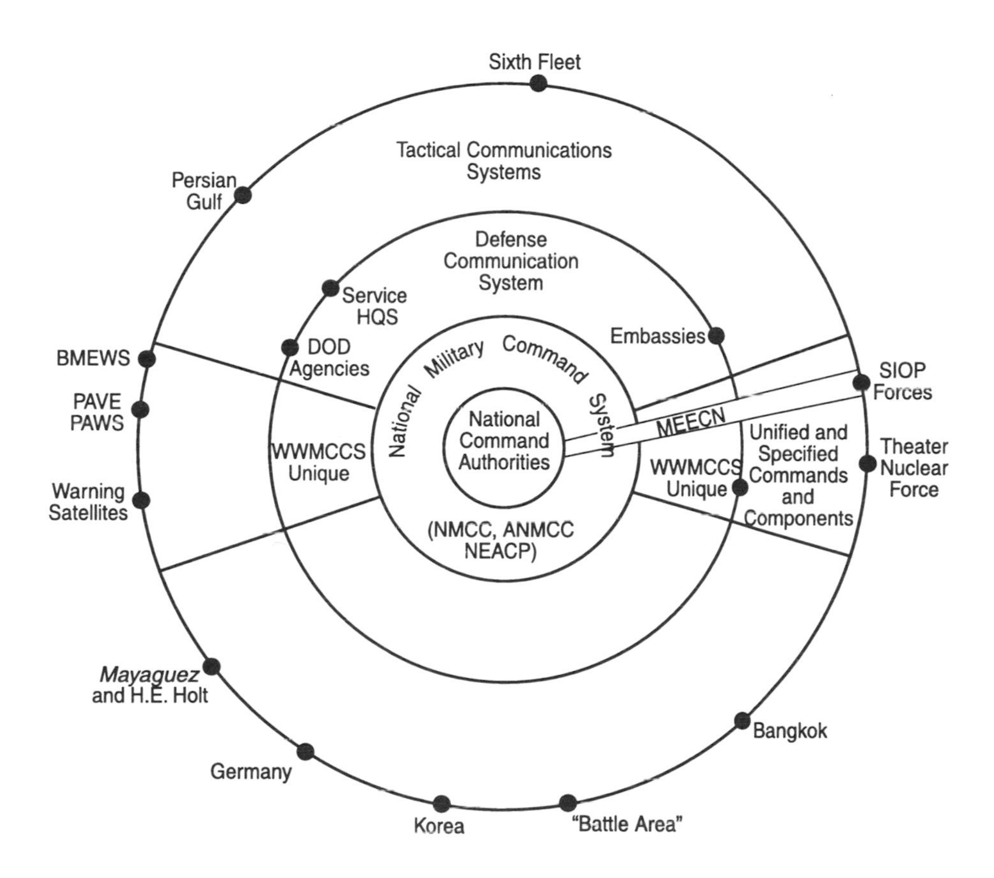

Source: C. W. Borklund, "Military Communication: The 'What' is Much Tougher than the 'How,'" *Government Executive,* June 1976, 25.

Figure 13. Conception of WWMCCS System Architecture, Given Revised DODD 5100.30

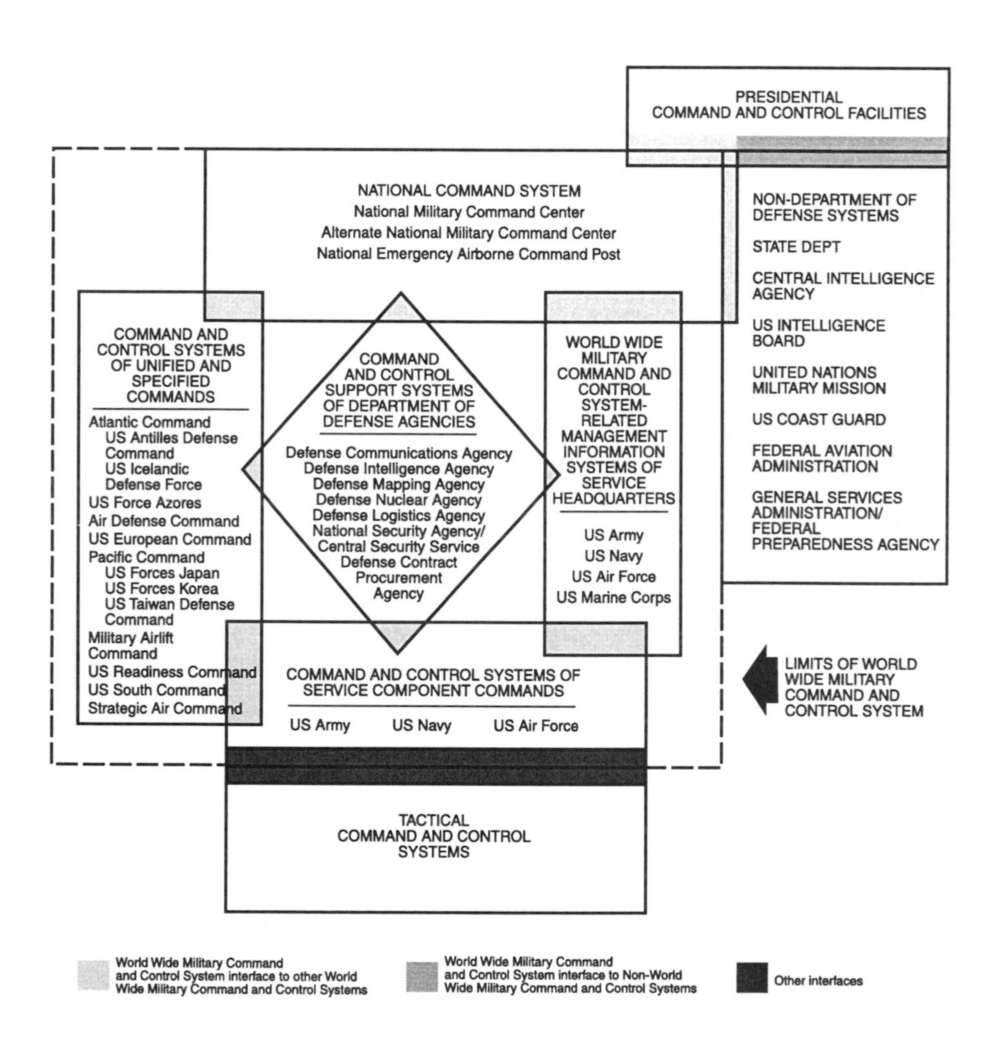

Source: *The World Wide Military Command and Control System—Major Changes Needed in Its Automated Data Processing Management and Direction*, LCD-80-22 (Washington, D.C.: General Accounting Office, 14 December 1979), 3.

Figure 14. WWMCCS System Relationships

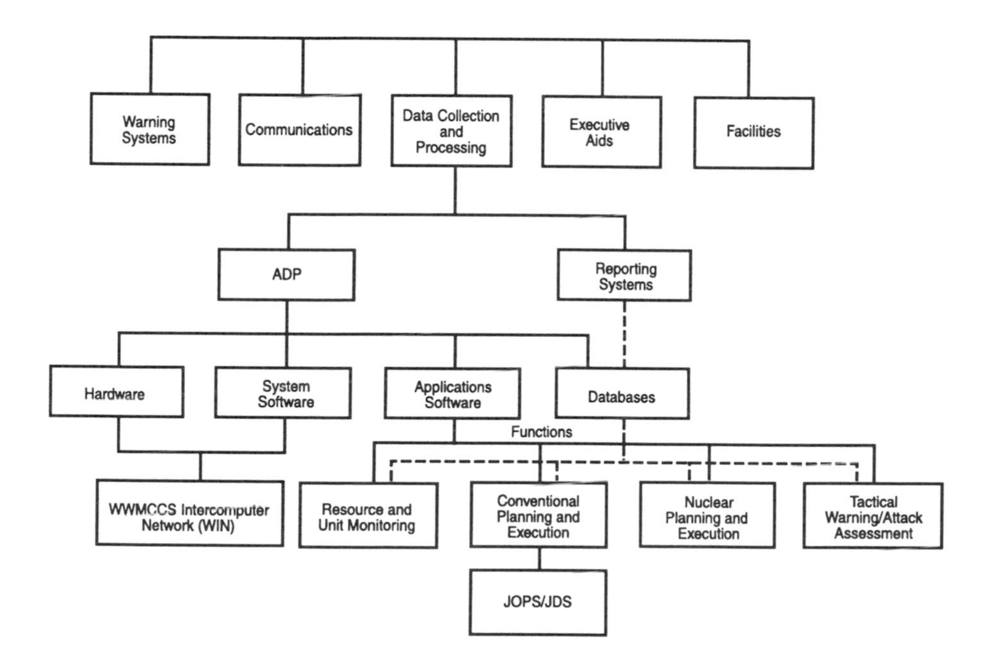

Source: The Marine Joint Staff Officer (Quantico, Va.: USMC Development and Education Command, January 1987), 5–20.

Figure 15. The Elements of WWMCCS

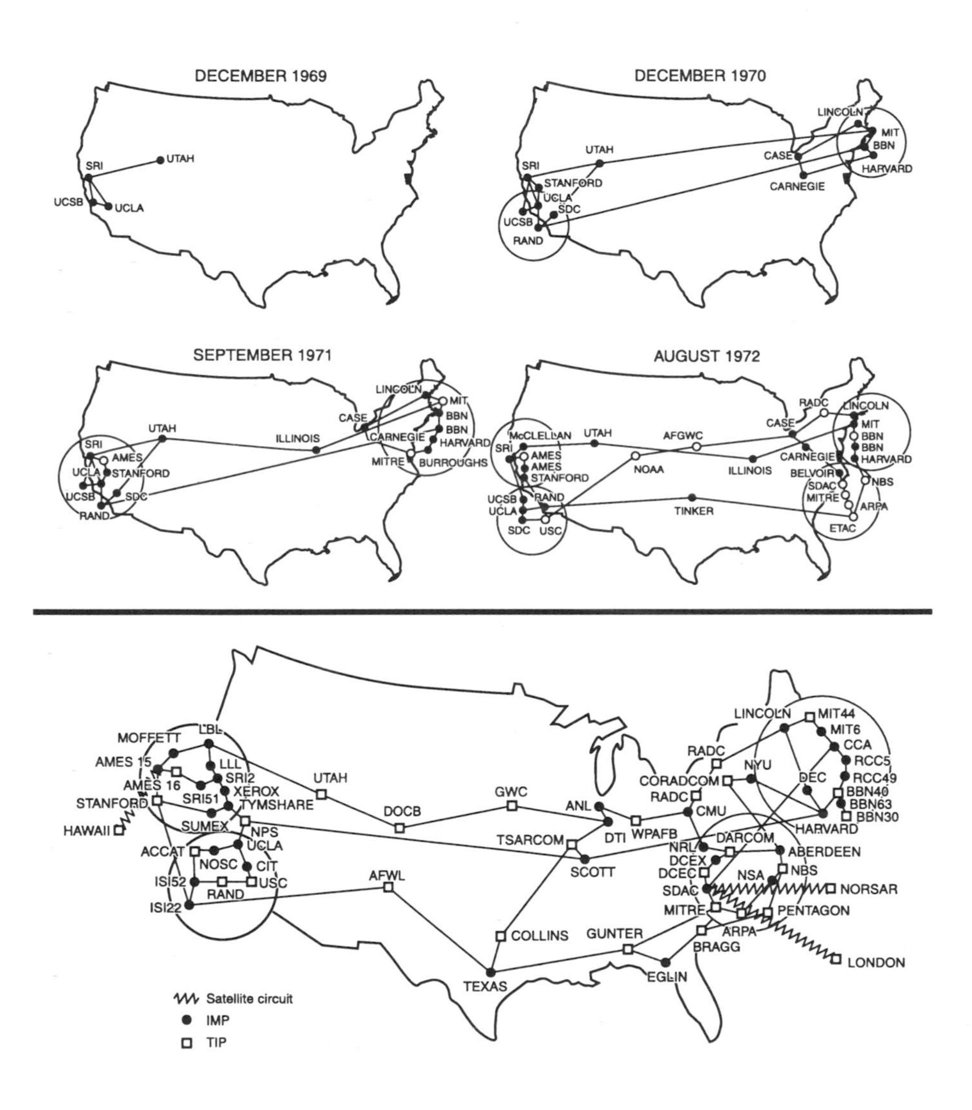

Source: Lee M. Paschall, "Command, Control, and Technology," *Countermeasures,* July 1976, 43.

Figure 16. PWIN Initial Three-Node Configuration

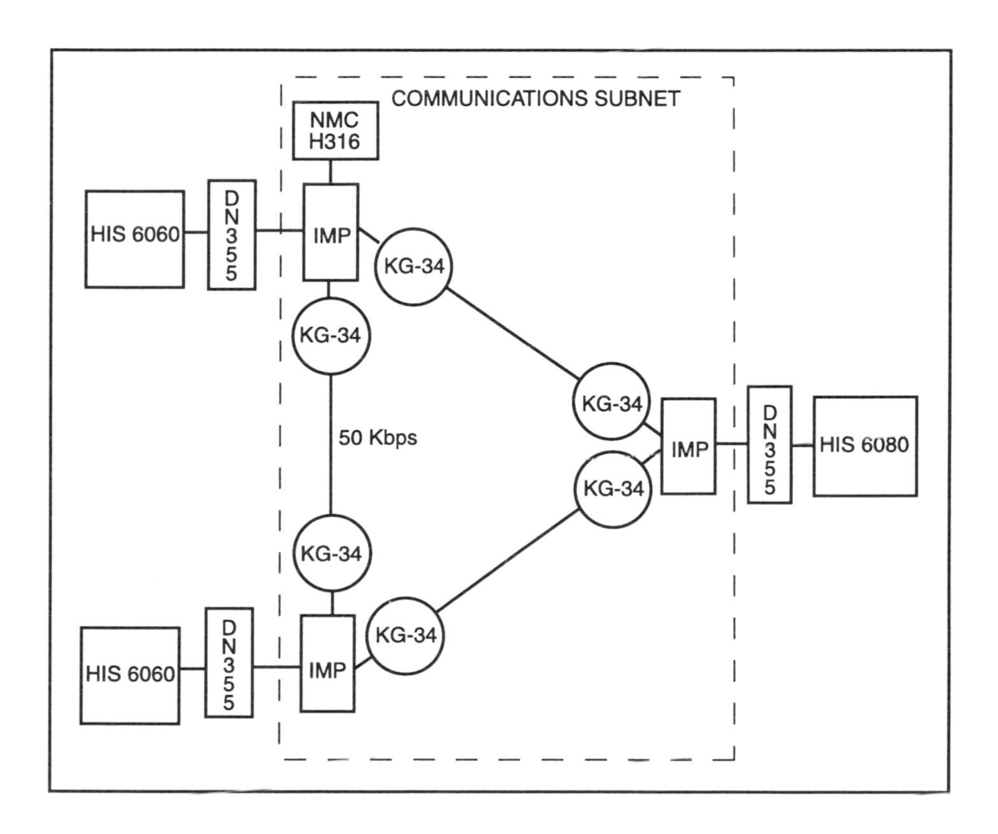

Source: Comptroller General, *The World Wide Military Command and Control System—Major Changes Needed in Its Automated Data Processing Management and Direction,* LCD-80-22 (Washington, D.C.: General Accounting Office, 14 December 1979), 49.

Figure 17. WIN Configuration in the Early 1980s

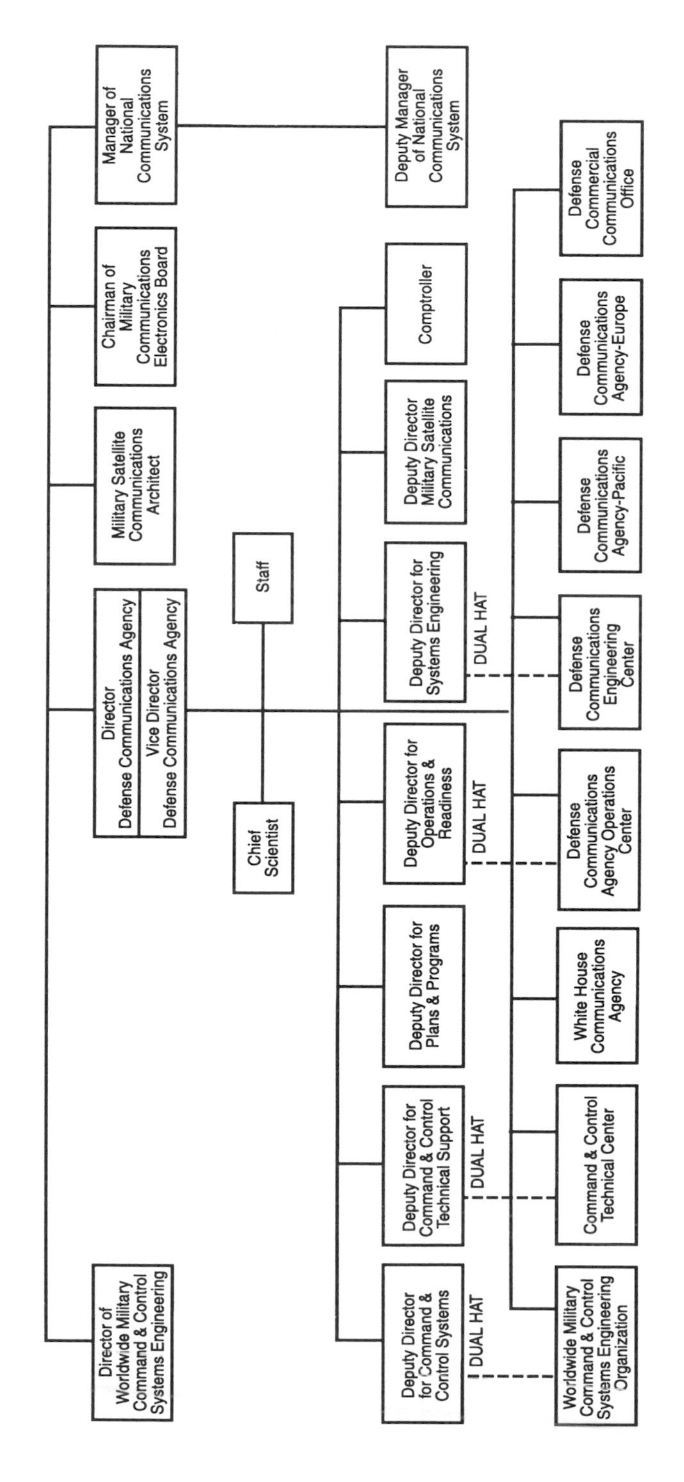

Source: Gerald P. Dinneen, "C² Systems Management," *Signal*, September 1979, 18.

Figure 18. New Organizational Structure for Defense Communications Agency

Satellite
Satellite
BMEWS
Site 1
BMEWS
Site 3
BMEWS
Site 2
COBRA
Dane
PARCS
PAVE PAWS
East
ANMCC
PAVE PAWS
West
SAC
NMCC
NORAD
FS8-7
DSCS
Satellite
DSCS
Satellite

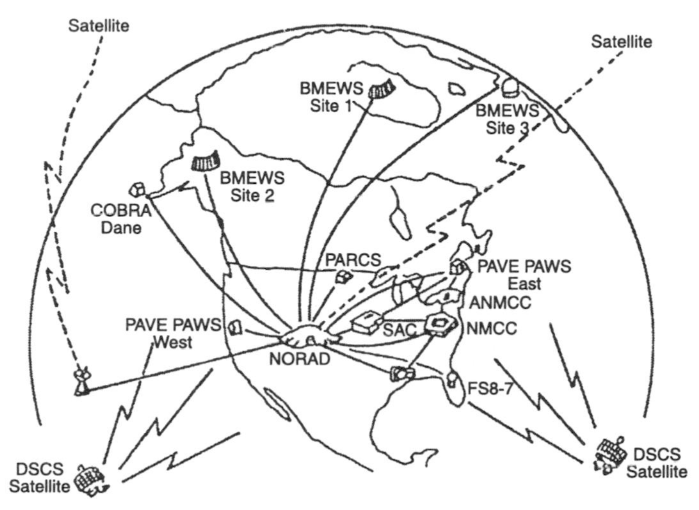

Source: Senate Committee on Armed Services, *Recent False Alerts From the Nation's Attack Warning System* (Washington, D.C.: General Accounting Office, 9 October 1980), 2.

Figure 19. NORAD's Missile Warning System

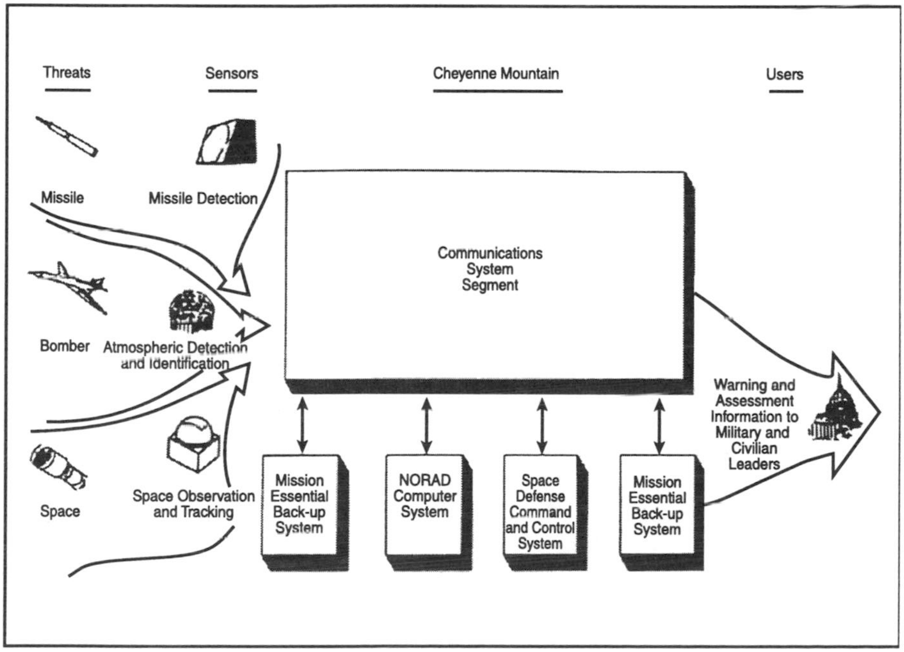

Threats
Sensors
Cheyenne Mountain
Users
Missile
Missile Detection
Communications
System
Segment
Bomber
Atmospheric Detection
and Identification
Warning and
Assessment
Information to
Military and
Civilian
Leaders
Space
Space Observation
and Tracking
Mission
Essential
Back-up
System
NORAD
Computer
System
Space
Defense
Command
and Control
System
Mission
Essential
Back-up
System

Source: General Accounting Office, *NORAD's Communications System Replacement Program Should Be Reassessed,* GAO/IMTEC-89-1 (Washington, D.C.: General Accounting Office, November 1988), 10.

Figure 20. Communications System Segment Connections

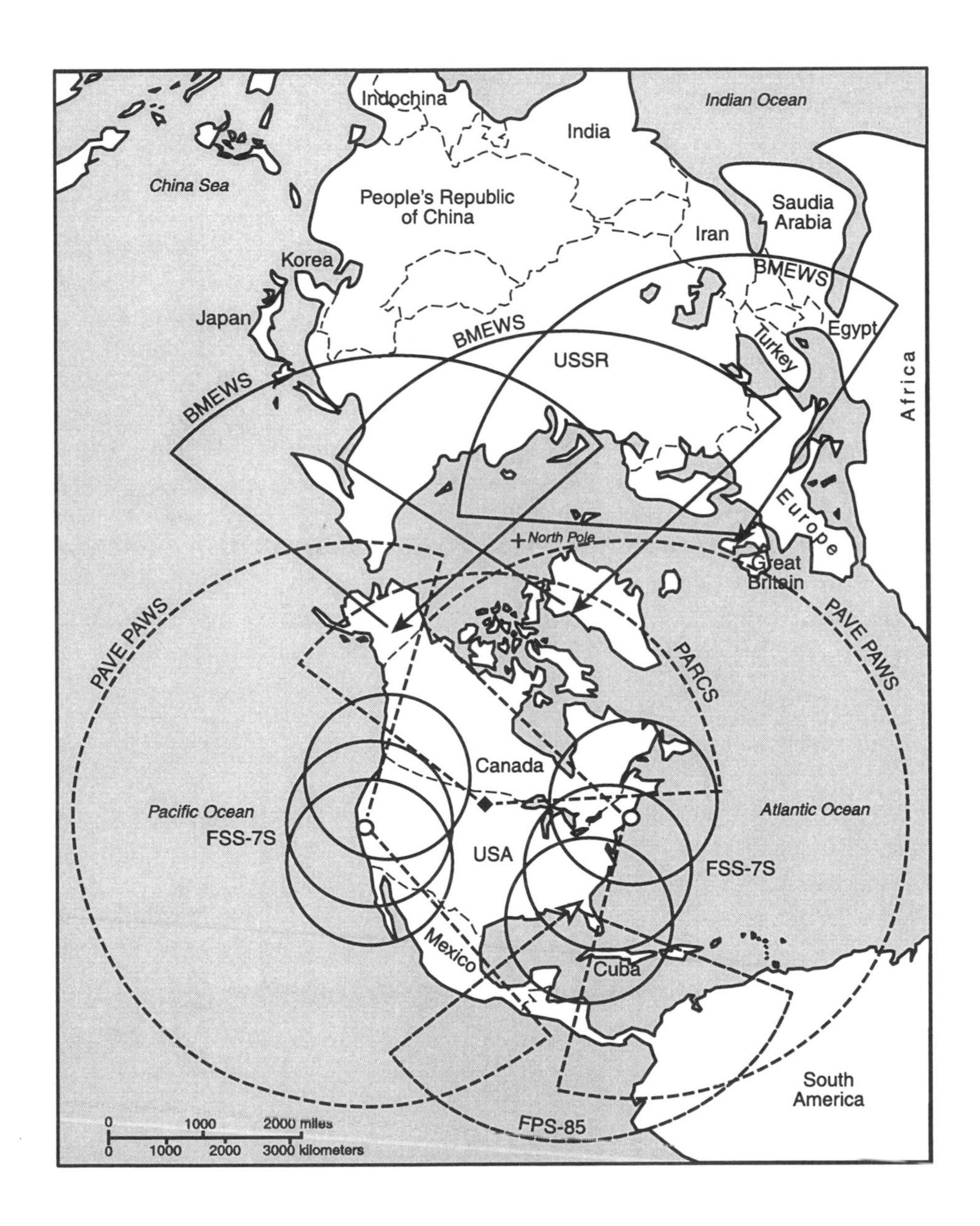

Source: Harry V. Martin, "Communications Vulnerability," *Military Science & Technology,* June 1982, 46.

Figure 21. Land-based Ballistic Warning and Detection

Chapter 15

The C^3I Triad: Programs

If the 1960s had been a decade of WWMCCS conceptualization and the 1970s one of system formalization, the 1980s represented a decade of realization, a time for bringing to fruition many of the programs judged necessary for a World Wide Military Command and Control System in fact as well as in name. The Strategic Modernization Program identified command and control improvement as its highest priority item, and the programs that fell beneath this rubric, the majority of them already under development, were overwhelmingly intended to enhance control over the nuclear forces. But in one sense most of the efforts took place at the periphery, focused on sensor systems and other assets controlled by the military services, rather than on core capabilities designed to merge those assets into a more cohesive and effective system. Not surprisingly, perhaps, given that key administration officials, Defense Secretary Weinberger among them, believed that centralization had gone too far and that programmatic authority should reside at the subunit level, with the military services. But despite these caveats, and after years of second-class status, it was during the Reagan administration that WWMCCS finally came of age. Below are considered in brief a number of the major WWMCCS initiatives, presented within the compass of each of the three legs of Latham's C^3I Triad.

Early Warning and Attack Assessment

One key focus of the command and control modernization program was to upgrade and enhance the capabilities of the US radars and satellites that provided early warning of Soviet attack. Broadly speaking, that attack could come either by way of ballistic missiles or through such air-breathing threats as cruise missiles and aircraft, and specific systems were intended to deal with each type of threat. For the detection of ballistic missiles, potential approach corridors were covered by at least two different types of warning sensors, designed to detect different types of physical phenomena. Radars such as those of the venerable Ballistic Missile Early Warning System were for detecting the presence of missiles by reflected radio energy. Depending upon the radar's power and capability, analysis of the returned signal could provide information about the location, trajectory, and speed of incoming missiles. Additionally, sensors on board early warning satellites would register the infrared energy given off by burning rocket engines, providing similar information. Working together, these radars and satellites provided "dual phenomenology" coverage, a critical redundancy for avoiding errors in the assessment of ballistic missile attack against the United States. Three ballistic missile early warning-and-attack characterization systems were identified as key elements of the Strategic Modernization Program, namely, BMEWS, the Precision Acquistion of Vehicle Entry Phased Array Warning System, and DSP satellites, the first two of which are considered here.

The Strategic Modernization Program additionally focused attention on upgrading the capabilities of the radar systems deployed around the North American periphery for providing early warning, attack assessment, and tracking of air-breathing threats such as aircraft and cruise missiles. Two programs of this sort were already in progress when the Reagan administration arrived in Washington, specifically, the Over-the-Horizon Backscatter (OTH-B) radar and the North Warning System, an upgrade for the antiquated distant early warning line (DEW LINE) radars. These were accorded a high priority in the command and control modernization efforts of the eighties.

PAVE PAWS

The original US system for detection of submarine-launched ballistic missiles was constructed during the early 1960s, a series of six FSS-7 height-finding radars (the so-called fuzzy sevens) deployed along the east and west coasts in the early 1960s as part of the Air Force's Back-up Interceptor Control System. Coverage to the south came later in the decade as the original sites were supplemented by facilities in Texas and Florida. Overall it was a limited system, capable of providing only minimal warning in case of a mass SLBM attack. Most of the facilities had very restricted data-processing capabilities, rendering them unable to provide detailed information about incoming missile trajectories.

The replacement for the original SLBM detection system was PAVE PAWS, whose radars were designed to detect SLBMs flying minimum-energy ballistic trajectories at ranges of approximately 2,000 nautical miles (NM), and at 3,000 NM for missiles in higher, lofted trajectories. On top of this, the radars' computer-based attack characterization capabilities would permit extremely precise prediction of launch and impact points of multiple targets, with warning data automatically forwarded by way of redundant ground communications links to the NMCC and ANMCC, SAC and NORAD headquarters, and other command centers.[1] A secondary function of PAVE PAWS was to support NORAD's spacetrack mission, and it was for this purpose that the radars would be most frequently used, continuously cataloging the position and velocity of satellites and other objects in low Earth orbits, all the while maintaining a constant watch for SLBMs.

The first PAVE PAWS radar, located at Otis AFB in Massachusetts, came on line in early 1979. The second site, at Beale AFB in California, became operational later that year. In the Strategic Modernization Program, one of the major priorities was a PAVE PAWS expansion. Two new facilities in the southeast and southwest were to be added to the two already in operation, providing dual phenomenology coverage toward the south. Located at Robins AFB, Georgia, and Goodfellow AFB, Texas, these two sites became operational in 1987. Even while the new facilities were under construction, upgrades to the original PAVE PAWS sites commenced to bring them into conformity in the areas of computer processing, displays, and radar technologies.[2] As the 1980s drew to their close, the four PAVE PAWS radars, along with the Perimeter Acquisition Radar Attack Characterization System in North Dakota, provided complete coverage of potential ocean launch areas around the continental United States, closing earlier gaps and significantly improving the US early warning capability.

Ballistic Missile Early Warning System

The second ballistic missile program accorded a high priority by the Reagan administration involved modernizing the three sites of the venerable BMEWS, designed to detect and provide attack characterization information about incoming intercontinental ballistic missiles from the north. The BMEWS project began in 1958, when the Air Force contracted for the development and installation of three separate radar sites, to be located in far-northern latitudes to maximize tactical warning time of Soviet attack. Additional contracts were awarded to develop a dedicated communications network to link these remote sites to NORAD headquarters in Colorado.[3] Yet while powerful, all of these BMEWS radars rapidly became outdated as defense requirements shifted from detection of relatively small scale attacks to detecting and tracking far larger numbers of incoming missiles with a high degree of precision.

Even before the Reagan administration arrived in Washington, some efforts had been made to improve BMEWS, although funding had proved a perennial problem. The Strategic Modernization Program intended to accelerate the modernization effort, calling for the old BMEWS radars to be replaced with far more powerful state-of-the-art phased array radars of the type used in PAVE PAWS, and for the facilities' 1950s-vintage vacuum tube computers to be replaced with more modern and powerful ADP equipment. The new electronically steered radars would provide far better detection and tracking of ICBMs than the radars they replaced, permitting objects 10 square meters in size to be tracked at a range of 3,000 nautical miles. Initial operational capability for the upgraded BMEWS would come in the early 1990s.

Over-the-Horizon Backscatter

To enhance America's ability in detecting air-breathing threats, a key initiative involved the deployment of a comprehensive OTH-B radar defense perimeter around the United States, capable of detecting airborne objects at distances beyond the range of conventional line-of-sight radars; that is, between 550 and 2,000 miles from the transmitter sites at altitudes from 100,000 feet right down to Earth's surface. OTH-B would additionally provide attack assessment information and permit airspace control over both the United States and Canada.[4]

The origins of the program extend back to the mid-1960s, but with the Reagan administration's strategic modernization program, developing an operational OTH-B capability became a front-burner priority. Plans called for a comprehensive OTH-B defense perimeter around the United States consisting of four separate radars—an East Coast, West Coast, Central, and Alaskan facility—each containing two or more integrated sectors of 60-degree radar coverage. The way the system worked, powerful high-frequency signals would be reflected off the ionosphere back toward Earth beyond the horizon from the transmitter. They would strike any airborne targets they encountered, bounce back to the ionosphere, and be reflected back in the direction of the radar's receiving antennas, which were located some one hundred miles from the transmitting site to minimize interference. The raw radar returns would then be subjected to advanced digital processing at the system's operations control center, moving targets would be isolated, and the results would be transmitted to NORAD and other locations.[5]

That, at any rate, was the intention. But as a series of tests conducted during 1988 made clear, the radars did not perform particularly well at night or under poor atmospheric conditions. Worse still, their discrimination was poor, meaning that they had difficulty detecting and tracking small targets at long distances.[6] When these problems were coupled with a declining defense budget toward the end of the decade, it was clear that the number of deployed radar sectors would have to be scaled back. But perhaps even more importantly, the end of the cold war dramatically reduced the need for OTH-B, and NORAD began looking for ways to justify funding for the not-yet-completed portions of the system. Specifically, they began emphasizing the radars' ability to provide coverage over Mexico, a primary staging area used by drug traffickers.[7] Underscoring this new role for OTH-B in the post-cold-war era, the program's entire 1991 budget was transferred to the Pentagon's Drug Interdiction and Counterdrug Activities budget.

North Warning System

To complement the OTH-B coverage and to fill that system's gaps to the north (a direction where high-frequency signals are susceptible to auroral disruption), a second initiative for improving detection of air-breathing threats was the North Warning System (NWS), a series of short- and long-range radars, which, along with their operations centers, support facilities, and communications links, would replace both the aging SAGE radars in Alaska and the radars of the stretching across Alaska and northern Canada.

Two programs were involved in NWS, aptly named SEEK IGLOO and SEEK FROST. The first to get under way was SEEK IGLOO, which called for the replacement of 13 SAGE air defense radars in Alaska, operated by the Alaskan Air Command, with modern, highly reliable, long-range radars. The replacement radar of choice was the FPS-117, which integrated into a single antenna a long-range surveillance radar, a heightfinder radar for medium- to high-altitude tracking of targets, and an Identification Friend or Foe System for automatically querying the transponders of aircraft. Unlike its predecessors, the new rotating radar was solid-state and designed to be frequency-agile, capable of operating at 20 different frequencies and capable of distinguishing targets among the calving ice, flocks of migratory birds, and the other radar clutter of the Arctic region.[8]

The second NWS program, SEEK FROST, would replace the antiquated DEWLINE with two different types of radars. First, 13 strategically located DEW LINE sites would be equipped with minimally attended FPS-117s of the same type used in SEEK IGLOO.[9] But because these were designed for long-range, medium- to high-altitude coverage, line-of-sight constraints meant that there would be frequent gaps in coverage. To fill those gaps, the 13 radar sites would be complemented by another 39 short-range unmanned radars for low- to medium-altitude coverage.[10] Depending upon terrain restrictions, between two to six of these "gap-filler" radars would be located between pairs of the FPS-117s, creating an unbroken electronic fence to the north.

The North Warning System was declared operational in the early 1990s, scanning the sky ceaselessly for a threat that in many people's view no longer existed. But system proponents put the best possible spin on the situation: "Just because there are fewer burglaries in the neighborhood doesn't mean you throw away the burglar alarms,"[11] one NWS operator remarked.

Communications Initiatives

The second leg of Donald Latham's C3I Triad involved the upgrade or deployment of a series of communications systems to permit the leadership to remain in contact with the strategic forces on a continuing basis under all conditions, up to and including a strategic nuclear exchange. A number of ground-, air-, and space-based communications initiatives were pursued under the aegis of the Strategic Modernization Program, many of them elements of WWMCCS's Minimum Essential Emergency Communications Network (MEECN). For communicating with the ballistic missile submarine force, plans called for upgrading the Navy's fleet of TACAMO planes, and for deploying a ground-based extremely low frequency (ELF) communications system. Also called for was the development of the Air Force's Ground Wave Emergency Network (GWEN). Finally, attention was given to the deployment of the third-generation Defense Satellite Communications System and to the development of the exceedingly ambitious Military Strategic and Tactical Relay Satellite. Several of these systems are discussed below.

Extremely Low Frequency Communications

A number of independent, redundant systems have long existed to provide communications to the strategic submarine force, but all imposed substantial limitations in operational flexibility. What was needed, in the Navy's judgment, was a system that was not only survivable and secure, but one that was jam resistant and whose signals could be received even in a wartime environment of multiple nuclear bursts, thus guaranteeing the submarines' usefulness as a deterrent force.

Since the late 1950s, the Navy had been examining the feasibility of a communications system utilizing the portion of the electromagnetic spectrum between 10 and 100 Hertz, known as the extremely low frequency or ELF band. On the theoretical level, ELF offered many benefits. ELF signals are characterized by low-propagation attenuation, meaning that they can be received over great distances, indeed worldwide. The signals have low attenuation in seawater, such that they can penetrate to depths some 20 times deeper than very low frequency signals.[12] With ELF, submarines could operate well below the 200-foot maximum detection depth for ocean surveillance satellites and below noise-reducing thermal layers in the ocean, making them less susceptible to antisubmarine warfare activity. Additionally, the way ELF signals are transmitted makes them highly resistant to jamming and other forms of interference, including nuclear detonations.[13] Finally, it was felt that an ELF system could be survivable, since its transmitters and antenna arrays could be geographically dispersed and hardened.[14]

The original system design advanced by the Navy, known as Sanguine, was exceedingly ambitious, covering approximately 41 percent of the state of Wisconsin. Over six thousand miles of antenna wires would be buried six feet deep, checkerboard fashion, on state and federal property and along public rights-of-way. The wires would be grounded at each end by an elaborate grounding configuration that, depending upon local conductivity conditions, could extend for distances of up to two miles. Two hundred and forty underground transmitters, each installed approximately at the center of an antenna cable, would power the system. To ensure their survivability, the transmitters would themselves be buried about 35 feet below ground in reinforced concrete capsules.[15] The Navy estimated that some 800 million watts of power would have to continually flow through Sanguine's wires to maintain operational readiness.

In the years that followed, unremitting criticism of the proposed system by various state and federal officials, concerned citizens, and environmental groups resulted in an ongoing search for alternative host sites and a progressive scaling-back of the system's dimensions. To increase the system's acceptability among cost-conscious members of Congress, during the mid-1970s the Navy shifted its preferred ELF system from a downsized, survivable Sanguine to a new nonsurvivable system called Seafarer. Despite many changes and adjustments, however, the criticism continued unabated. The Navy's response was to keep searching for new ELF designs that would prove more publicly palatable, resulting in even further reductions in the proposed system's size. The most promising alternative settled upon by the Navy involved linking its small Wisconsin ELF test facility to another small facility to be constructed at Sawyer Air Force Base in Michigan. The Navy called this far more restrictive system Austere ELF, and President Carter approved the plan in January 1979.

But congressional resistance to the cost of even an "austere" ELF necessitated a further reduction in the size of the Wisconsin-Michigan system, halving the amount of antenna wire to be used. As part of the Strategic Modernization Program, the Reagan administration gave the go-ahead for this downsized Austere ELF, which was renamed simply Project ELF. Work proceeded throughout the decade, and in May 1989 the two linked facilities went to full power. Full operational capability was achieved the following October.[16] So even as the cold war ended and the 1990s dawned, the Navy had at last assured itself of a real-time submarine communications capability.

Military Strategic and Tactical Relay Satellite

Given the substantial and increasing US reliance upon space-based assets for communications, a major concern of defense planners during the 1980s was satellite vulnerability. The problem seemed particularly pressing, because in many families of defense satellites, toughness had been deliberately sacrificed to pack on board the maximum mission capability. The result was highly capable but inherently fragile spacecraft that were vulnerable, given the active pursuit of antisatellite technologies by both superpowers. Consequently, much thought was given to enhancing satellite survivability through such means as hardening, maneuverability, the use of deceptive technologies, and proliferation.[17]

These concerns found expression in the Military Strategic, Tactical, and Relay (MILSTAR) satellite communications system.

Representing the most ambitious and technically difficult communications program ever undertaken by the Pentagon, MILSTAR also represented the highest-priority command and control element of the Strategic Modernization Program. Intended to employ an internetted constellation of satellites both in synchronous equatorial orbits and elliptical polar orbits, MILSTAR was intended to link the National Command Authorities to surface ships, submarines, aircraft, ICBM forces, Army ground mobile forces, and theater nuclear forces around the globe.

Plans called for MILSTAR to incorporate a number of features to ensure its continued availability during crises up to and including general nuclear war. The satellites would be hardened against radiation from nuclear detonations and laser attack. They would be maneuverable and incorporate on-board decoys to enhance their survival against physical attack. The satellites would offer enhanced resistance to jamming, with much of the improvement arising from their use of the extremely high frequency (EHF) band. EHF signals travel in extremely straight lines and are relatively unaffected by passage through the atmosphere. This means they can focus more power on ground receivers, allowing for the use of smaller antennas and avoiding jammers and interception.[18] In addition, the use of rapid frequency-changing burst transmissions and antenna-hopping techniques would make the transmissions even less vulnerable. Finally, the spacecraft would be outfitted with satellite-to-satellite cross-links (hence the term *relay* in MILSTAR), eliminating dependence on ground stations for communications relay and minimizing the risk of eavesdropping by hostile terrestrial terminals.

When fully operational, MILSTAR was intended to enhance or assume the functions of a number of existing satellite systems. Critical emergency action message services provided by UHF systems like AFSATCOM would be shifted to MILSTAR. MILSTAR would augment and possibly replace the Satellite Data System, serve as a relay for data from intelligence satellites such as the KH-12, and assume at least some of the broadband communications functions performed by the Defense Satellite Communications System.[19]

That, anyway, was the plan. But it was at the moment when MILSTAR's requirements had to be translated into reality that the problems began. The ambitious and technically complex nature of the undertaking was immediately apparent, and within a very short time the program began running into a series of technological roadblocks. These included the need for reliable traveling wave tube amplifiers, fault-tolerant spaceborne computers, advanced adaptive antennas, nulling antennas, hardened electronics and high-speed data processors, and low-cost EHF terminals.[20] The Air Force assured critics that the problems were not insuperable, but their large number virtually guaranteed cost overruns that would in turn force the program's timetable to be stretched out. Complicating matters further was the catastrophic loss of the space shuttle *Challenger* in January 1986, freezing shuttle launches until the cause of the *Challenger* mishap could be ascertained and corrected.

On top of all this, almost continuous interservice skirmishing over technical requirements and bureaucratic prerogatives beset the MILSTAR program. On the one hand, this led to one of the most heavily gold-plated programs the defense establishment had ever seen. The ability to satisfy everyone by adding capabilities was not unlimited, however—given the irreducible fact that only so much weight could be lifted into space by existing launch vehicles, and any decision to enhance one capability (such as hardening) meant that sacrifices had to be made in other areas. As the tradeoffs were made, it quickly became apparent that MILSTAR would likely have a low data rate, and might have to limit simultaneous access to as few as 15 users.[21] Such low capacity and flexibility strongly suggested that MILSTAR would never achieve its promise as a war-fighting "switchboard in the sky," nor would it be able to assume the functions of such satellite communications programs as the Defense Satellite Communications System or Fleet Satellite or meet intelligence relay requirements. By the mid-1980s program officials were confirming these limitations to Congress, even as projections of program costs were ballooning.[22]

Perennially over budget, in 1988 the plans for MILSTAR were pushed back. Additional stretch-outs appeared inevitable as the new Bush administration's defense secretary, Dick

Cheney, tried to meet near-term budgetary targets by cutting $1 billion from the defense budget submitted by the outgoing Reagan administration.[23] In 1989 the House Appropriations Committee called for termination of MILSTAR. Recognizing a sinking ship when they saw one, the services began abandoning MILSTAR. The axe finally fell in July 1990, when Democrats on the Senate Armed Services Committee overrode the unanimous objections of committee Republicans and voted to kill MILSTAR. The reasons were clear enough: MILSTAR's sticker price had tripled, but for a less capable system that would serve far fewer users than had been originally envisioned.[24] But perhaps most important of all was the increasing inappropriateness of the system. As congressional critics were quick to point out, MILSTAR had been conceived in a period of resource munificence and intense Soviet-American hostility, when great emphasis was placed on fighting and winning a nuclear war. But in the altered national security landscape of the 1990s, it was said, neither of these concerns applied any longer.

Ground Wave Emergency Network

The prevailing logic during the 1980s was that satellites were inherently fragile, and hence a nuclear war might quickly prove fatal to many key space-based communications systems, including MILSTAR itself, and protecting all of the satellites would almost certainly prove prohibitively costly.[25] The response to this was to put in place additional MEECN systems utilizing the various frequencies of the electromagnetic spectrum, so that the NCA could relay instructions to the strategic nuclear forces should satellite communications be disrupted. ELF was one such system. Another major system was GWEN, which commanded a full two-thirds of the funding devoted to MEECN improvements during the Reagan years.

GWEN was intended to provide assured transmission of force direction messages in case of a surprise, "bolt from the blue" Soviet nuclear attack and to provide an enduring communications capability thereafter. The characteristics that permitted GWEN to accomplish its mission were several. First, there was its use of ground-hugging, low-frequency radio, which resists disruptions of the ionosphere caused by nuclear weapons. Second, there was its use of packet-switching, the technique developed during the 1970s for use in the ARPANET and later slated for use in several other WWMCCS-related networks. Third, there was the system's nodal redundancy, enhancing the probability that messages would reach their destinations.[26] Plans called for developing GWEN in stages. The first phase, called initial communications connectivity, was intended simply to demonstrate GWEN's technical viability. Phase 2 called for the construction of a thin-line communications connectivity (TLCC), consisting of 56 GWEN relay stations arranged in a giant figure eight with SAC headquarters in Omaha at its center. Linking together a total of eight command centers and 30 military bases, the TLCC by design incorporated as many of the desired GWEN features as were then within the state of the art, and it was intended to serve as a foundation for the much larger Phase 3 system to come.[27]

Even as the TLCC was moving toward completion, the number of relay towers to be included in the Phase 3 GWEN system was undergoing a process of incremental diminution. Plans called for between 400 to 500 towers at the time the system was announced in 1982; but well before the TLCC came on line, this figure had been reduced to 236 and then to 127. Assistant Secretary of Defense for C^3I Donald Latham noted that for the current GWEN mission, 127 GWEN nodes were sufficient but that "other missions and a more realistic threat could require additional nodes."[28] The Phase 3 GWEN was thus explicitly viewed as an interim system, an unsatisfactory resolution to the dilemma that an appropriately redundant system was prohibitively costly, while anything less could be easily destroyed, rendering it useless for its intended purpose.

These concerns with cost, coupled with strong public opposition to the construction of relay towers in some localities, had the effect of slowing the construction of the GWEN. It was not until the fall of 1987 that RCA, the prime system contractor, began testing the TLCC.[29] By early 1988, 49 of the 56 TLCC nodes were undergoing initial operational testing and evaluation. Work on Phase 3 was simultaneously in progress, and in mid-1988 the Air Force approved an expansion of GWEN from the TLCC's 56 nodes to 96.[30] Despite the substantial reduction, Air Force officials claimed that the much-diminished GWEN would still constitute a "final operational capability" that would be operational within a few years.[31] As it turned out, those few years brought about not the completion of the Ground Wave Emergency Network but rather the end of the cold war, raising questions as to whether this element of MEECN was simply an expensive anachronism.

Command Initiatives

The final leg of Latham's C^3I Triad involved improvements in the command structure. This included substantial upgrades to the capabilities of the ground-based command centers, most prominently the Alternate National Military Command Center, the National Military Command Center, and NORAD's Cheyenne Mountain headquarters. Major improvements were also planned for various air-based command posts, including SAC's "Looking Glass" aircraft, the EC-135 airborne command posts used by the nuclear CINCs, the National Emergency Airborne Command Post, and the president's personal plane, Air Force One. Also falling beneath the rubric of command initiatives were various initiatives intended to enhance the national leadership's ability to direct the forces: programs such as the NAVSTAR Global Positioning System (GPS) and the WWMCCS Information System and research into such decision-making aids as artificial intelligence and expert systems. One of these command initiatives, NAVSTAR, is examined in greater detail below. The WWMCCS Information System is discussed in the next chapter.

NAVSTAR Global Positioning System

The Navigation Satellite Timing and Ranging (NAVSTAR) Global Positioning System is a satellite-based radio navigation system designed to provide highly accurate altitude, latitude, longitude, velocity, and time data to properly equipped users anywhere in the world. The global positioning system was long recognized by the Pentagon as a research priority. The first major milestone in the evolution of the NAVSTAR GPS came in 1967 when the Joint Chiefs of Staff conducted a comprehensive review of all navigation systems then in use or at the stage of advanced development to determine which would be most appropriate for meeting the criteria of worldwide coverage, redundancy, instantaneous response, continuous availability, and the ability to resist enemy countermeasures. The study concluded that none of the systems then in use could meet all of the military's requirements and that of those options then being considered, a satellite-based system for three-dimensional navigation offered the greatest potential to do so. The result of the study was a JCS master navigation plan, a general guidance document directing that a program to orbit such a system be established on a priority basis.[32]

As originally envisioned, the system would serve over 27,000 individual American and allied users, which included fixed-wing aircraft, helicopters, surface ships, submarines, land vehicles, and ground troops. Civilian use was also projected, although likely more for the large external constituency it would rally behind the program than out of any Pentagon desire to improve the navigation systems then in use by the maritime and airline industries. A major problem with the JCS plan from the outset, however, was that no preliminary studies were conducted to identify user needs; thus, there was no way to know whether the capabilities proposed for the new system would address specific unmet needs or identified deficiencies of prospective users.[33]

NAVSTAR contained three distinct elements: a space segment, a control segment, and a user segment. The space segment consisted of 24 satellites, of which 21 were active and three were on-orbit spares. They were divided into three planes of eight satellites each, in circular orbits at an altitude of approximately 10,900 nautical miles, that provided continuous worldwide coverage. Each satellite was designed to transmit two ultra-high frequency signals that could be captured by receiving sets, one for positioning and one for timing. Receiving the "composite signal" from four satellites was required for achieving the expected level of accuracy in navigation, determination of velocity, and three dimensional positioning, by some estimates to within 11 feet laterally and 12 feet in altitude.[34] The control segment, consisting of a number of ground facilities scattered across the globe, was intended to track the satellites and provide updates to their position and timing information. The user segment of the system included the various terminals and devices that would capture the satellite signals and convert them into navigation, time, and position information, which in turn could be used for a variety of military and civilian purposes—including positioning air, sea, land, and space platforms prior to weapons launch.[35]

From the perspective of command and control, several significant changes in NAVSTAR's mission occurred during the 1970s. Under the original system concept, NAVSTAR signals would be available to ships, aircraft, and some land vehicles. Missiles were not considered prospective users because the Pentagon was planning a dedicated satellite constellation for in-flight missile guidance. But a series of technological advances resulted in the dedicated system being scrubbed and its functions assigned to NAVSTAR. Most notably, these were advances in the area of miniaturization, coupled with energetic efforts by the director of defense research and engineering to have the new navigation satellite support the missile accuracy improvement program. The implications of such a change were far reaching. A traditional weakness of submarine-launched ballistic missiles had always been their accuracy, a problem arising from the fact that the launch platform itself is continuously moving, if only slightly, with deleterious effects on gyroscope-based guidance systems. The possibility for midcourse flight corrections changes things completely, giving the strategic submarine force unprecedented hard-target capability.[36]

Of perhaps greater relevance from the perspective of command and control was the decision to put Nuclear Detection System sensors on board NAVSTAR satellites. These sensor packages were designed to detect the visible light, X-rays, gamma rays, and electromagnetic pulse associated with a nuclear detonation, allowing for determination of bomb yield and type.[37] Since the energy from detonations would arrive at different satellites at slightly different times, the locations of these satellites could be precisely determined by NAVSTAR's atomic clocks, accurate to a few billionths of a second.[38] With this information, the status of enemy military forces could be ascertained, allowing for the redirection of forces, the rendezvous of nuclear-equipped aircraft with tankers without breaking radio silence, the reconstitution of command and control assets, and a variety of other adjustments during the trans- and post-attack stages of conflict. To ensure their usefulness in a protracted war situation, the satellites would be hardened against jamming, laser and nuclear weapons effects, and other types of electronic interference.[39]

Operational testing of NAVSTAR began in 1985. Acquisition commenced in 1986, and NORAD's Consolidated Space Operations Center took over operational control of the system that same year. But a series of production difficulties, coupled with the problems afflicting the space shuttle program following the *Challenger* disaster, resulted in major program delays. It was not until February 1989 that the launch of the first production NAVSTAR satellite took place, and the cold war drew to its close with the fielding of a fully operational global positioning system still somewhere in the indeterminate future. When it finally came on line, NAVSTAR would serve a somewhat different set of constituencies than had been envisioned. SAC decided not to install GPS receivers on board its aircraft, cutting the total number of Air Force planes slated for receivers by more than half. Additionally, it turned out that a whole class of small, instrument-stuffed aircraft like the F-16 had insufficient space for the receivers. In total, some 3,600 GPS receiving units were slated for platforms that could not accommodate them.[40] The most visible benefits of NAVSTAR were, rather, in the civilian sector. For example, when France and Great Britain were constructing the so-called Chunnel beneath the English Channel to connect the two countries, they turned to NAVSTAR. Construction of the tunnel's three tubes proceeded simultaneously from the French and English coasts; without NAVSTAR to correct errors in the measuring instruments used, the tubes would not have been able to meet at the channel's mid-point at the same altitude.[41]

These were some of the major command and control initiatives advanced by the Reagan administration under the rubric of the Strategic Modernization Program. The importance and centrality to the program of one additional command initiative, the WWMCCS Information System, is such that it is given a more detailed treatment in the next chapter.

Notes

1. Edgar Ulsamer, "The Growing, Changing Role of C^3I," *Air Force Magazine* 62 (July 1979): 36.

2. "Pave Paws, BMEWS Radar Site Upgrades Will Broaden Missile Threat Coverage," *Aviation Week & Space Technology*, 9 December 1985, 52.

3. "Surveillance and Control Systems," *Signal* 20 (April 1985): 22.

4. House, Committee on Appropriations, *Department of Defense Appropriations for 1987*, 99th Cong., 2d sess. (Washington, D.C.: Government Printing Office [GPO], 1986), 441.

5. Jerry Mayfield, "Air Force Upgrading Radar Network," *Aviation Week & Space Technology*, 16 June 1980, 232.

6. House, Committee on Appropriations, *Department of Defense Appropriations for 1989*, pt. 6, 100th Cong., 2d sess. (Washington, D.C.: GPO, 1988), 545.

7. General Accounting Office, *Over-The-Horizon Radar: Better Justification Needed for DOD Systems' Expansion*, NSIAD-91-61 (Washington, D.C.: General Accounting Office [GAO], January 1991), 2, 8.

8. Bruce Gumble, "Air Force Upgrading Defenses at NORAD," *Defense Electronics* 17 (August 1985): 106.

9. Clarence A. Robinson Jr., "Soviets Test New Cruise Missile," *Aviation Week & Space Technology*, 2 January 1984, 15.

10. "What's Happening in Electronics at ESD," *Air Force Magazine* 66 (June 1983): 48.

11. Jon Bowermaster, "The Last Front of the Cold War," *The Atlantic Monthly*, November 1993, 44.

12. Stockholm International Peace Research Institute, *World Armaments and Disarmament: SIPRI Yearbook 1979* (London: Taylor & Francis, 1979), 397.

13. William J. Ruhe, "ELF (Extremely Low Frequency) in Wartime," *Signal* 33 (January 1979): 19.

14. John Merrill, "Some Early Historical Aspects of Project Sanguine," *IEEE Transactions on Communications*, 4 April 1974, 359.

15. Lowell L. Klessig and Victor L. Strite, *The ELF Odyssey: National Security Versus Environmental Protection* (Boulder, Colo.: Westview Press, 1980), 16.

16. Bruce Borcherdt, "Woodpeckers Tapping U.S. Submarine Radio Network," *Chicago Tribune*, 1 June 1989, 4.

17. William E. Burrows, *Deep Black: Space Espionage and National Security* (New York: Random House, 1986), 321–23.

18. "Critical C^3I Programs Emphasize Survivability, Jam Resistance," *Defense Electronics* 18 (July 1986): 92–94.

19. Edgar Ulsamer, "C^3 Survivability in the Budget Wars," *Air Force Magazine* 66 (June 1983): 38.

20. James B. Schultz, "Milstar to Close Dangerous C^3I Gap," *Defense Electronics* 15 (March 1983): 52–53.

21. James W. Rawles, "MILSTAR Soars Beyond Budget and Schedule Goals," *Defense Electronics*, February 1989, 67.

22. House, *DOD Appropriations for 1987*, 615.

23. "Cheney Stretches Milstar?" *C^3I Report*, 1 May 1989, 1.

24. R. Jeffrey Smith, "If The Cold War Has Ended, Why Is It So Chilly in Here?" *Washington Post National Weekly Edition*, 19–26 August 1990, 15.

25. House, *Department of Defense Appropriations for 1984, Hearings before a Subcommittee of the House Committee on Appropriations*, 98th Cong., 1st sess. pt. 8 (Washington, D.C.: GPO, 1983), 504.

26. Ulsamer, "C^3 Survivability in the Budget Wars," 54.

27. *Defense Electronics*, 99.

28. House, *DOD Appropriations for 1987*, 496.

29. James W. Canan, "Steady Steps in Strategic C^3I," *Air Force Magazine* 70 (June 1987): 46.

30. *Defense Electronics*, 96–99.

31. House, *DOD Appropriations for 1989*, 468.

32. Comptroller General, *Defense Acquisition Programs: Status of Selected Systems*, NSIAD-90-30 (Washington, D.C.: GAO, December 1989).

33. Comptroller General, *The Global Positioning System—A Program with Many Uncertainties* (Washington, D.C.: GAO, 17 January 1979), 1.

34. D. Douglas Dalgleish and Larry Schweikart, *Trident* (Carbondale, Ill.: Southern Illinois University Press, 1984), 254–55.

35. Comptroller General, *Should NAVSTAR be Used for Civil Navigation? FAA Should Improve Its Efforts to Decide*, LCD-79-104 (Washington, D.C.: GAO, 1979), 1.

36. Dalgleish and Schweikart, 254–55.

37. John C. Toomay, "Warning and Assessment Sensors," chap. 8 in *Managing Nuclear Operations*, eds. Ashton B. Carter, John D. Steinbruner, and Charles A. Zraket (Washington, D.C.: Brookings Institution, 1987), 310.

38. Gerald Green, "C^3I: The Invisible Hardware," *Sea Power* 26 (April 1983): 113.

39. "Production GPS Satellites in Ground Tests," *Aviation Week & Space Technology*, 9 December 1985, 73.

40. House, *DOD Appropriations for 1989*, 548.

41. "Gran Bretaña Deja De Ser Una Isla," Madrid, *El Pais*, 5 Mayo 1994, 6.

Chapter 16

WWMCCS Information System

By the middle of the 1970s the initial 35 WWMCCS standard computer systems had been successfully installed in 26 centers around the world.[1] Users could access the computers by way of local terminals in the immediate vicinity of the host computer or by using remote terminals located hundreds, even thousands, of miles away. The growth of the remote terminals had been extremely rapid, such that by the beginning of the 1980s almost every major American installation in the continental United States, Korea, and Europe was connected to WWMCCS. It was the culmination of years of effort to link the major command centers into a coordinated and responsive whole. While the underlying concept was sound, the practical deficiencies of WWMCCS automatic data processing were myriad, many of them directly attributable to the Honeywell 6000-series computers. Additions of various sorts were made as users endeavored to enhance their performance, decrease response time, provide backup capability, and improve compatibility with the new peripheral equipment that continued to appear.[2]

Software problems also abounded. The Honeywell-provided hardware architecture and database management system software could not provide a full range of user support services, especially in crisis situations. And since the software was owned by Honeywell rather than the government, the Pentagon was constrained in the modifications it could undertake to correct deficiencies and tailor the software to specific defense needs.[3] The result was a proliferation of software applications as users tried to find ways to make the system perform more effectively.

The Prototype WWMCCS Intercomputer Network—subsequently transitioned to operational status as WIN, the WWMCCS Intercomputer Network—had been a significant effort to address these concerns. Additional experience with networking had been gained, to be sure, but again a major impediment turned out to be the WWMCCS computers, which were incapable of performing the necessary packet-switching functions. Here too, additional hardware and software were necessary to work around the shortcomings and make all of the equipment play together. Nobody denied that there were problems, and the growing consensus was that what was required was a "total system," an intercomputer network capable of addressing all command and control requirements—one in which data could be widely shared and one that would degrade gracefully under conditions of stress rather than experience catastrophic failures. The result of these concerns was a congressional order to prepare plans for the replacement of the faulty computer network.

Recognizing that any satisfactory solution to the WWMCCS ADP problem would take a number of years to implement, an interim program was begun in 1980 to enhance the existing system's reliability and remedy some of its information-processing shortfalls. As part of this, an additional eight-year contract was signed with Honeywell to maintain and support the existing WWMCCS standard computers and associated software, and more than $100 million would be spent on these upgrades during the ensuing years. Nonetheless, the long-term prospects for such patchwork solutions appeared bleak. Advances in automatic data-processing technologies had been nothing short of spectacular since the time the Honeywells were acquired, and the number of people still using the antiquated 1960s-vintage computers was rapidly declining toward zero. This, in turn, had predictable effects upon system reliability; for as the number of systems dwindled, there remained no incentive for a commercial firm such as Honeywell to continue its investment in the outdated technologies. Resources were directed elsewhere, resulting in a lack of spare parts and trained personnel to service the older machines. It was a situation with serious implications for WWMCCS maintenance, the costs for which were rising steeply, and it was apparent to practically everyone that a definitive solution would have to be found soon.[4]

The search for that solution became more aggressive upon the arrival of the Reagan administration in January 1981. Immediately upon taking office, Assistant Secretary of Defense for C³I Latham began pointing out the many deficiencies in

WWMCCS's capabilities, especially its inability to support decision makers in time-sensitive situations and crises. He described how the rapid pace of technological advance had created a situation in which the Pentagon was locked into a spiral of escalating costs for maintaining obsolete hardware and software whose performance was in continual decline relative to the state of the art. The system lacked a data management capability. There was inadequate computer availability, response time was poor, capacity could not be expanded to support defined applications, and the system did not meet the requirement for multilevel security.[5] Latham consequently issued a call for a major modernization of WWMCCS ADP, in particular the pursuit of a reliable and secure, high-capacity intercomputer network. But how this was best accomplished was by no means clear. How should computing power be distributed within the system? Should ADP modernization take place at each individual user site, or should it be concentrated in fewer locations using a smaller number of more powerful central computers? Should identical modernization take place at all sites, or should hardware and software be tailored to users' specific needs? Should modernization occur simultaneously at all sites, or should attention be directed first to those elements of the system considered most deficient? For questions such as these, there was often no obvious or easy answer.

As in other instances where efforts had been undertaken to strengthen multiuser command and control systems, the services were less than enthusiastic supporters of Latham's proposal. DCA Director Samuel Gravely pointed out how service critics were charging that an upgraded WWMCCS ADP capability was not in their interests since it contributed mainly to support of the unified and specified commanders, not the services. "I guess I do not understand service interests," Gravely acidly remarked, "if they are not coincident with CINCs' interests."[6] But the services' resistance was actually not at all difficult to understand. Whereas existing vertically oriented systems were designed for service-specific purposes, an upgraded WWMCCS ADP capability of this sort promised greater horizontal integration, and with it the possibility for increased centralization of control.[7] What this meant for reformers was that organizational dynamics and concerns—bureaucratic inertia, organizational loyalties, suboptimization, resistance to change, and limited rationality—were issues that had to be dealt with on top of the technological issues involved in the construction of an upgraded WWMCCS ADP capability. With money and bureaucratic turf up for grabs, the perhaps inevitable result was contentiousness and confusion over the requirements the new system should address, the hardware and software necessary to do the job, and precisely who would make these judgments.

The Reagan administration was committed to command and control modernization, however, and the "total system" it had in mind was called the WWMCCS Information System (WIS), an effort that would implicate to varying degrees the Defense Communications Agency, the Joint Chiefs of Staff, the Office of the Secretary of Defense, and the three services. What planners had in mind was something more ambitious than WIN; they envisioned an interactive computer network that would tie together a series of central computers and local area networks, permitting the sharing of databases and workloads between command centers. It would be easier to use than its predecessor and offer improved means for protecting classified information.[8] The network would be reliable and redundant, capable of functioning under a range of peacetime and wartime conditions, up to and including nuclear war. The development of WIS represented the most important aspect of the command and control modernization effort, itself hailed as the most important element of the administration's vast and ambitious Strategic Modernization Program.

With respect to the new system's structure, or architecture, WIS system engineers opted for modernization of individual sites as opposed to concentrating ADP functions in fewer locations, the operative assumption being that a larger number of system nodes offered greater reliability and survivability than putting more figurative eggs in fewer nodal baskets. It was next decided that there was no need for similar modernization at all sites. While hardware and software standardization was considered essential for common functions among WWMCCS sites, it was acknowledged that custom-tailored modernization could take place where command-unique ADP functions were concerned. This was termed a *functional family* approach, and the hoped-for result would be a flexible, modular system in which smaller computers dedicated to specific functions could be tied together into an integrated network, providing the possibility of graceful rather than wholesale degradation in the event of failures. Finally, it was decided that the modernization effort should be phased in gradually, both to prevent disruption of operations at the various WWMCCS sites and to spread out the costs.[9]

As conceived, WIS would have two main emphases, the first of which involved hardware. Here, the most important task was the replacement of the outdated WWMCCS standard computers with more modern state-of-the-art ADP equipment, as well as the upgrade of associated terminals, displays, and peripherals. At the time, 83 of the Honeywell 6000-series computers were in operation at 49 locations. Of these, the WIS modernization would affect 78 computers at 46 locations. (The other five computers were operated by the Defense Intelligence Agency (DIA) as part of the Intelligence Data Handling System and would be upgraded by DIA as part of a different program.) Also included under the rubric of hardware was the upgrading of a variety of non-Honeywell computers then in use throughout WWMCCS: for example, the UNIVAC 1100/42s at SAC, NORAD, NMCC, and ANMCC.[10]

The second WIS component involved upgrading existing software and developing new software so that WWMCCS could more effectively perform its tasks of situation assessment, crisis management, rapid force deployment, and support. The need for the upgrade was pressing. First there was the system software. Because it was based on the batch-processing concepts of the 1960s, it simply could not provide a full range of user support services, including adequate on-line software development and data management tools. Moreover, since the system software was not owned by the government, the Department of Defense was constrained in the modifications it could undertake to correct deficiencies and tailor the programs to specific defense needs. Next was the WWMCCS standard applications software, which was limited and in need of redesign and reorganization. Finally, modernization of much of the command-unique software was urgently required. Many of the difficulties derived from the fact that commercial software had slowly begun to replace software written specifically for

WWMCCS as users tried to meet their site-unique requirements, leading to a proliferation of such applications within WWMCCS.[11] To provide the sorts of functions required by WWMCCS, it was clear that the operating system software was in need of complete replacement, while between one-quarter and one-half of all standard applications software and command-unique software was in need of upgrading.[12]

The movement toward WIS was given additional impetus when, in early 1982, Honeywell announced that it had decided to phase out maintenance and support for the operating system software used on the WWMCCS computers. What this meant in practical terms was that by the time the phase-out was concluded, in January 1986, the Pentagon itself would have to bear the full costs for software maintenance and modifications. A study was promptly initiated to compare the costs and benefits of developing entirely new operating system software as opposed to maintaining and enhancing the existing software, and the conclusion was that pursuing entirely new software would be more cost effective. With the need clear for both new hardware and software, Secretary of Defense Weinberger approved the development of WIS in July 1982.[13]

Perhaps not surprisingly, new procedures and management practices were put into effect to implement the new WIS, making the program an initiative for organizational as well as technological change. The Air Force chief of staff was designated the executive agent for WIS modernization. To raise the visibility of WIS and provide strong centralized management for all associated activities, a WIS joint program manager (JPM) was created and would report through the joint chiefs to the assistant secretary of defense for C^3I. To oversee the acquisition of new WIS hardware and software ordered by the JPM, a system project office was established at the Electronic Systems Division in Massachusetts. Finally, and to facilitate the modernization of the high-priority areas of tactical warning and attack assessment, a systems integration office was established at the Aerospace Defense Command in Colorado.[14]

In October 1983 the Air Force awarded GTE's Government Systems Division the contract to serve as the prime integration contractor and systems architect for WIS. GTE's responsibilities included system definition and development, including establishing standards for terminal-to-terminal, terminal-to-host, and host-to-host communications. The contract also called for testing and integrating subsystems supplied by subcontractors and for modernizing WWMCCS's extensive supporting software with a high-order programming language. What was perhaps most noteworthy about GTE's approach was that, at ESD's insistence, hardware selection would take place after software development. The point was to select a software approach that would allow the software to be machine-independent and portable across different types of computing hardware, rather than locking it into specific types of hardware as had been the case in the WWMCCS ADP Upgrade Program. For this purpose all old software (written in COBOL) as well as all new software would be rewritten in Ada, a language implemented by order of the undersecretary of defense for research and engineering and explicitly designed to run on different types of computing hardware.[15]

Ada

The need for a programming language such as Ada goes way back. By the middle of the 1970s, approximately five hundred programming languages or different versions of languages were in use throughout the DOD. A veritable electronic Tower of Babel, in practical terms it was a severe limitation for support maintenance and training, since moving applications programs among computer systems required different tools and expertise for each language.[16] It was hardly a recipe for effective performance, not to mention that it was extremely expensive. One obvious way to deal with the problem of software proliferation was to use a single data format, or computer language, to meet the requirements of the whole range of DOD computer systems, including those employed by WWMCCS. Nothing like this existed. While such defense organizations as ARPA had been interested in computer science research for years—in such areas as networking, artificial intelligence, computer image processing, speech recognition, and the like—the vast bulk of software produced was emphatically system-specific. There was no agreement, either within ARPA or externally, as to precisely what elements a common computer language should include, and for good reason: there was no unambiguous way to assess whether any given language was better than another. Was ease of programming the criterion? Was it the ease with which the program could be modified later? How about ease of documentation or transfer? There was no easy answer, no obvious way to strike a balance between various concerns and criteria. What was clear was that the number of languages then in use throughout the Defense Department was far too large, making the pursuit of a single high-order programming language a reasonable goal to pursue.[17]

And so in mid-1975, the Institute for Defense Analysis established a high-order working group whose mission was to draft a series of initial requirements for a programming language that could be used on computers built by different manufacturers and that could be transferred among them. These requirements were reviewed by experts from the military services, defense agencies, industry, and the academic world, and further modifications and revisions were made. This iterative, multistage process was sufficient to convince key officials that it was in fact possible to develop a single programming language to meet defense needs.

A final set of requirements was approved in January 1976, and contracts to develop a prototype standard language were competitively awarded to four contractors the following year. The contractors' preliminary designs were then widely distributed within the defense community, and based on the comments received, the requirements for the new language were finalized. Two contractors were chosen to continue the design work to meet these requirements, and their designs were again distributed for comment. In 1979 the Pentagon selected the language designed by a team from Cii-Honeywell Bull. The language, called Ada, was named for Augusta Ada Byron, the daughter of poet Lord Byron and the world's first computer programmer, who had worked on Charles Babbage's mechanical computing engine in the early 1800s. The Pentagon approved Ada as a military standard programming language in 1980.[18] Standardization was the goal, but unlike in the past when it had taken place through the use of standard computing hardware, it would now be accomplished with Ada.

To effect the transition to Ada and support it thereafter, three organizations were established, the first of which was the Ada Joint Program Office. Housed within the Office of the Deputy Director for Defense Research and Engineering (Research and Advanced Technology), it had several responsibilities, the most important of which was to ensure that Ada was implemented, maintained, and used throughout the Department of Defense. Another task was to support the development of further Ada tools to improve productivity, and to that end, an Ada Information Clearinghouse was set up to make available information on all Ada-related projects, tools, conferences, seminars, and training activities.[19] Software module libraries were also established at a number of locations and training films were released for program managers and software engineers.

The other two Ada-related organizations were the Software Technology for Adaptable, Reliable Systems (STARS) Joint Program Office, and the Software Engineering Institute, both of which were managed by DARPA. The goals of STARS, which was supported by the military services and defense agencies, were to improve the quality and reliability of computer applications programs, promote the development and reuse of software modules, and reduce the time and cost necessary for software development. The Software Engineering Institute, a federally funded research and development center set up at Carnegie-Mellon University in Pittsburgh, would focus on general software engineering issues, using Ada as the primary programming language.[20]

The use of Ada on a WWMCCS-wide, even defensewide, basis promised enormous advantages. With Ada, it would be possible to capitalize on hardware advances made in the commercial sector while at the same time avoiding problems created by the use of multiple-software applications. Ada would be portable, meaning that software modules written in the language would be reusable in different applications. It would reduce the costs of modifying and maintaining software and for training personnel.[21] By all indications, Ada would be the last new major language to be developed by the DOD prior to the advent of automatic programming, and studies conducted during the early 1980s indicated that by obviating the need for a large number of individual programming languages, Ada would result in substantial cost savings.[22]

The Ada writing on the wall was clearly visible to all, and with few Ada tools available from commercial sources initially, the services initiated their own projects to develop the necessary tools for writing Ada application programs. The Air Force was first, in 1979 initiating a project known as the Ada Integrated Environment. The following year the Army made its move, initiating its Ada Language System project to support software development, improve programming, and improve management control over the software life cycle. Last came the Navy and its Ada Language System project, the expressed purpose of which was to limit the proliferation of service-unique Ada language support systems and reduce the costs associated with implementing Ada.

In June 1983 the Defense Department proposed a revision to DOD Instruction 5000.31 (originally issued in November 1976), which had limited the number of DOD-approved computer languages to seven. The proposed revision, *Interim List of DOD Approved High-Order Programming Languages*, went even further, stating that Ada would become the single common computer programming language for critical mission applications. In a subsequent memorandum, the undersecretary of defense for research and engineering directed the services and defense agencies to implement the proposed revision immediately, and they did so shortly thereafter.[23] The movement to Ada was ultimately formalized by DOD Directive 3405.1, *Computer Programming Language Policy*, which designated Ada as the single defense programming language for general purpose ADP systems. At the same time, DOD Directive 3405.2 established the use of Ada for computer systems that were integral to weapons systems. According to the directives, Ada would be used for all intelligence, command and control, and other general purpose computer applications, except in those instances where another approved language was demonstrably more cost effective.[24]

While in principle the widespread use of Ada represented a considerable improvement over the current situation of software profligacy, it did not guarantee a complete solution to all DOD's automatic data-processing problems. Part of the reason derived from the fact that higher-order languages such as Ada themselves depended to an extent upon instruction-set architectures; that is, software programs, usually commercial proprietary ones, associated with a specific family of computers. Depending upon the instruction-set architecture used, a high-order language might require a substantially greater amount of computer memory, produce different results, or in some instances even fail altogether. For this and other reasons, the services fought hard and successfully for the right to be granted a waiver for Ada use in specific programs, and a number of such requests were subsequently granted. But by no means were the exemptions automatic. Criteria for being granted a waiver included conducting a developmental risk analysis, which included technical performance, cost, and schedule impact as well as a complete life-cycle cost analysis. To meet that challenge, each of the services designated an Ada executive.[25]

There were other potholes in the road to an all or mostly Ada programming world, including the "culture shock" resulting from the transition to Ada. As one industry writer described things, Ada was not simply a computer programming language but rather the fount of a new and entirely different way of approaching programming—a whole different software culture. Whereas in the past software development had taken place in an "artistic culture,"—a context wherein innovative approaches were applauded and rigid development standards were the exception—Ada emphasized the reverse: an "engineering culture" in which software development was subject to rigorous controls in much the same way as in any engineering project. One consequence of this development was that Ada was an exceedingly difficult language to learn, often requiring well over a year for personnel to achieve full proficiency.[26] Still another area of concern was the use of Ada for command and control applications requiring real-time data processing. Earlier programming languages such as FORTRAN and JOVIAL had serious deficiencies when used for real-time applications; but while unfortunate, this was perhaps forgivable since they had not been explicitly designed for that purpose. But Ada *had* been so designed, and it did not work well in time-sensitive situations either. While some software experts suggested that the problems were attributable to the compilers that implemented the language, others suggested that they might be inherent in Ada itself, and opinion was divided as to whether the problems could ultimately be resolved as the language matured.[27] Throughout the remainder of the cold war, a number of issues concerning Ada would remain unresolved. But the basic elements of the WWMCCS Information System were now in place.

Notes

1. General Accounting Office (GAO), *Command and Control: Upgrades Allow Deferral of $500 Million Computer Acquisition,* GAO/IMREC-88-10 (Washington, D.C.: General Accounting Office, February 1988), 8.

2. Department of Defense (DOD), *Modernization of the WWMCCS Information System (WIS),* Assistant Secretary of Defense for Communications, Command, Control, and Intelligence, 19 January 1981, 5.

3. Ibid., 8–10.

4. Ibid., 13.

5. GAO, 13.

6. Samuel L. Gravely Jr., "The DCA—A Rock and a Hard Place," *Signal* 34 (April 1980): 9.

7. Julian S. Lake, "C^3 Software Myopia," *Countermeasures* 2 (8 December 1976): 62.

8. GAO, 8.

9. DOD, 20–21.

10. Ibid., 12.

11. Carol Hamilton, "Worldwide C^2 System Networks Strategic Data for Joint Chiefs," *Defense Electronics,* June 1988, 57.

12. DOD, 8–10, 14.

13. GAO, 2, 14.

14. DOD, vii–viii.

15. David A. Boutacoff, "WWMCCS (Worldwide Military Command and Control System) Evolves to Meet Expanding Requirements," *Defense Electronics* 16 (November 1984): 84–85.

16. Len Famiglietti, "Ada 'May Never Bring Savings,'" *Jane's Defence Weekly,* 29 April 1989, 761.

17. Senate, Committee on Armed Services, *Fiscal Year 1976 and July-September 1976 Transition Period Authorization for Military Procurement, Research and Development, and Active Duty, Selected Reserve, and Civilian Personnel Strengths,* pt. 6, *Research and Development,* 94th Cong., 1st sess. (Washington, D.C.: Government Printing Office [GPO], 1975), 3349–50.

18. GAO, *Programming Language: Status, Costs, and Issues Associated with Defense's Implementation of Ada,* IMTEC-89-9 (Washington, D.C.: GAO, 1989), 8–9.

19. Ibid., 16.

20. Ibid., 11.

21. House, Committee on Appropriations, *Department of Defense Appropriations for 1989*, pt. 6, 100th Cong., 2d sess. (Washington, D.C.: GPO, 1988), 494.

22. House, House Committee on Government Operations, *DOD Should Change Its Approach to Reducing Computer Software Proliferation: Report to the Chairman*, MASAS-83-26 (Washington, D.C.: GPO, 26 May 1983), 5.

23. GAO, *Programming Language*, 18–20.

24. Ibid., 11–12.

25. House, *DOD Appropriations for 1989*, 494.

26. Susan Lorenz, "Ada Shock: A Computer Cultural Transition," *Signal*, May 1987, 147–48.

27. GAO, *Programming Language*, 3, 30.

Chapter 17

Defense Centralization

Even while WIS and other C^3I Triad programs were advancing the cause of centralization through new technologies, events were taking place that would influence centralization on the organizational and managerial fronts. Two incidents during the early 1980s underscored in a dramatic and public way the disconcerting fact that the existing command structure was imperfect and that despite a series of efforts over the years to improve jointness in military operations, an effective command and control capability had yet to be realized. The first of these was Operation Eagle Claw, the failed effort to rescue the American hostages in Iran. The second and more influential incident took place on 25 October 1983, when a multinational force led by the United States invaded the Caribbean island of Grenada, ousted its leftist government, and demonstrated again the difficulties the services had in mounting joint operations.

Operation Urgent Fury

Grenada had been a thorn in the side of the United States since 1979, when Maurice Bishop ascended to power in a bloodless coup. When the Reagan administration took office in 1981, Grenada was promptly grouped with Nicaragua and Cuba as a threat to vital US interests, and the pressure on the Bishop regime was turned up. Part of the pressure came by way of a series of military exercises, practically dry runs for a Grenada invasion.[1] Other pressures were political. For example, in announcing his Caribbean Basin Initiative in February 1982, President Reagan tried to exclude Grenada from participation. By early 1983 Pentagon officials were publicly declaring that Cuban influence in Grenada had reached such a high level that the island could now be considered a "Cuban prótegé."[2]

On 14 October 1983 Grenada's deputy prime minister, Bernard Coard, staged a coup against Bishop, placing him under house arrest. But Coard quickly lost the support of the military, and power was seized by Army commander Gen Hudson Austin, a self-described Marxist.[3] On 19 October a crowd of several thousand people freed Bishop, who promptly marched to Army headquarters to try to persuade the soldiers to rally behind him. Austin responded by sending an armored troop carrier to the scene. Gunfire broke out, a mob scene ensued, and Bishop and three members of his Cabinet were separated from the crowd, lined up against a wall, and shot.[4] With the island in a state of political turmoil and a Marxist even more extreme than Bishop now in charge of its government, the Reagan administration was spurred to action. On 21 October a 10-ship, fifteen thousand-man Navy task force bound for Lebanon was diverted toward Grenada. Operation Urgent Fury, the invasion of Grenada, began four days later. More than a thousand Marines and Army Rangers made the initial landings on 25 October; the Rangers' objective was the airport at Point Salines on the southern part of the island, while the Marines concentrated their attention on Pearls Airfield north of the capital city of St. George's. Accompanying the invasion force, known as Joint Task Force 120, was a token 300-man Caribbean peace force from seven Caribbean countries.[5]

Much has been written about Operation Urgent Fury, its justification, and the extensive press censorship that accompanied it. But the relevant point here is how well the joint task force performed, and despite the rapid and complete military victory, that performance was considerably less than optimal. As with Operation Eagle Claw earlier, the Pentagon appeared to have subordinated the principle of unity of command, and hence maximum effectiveness, to give each service its share of the action.[6] The Navy was in overall charge of the operation, but coordination between the Navy and the Army was essentially nonexistent. Control of the air units participating in the operation was divided between the Navy and the Air Force. On the ground a similar division of responsibilities was made between the Army and the Marines. In defense of these arrangements, the commander in chief of Atlantic forces, Adm Wesley McDonald, pointed out how dividing a command is "not unique."[7] But unique or not, the lack of joint air and land commanders resulted in delays and serious problems of coordination.

Communications between the various elements of the invasion force were highly problematical, representing perhaps the most serious command and control problem. Army forces frequently found themselves unable to communicate with the Marines because they used different equipment and radio frequencies. Army troops were unable to contact Navy ships for fire support because of incompatibilities in the communications security equipment used by each service. In what was perhaps the most notorious incident, an exasperated Army officer reportedly went to a pay phone and, using his AT&T calling card, phoned 82d Airborne headquarters at Fort Bragg, North Carolina, to ask them if they could raise the ship. On several occasions Army officers flew by helicopter to the USS *Guam*, the flagship for the invasion force, in an unsuccessful attempt to coordinate naval gunfire. Even where communications were possible, Army officers found it difficult to request fire support from Navy ships because they could not authenticate these requests using Navy codes.[8]

The upshot was that despite all of the high technology equipment available, problems of compatibility and procedure produced a serious lack of communications during critical stages of the operation. "The elite units and pilots sent in to provide air cover may as well have been from different countries and speaking different languages," one observer lamented.[9] Assistant Secretary of Defense for C^3I Latham agreed, noting that it was fundamental to "have a good communications plan before you mount an operation, and that it's an adequate plan."[10] Sensitive to the criticism and to the fact that much of it was directed toward their stubborn parochialism, the services quickly initiated a number of corrective actions. But so serious had the problems been during Operation Urgent Fury that they would soon be cited by Congress as a key reason for reorganizing the Department of Defense.

Defense Centralization in the 1980s

In fact, the bureaucratic movement toward that reorganization was already under way. In 1982 Joint Chiefs of Staff chairman David C. Jones testified before Congress regarding the many shortcomings of the existing JCS system, setting in motion a series of congressional hearings and reviews that were still in progress when the invasion of Grenada took place. Because of the many serious interoperability problems highlighted by Operation Urgent Fury, several lines of action were initiated. The joint chiefs organized a high-level Joint Requirements Management Board, composed of the vice chiefs of each service, to review joint aspects of service acquisition programs and enhance interoperability of service assets. More important for the shape of the defense future, however, were two staff studies launched by the Armed Services Committees of both branches of Congress: in the Senate under the chairmanship of Barry Goldwater and in the House under that of Bill Nichols.[11] Of the two, that of the Senate Armed Services Committee, titled *Defense Organization: The Need for Change,* was especially critical. Made public in October 1985, the report laid out the many organizational problems that had plagued the DOD and continued to do so.

The first problem identified was limited mission integration at the policy-making level. The three major Pentagon elements—the Office of the Secretary of Defense, the Joint Chiefs of Staff organization, and the military departments—tended to emphasize functional efficiency rather than substantive goals: "In colloquial terms, material inputs, not mission outputs, are emphasized." The specific problems issuing from this efficiency orientation included the domination of program decisions by service interests and the neglect of those functions not central to the services' missions. They included the inadequate development of joint doctrine and the suborning of the needs of the unified commanders to service needs. They also included deficiencies in interoperability between service assets and an inability to make trade-offs between competing service programs when both contributed to a specific mission.[12] It was an old story, of course, dating back at least to the time of the 1958 defense reorganization.

The impetus behind this functional emphasis becomes clearer when a second and related problem is considered—the imbalance between service and joint interests. The committee report succinctly summarized the decades-old problem, "Under current arrangements, the Military Departments and Services exercise power and influence which are completely out of proportion to their statutorily assigned duties." They did this by dominating the Joint Chiefs of Staff system, exercising a de facto veto power over virtually all JCS decisions and actions. They did it by dominating the unified command system, such that the unified commanders remained dependent upon their service components. They did it because the OSD lacked the ability to effectively integrate service capabilities and programs.[13] There could hardly have been a better description of a decentralized decision-making structure and the suboptimization it fostered.

The overriding of joint interests by service interests created a third problem: excessive emphasis on ongoing modernization to meet hypothetical future needs at the expense of present operational readiness. The continual upgrading of capabilities is, of course, the essence of the evolutionary approach, and because it served their interests and enhanced their autonomy, the services were its strongest proponents. The services adopted the evolutionary approach because they themselves operated within a context of formal rationality. That is, they were concerned with efficiency and operative goals and viewed their modernization efforts as an ongoing process with no clearly defined end state. Focusing on process, evolution itself was the goal. To the contrary, the proponents of centralization were more solution-driven, concerned with discrete end states of capability and readiness. The problem was that these effectiveness criteria were often subordinated to service interests.

If it seems that the major dynamic within the DOD was less the frequently mentioned rivalries between the services and more the conflict between the services and the proponents of centralization, it was precisely this point that constituted the fourth problem identified in the staff report; what was characterized as "inter-service logrolling." It is not that competition and strife between the services was denied. To the contrary, secretiveness, duplication, and lack of understanding between the services were acknowledged as continuing problems. But a more important problem, one whose origins seem to date rather precisely to the centralizing fever of the early 1960s, was the services' tendency to provide a united front in their dealings with the civilian leadership. The report suggested that the intention as well as the effect of this tendency—one is tempted to say of this collusion—was precisely to stay the juggernaut of centralization, isolating the OSD and weakening civilian control over the military establishment.[14]

A final problem identified in the committee report involved the inadequacies of the existing Joint Chiefs of Staff system. Dominated by the services, with a bureaucratic structure that emphasized committee decision making and consensus views, the advice offered by the JCS to the civilian leadership was all too often inadequate, irrelevant, untimely, or unclear. A key reason underlying this was the conflict of interest that was inherent in the "dual-hatting" of officers assigned to the Joint Staff. Unable to subordinate the interests of their own services to the larger national interest, the JCS was never able to evaluate objectively the appropriate missions and division of responsibilities among the services. What resulted was tepid and cautious advice, reflecting "whatever level of compromise is necessary to achieve the four Services' unanimous agreement."[15] It was indeed a heavy dose of criticism and an expansive call for reform.

The Packard Commission Report

In July 1985, even as the staff of the Senate Armed Services Committee was preparing its report, President Reagan established a blue ribbon commission on defense management to study current defense organization and management. Headed by David Packard, perhaps the most energetic proponent of enhanced centralization of defense management during the years he served as deputy secretary of defense, the Packard Commission's findings and recommendations were almost equally sweeping. Many of those recommendations were first released in an interim report, dated 28 February 1986. On 1 April President Reagan issued National Security Decision Directive 219, instructing the Department of Defense and other relevant executive agencies to implement all of the commission's recommendations that did not require congressional action. In a special message to the Congress three weeks later, the president endorsed the remaining recommendations and requested their prompt implementation. By the time the commission's *Final Report to the President* was released in July, both the House and Senate had already passed some of the requested legislation.[16]

The Packard Commission's recommendations for change fell into three broad areas: planning and budgeting, defense acquisition, and military organization and command. In the planning and budgeting area, the commission took direct aim at the prevailing means-oriented rationality, what it described as the excessive attention focused on the question of "how much," with inadequate attention paid to the more substantive questions "what for, why, and how well." To combat this misplaced emphasis, planning would have to start with a clear statement of defense objectives and priorities. After receiving advice and reviewing options, the JCS chairman then would frame broad military options, making explicit trade-offs among various defense elements in the process. The president would then select a specific option with associated spending. Budgets would move from annual to biennial, eliminating the current situation in which defense programs were in continual flux, being constantly adjusted to shifting budgets irrespective of the sense that the changes might make in terms of overall military strategy.[17]

The second major area was defense acquisition, a process described as "overwhelmingly complex," burdened under an immense weight of regulations, and suffocating under myriad unproductive layers of management. Acquisition was fragmented, with no single OSD official responsible for overall supervision of the process; in the absence of such an official, policy responsibility tended to devolve to the services, whose own interests almost always predominated over national-level concerns. Authority was diluted and accountability rendered less precise as a result. To counter this, it was recommended that a new three-tiered acquisition management chain be implemented.[18]

In the area of military organization and command, many of the areas of change recommended by the commission involved the unified and specified commands and their commanders in chief. The commission recommended that the unified commanders be released from the service restraints under which they had previously been operating, giving them greater latitude to structure subordinate commands and joint task forces in ways consistent with their missions. To give the commanders greater flexibility in operations that overlapped the geographic boundaries of other commands, it was recommended that the Unified Command Plan should be revised. To give the CINCs greater voice in JCS decision making, it was recommended that the position of vice chairman be created, a sixth JCS member whose key function would be to represent the interests of the CINCs. It was further recommended that the CINCs' reports and orders be channeled through the JCS chairman, to ensure their better incorporation into defense policy.[19]

Some of the most serious problems in the area of military organization and command, however, were directly attributed to the limited authority of the chairman of the Joint Chiefs of Staff. Where the civilian leadership required military advice integrating the views of the combatant commanders, no single military officer was responsible for providing such integrated advice. While in theory the JCS chairman did this, in practice under the current system he lacked the authority to do so. To give the position greater bureaucratic clout, it was recommended that the chairman be designated the principal military advisor to the president, the National Security Council, and the secretary of defense. And rather than simply a committee head, he should be able to present his own views in addition to the corporate views of the joint chiefs. Further, the Joint Staff and the JSC organization should be placed under the exclusive direction of the chairman and existing limits on the size of the Joint Staff should be removed so the chairman could better discharge his responsibilities. Finally, the secretary of defense, following the advice of the JCS chairman, should be given greater flexibility to shorten or bypass the established chain of command should he see fit—a deliberate move in the direction of the "White House to foxhole" model of centralized command and control universally derided by the services.[20]

Thus did David Packard, our "Mr. WWMCCS" of the early 1970s, lay the foundation for the most sweeping piece of defense legislation since World War II, the Goldwater-Nichols Department of Defense Reorganization Act of 1986. And if there was a single central theme to the vast centralization effort the act initiated, it was precisely to eliminate the lack of jointness and the problems in command and control interoperability that had plagued military undertakings throughout the course of the cold war.[21]

Goldwater-Nichols DOD Reorganization Act

The Goldwater-Nichols Department of Defense Reorganization Act of 1986 (Public Law 99-433) followed the recommendations of the Packard Commission in virtually all of its specifics. In the area of planning, the act required that an explicit statement of national strategy and an accompanying statement of military strategy be provided as a means of measuring the effectiveness of all defense programs. In this, the JCS chairman would be responsible for a number of key functions. These included, but were not limited to, preparing strategic plans; performing net assessments to determine the military capabilities of the United States, its allies, and potential adversaries; providing contingency plans that conform to the policy guidance of the president and secretary of defense; preparing joint logistic and mobility plans to support those contingency plans; advising the secretary regarding the priorities and requirements of the unified and specified commands; and formulating doctrine and policies for the joint training and employment of the armed forces.[22]

As the Packard Commission had recommended, Goldwater-Nichols completely revamped the functions of the JCS chairman, continuing the postwar trend of increasing the authority of that position. The act made the chairman a member of the National Security Council and designated him principal military advisor to the secretary of defense and the president. Now, rather than military advice coming from the JCS as a corporate body, it would come directly from the chairman, who would consult with and seek the advice of the other chiefs and the unified and specified commanders "as he deems appropriate." No longer would the JCS be constrained as it had in the past, a committee striving to achieve a single consensus viewpoint. Now it was the chairman's opinion that would weigh most heavily, although the other members of the JCS could submit dissenting opinions or advice differing from that of the chairman. The bureaucratic clout of the chairman was further increased by giving him personal authority and control over the Joint Staff, functions that had previously been in the hands of the JCS as a whole. To further aid the chairman in his work, and as the Packard Commission had recommended, the Goldwater-Nichols Act created the position of JCS vice chairman and designated the incumbent as the second highest-ranking military officer after the chairman himself.[23] The JCS chairman now had the authority to play a dominant role in the formulation and implementation of joint doctrine and in the resolution of any doctrinal disputes.[24]

Goldwater-Nichols also directed the secretary of defense to establish specific policies and procedures so that members of the armed forces could be trained as joint specialists. It was clear to congressional lawmakers that the joint area was not working out as efficaciously as hoped: low-quality officers were often assigned joint duty; it was used as a terminal assignment for officers prior to retirement; and those high-quality officers assigned such duty took few risks and left as soon as possible. Promotions came from within one's own service branch, after all, and given the traditions and strong cultural biases of the services, those assigned joint-service billets had little incentive to emphasize "jointness" over the interests of the services to which they would shortly return.[25] To put an end to this situation, Goldwater-Nichols directed that each service develop specific career tracks for joint specialty officers and promote them at a rate equal to that for officers of the same grade and category.[26] This would represent "a startling change to the historical prerogatives of the military departments," and in terms of breaking down the decentralized and subunit-dominated structure of the Department of Defense, the efforts to promote jointness were in the long run likely to prove the most potent agent of change.[27]

In the most fundamental of ways, then, the purpose of the 1986 Defense Reorganization Act was to improve and strengthen the command and control of joint forces. But at the same time, the passage of the act by no means heralded the demise of service influence or of the evolutionary approach which in various ways sustained it. In many key respects vestiges of the old power relationships remained, and to an important degree the defense establishment would continue to be bound by the earlier system of negotiation.[28] Significant in this regard was the fact that, with the exception of small operational budgets for the CINCs, budgetary control remained firmly with the services. Thus even the newest and most innovative of joint programs would continue to be funded the old fashioned way—by defense subunits whose interests were in a fundamental sense antithetical to the concept of jointness.

Implementing WIS

As these organizational changes were taking place, progress on the technological front for WWMCCS was also in evidence. The WIS modernization program was evaluated in May 1984 by the Defense Systems Acquisition Review Council, which determined that the system should be developed in three clearly defined phases, or blocks, each of which would require council review and approval to move into full-scale development. Block A would provide the system's technical foundation. This included an automated message handling system to improve controls over message receipt, preparation, and dissemination; computer workstations to provide data processing in user work areas; and a local area network to connect computer systems, automated message handling systems, and work stations at the various WIS sites.[29] Collectively, these improvements constituted an interim WWMCCS computer upgrade. Block B, which would begin once software development was sufficiently advanced, would involve competitive procurement of new state-of-the-art computers to replace the WWMCCS Honeywells, plus the development of new application software, procurement of a database management system, and the development of new security controls for data access. Block C would focus on improving joint mission planning and interfaces between the DOD, non-DOD agencies, and NATO systems. Gradually phasing in the new system in this fashion, it was believed, would allow the existing ADP capability to remain fully operational until the new capability was brought on line.[30]

In October 1984 the Electronic Systems Division awarded IBM's Federal Systems Division the contract to provide some thirty-five hundred of its PC-based workstations for the interim computer upgrade.[31] In September of the following year,

Defense Secretary Weinberger granted approval for the Air Force to proceed with system design for all WIS blocks and for full development of Block A. The Defense Acquisition Board (which had by this time replaced the earlier Defense Systems Acquisition Review Council) was expected to approve full-scale development of Block B in early 1988. If the acquisition of the new WIS computer systems was approved as planned, a contract would be awarded promptly, and the new systems would be installed during the 1989–91 time frame. But as decision time for Block B approached, it was by no means certain that new computer hardware would have to be procured for WIS, since the IBM computers involved in the interim WWMCCS upgrade had remedied many of the problems that had occasioned the initial decision to purchase new computing equipment. These computers were meeting or exceeding the JCS standards for availability during routine operations and simulated crises and were also doing quite well in response time. On top of this, they were seen as having substantial expansion capabilities, enough to meet most anticipated future needs.[32]

The upgraded computers allowed for additional benefits to be reaped as well. Where many of the operating system hardware features had once been WWMCCS-unique, especially in such areas as security controls, they were now part of the commercially available version of the software. That being so, DOD would not have to bear the entire cost of future software maintenance if, as proved to be the case, it could run the commercial version on the upgraded WWMCCS computers. For those unique features that remained, Honeywell was awarded a contract to incorporate them into future versions of the commercial software. Honeywell's contract to support the WWMCCS software ran through September 1991, with company officials indicating that they would be willing to negotiate an extension of the contract through the end of the century.[33]

Another shortfall that had been identified in the early 1980s was the lack of a data management capability that would allow users to deftly retrieve and summarize information. One of the key reasons for considering entirely new state-of-the-art computers for WIS had been precisely to support such a capability, which would also support high-order programming languages such as Ada. But as it turned out, the database management system for the upgraded WWMCCS computers not only could support Ada, but also in all probability it could support the Joint Operation Planning and Execution System (JOPES), a software application intended to improve all aspects of conventional joint operation planning and execution. JOPES was the primary application software to be developed during WIS Blocks B and C and was itself to be developed in two blocks, or increments. In Increment I, several WWMCCS applications would be integrated and modernized, including the Joint Deployment System, Joint Operation Planning System, and Unit Status and Identity Report System. By 1986 the WIS joint program manager concluded that JOPES Increment I could be fully supported on the upgraded WWMCCS computers. Since that increment could not provide automated support for joint mobilization plans and schedules, JOPES Increment II was intended to do so; by all indications, it too could be supported on the upgraded WWMCCS computers, although it was impossible to know for certain until the requirements for Increment II were defined—something that would not happen until the system itself was up and running.[34]

But as critics pointed out, basing the acquisition of new WIS computers on ill-defined JOPES requirements would also run the risk of acquiring computers with excess capacity and hence unnecessary costs, or, of acquiring inadequate machines that would require additional upgrades or replacements to meet mission requirements.[35] With the interim WWMCCS computers appearing increasingly attractive, the only remaining deficiency involved the requirement for multilevel security, whereby users with different security clearances could access authorized information while being denied an avenue to information for which they were not cleared. At the time, all WWMCCS sites practiced the "system high" approach to security, in which all users were simply cleared to the highest level of classified information used on the system. It was a costly and inefficient approach, and so the joint chiefs had mandated that the planned new WIS computer systems provide multilevel security. But although conceptually simple and eminently desirable, the software technologies necessary for multilevel security remained beyond the state of the art, meaning that neither the upgraded WWMCCS computers nor the planned new computers for WIS would be able to provide it. Once again, the upgraded WWMCCS computers appeared competitive.[36]

In short, most of the desired WIS capabilities were currently supported by the upgraded WWMCCS computers, while others, such as JOPES, most likely could be supported. The major remaining requirement, for multilevel security, was beyond the capabilities of both the upgraded WWMCCS computers or new WIS computers. All of this appeared to obviate the need for the new WIS computers, and, with potential savings on the order of $500 million to be had, the General Accounting Office recommended in February 1988 that the acquisition of the new WIS computers be postponed until later, when the needs of users and the need for entirely new computers to meet them had been clarified. The Pentagon concurred the following month, and for the remainder of the cold war, WIS would rely on the upgraded WWMCCS computers.[37]

By the end of the 1980s, this patchwork WIS involved more than one hundred mainframe computers and 65 remote processors, linking together hundreds of sites and more than three thousand individual workstations around the globe. System reliability was estimated to be on the order of 97 percent. It had as of that time cost some $800 million to create, with annual expenditures running in the range of $160–$200 million.[38] Things appeared to be working well, except perhaps with Ada. Toward the end of the decade, the GAO examined the use of that programming language in one hundred projects and found that it was not possible to determine whether its use was achieving many of its promised ends. How much had been spent on the transition to Ada? Were software development and maintenance costs being controlled as a result of its use? It was in fact impossible to know since the total number of projects using Ada was unknown and most of the known user organizations had kept inadequate records.[39] Moreover, no programs to assess Ada's costs and benefits over time had been established. While there was optimism that such information would gradually become available as the system further evolved, anecdotal accounts suggested that such information, if and when it finally came, might be unwelcome. For example, the undersecretary of the Army stated at an Ada exposition in Boston that he had yet to see convincing evidence that the new language would in fact realize the Pentagon's goal of reducing the ever-increasing costs associated with software development and maintenance.[40] Things had improved, certainly, but after three decades of development, the World Wide Military Command and Control System still remained part concept, part reality.

Notes

1. Alexander Cockburn and James Ridgeway, "America Invades Grenada: The Making of a Counterrevolution," *The Village Voice*, 8 November 1983, 1.
2. Philip Taubman, "U.S. Now Puts the Strength of Cubans on Isle at 1,100," *New York Times*, 29 October 1983, 6.
3. "New Details Are Given On Coup in Grenada," *New York Times*, 28 October 1983, A14.
4. House, Subcommittee on International Security and Scientific Affairs and on Western Hemisphere Affairs, Committee on Foreign Affairs, *U.S. Military Actions in Grenada: Implications for U.S. Policy in the Eastern Caribbean* (Washington, D.C.: Government Printing Office [GPO], 2, 3, 16 November 1984), 189.
5. Senate, Committee on Armed Services, *Organization, Structure and Decisionmaking Procedures of the Department of Defense*, pt. 7 (Washington, D.C.: GPO, 1983), 285–86.
6. John A. Wickham Jr., "Jointness and Defense Decision Making," *Signal*, February 1988, 17.
7. Richard Halloran, "U.S. Admiral Defends Dual Command," *New York Times*, 4 November 1983, A17.
8. Senate, Committee on Armed Services, *Defense Organization: The Need for Change* (Washington, D.C.: GPO, 16 October 1985), 365.
9. Richard C. Gross, "C^3: Fewer Mixed Signals," *Military Logistics Forum* 3 (June 1987): 20.
10. Ibid.
11. Ibid., 17–18.
12. Senate, *Defense Organization*, 3.
13. Ibid., 3–4, 6.
14. Ibid., 5.
15. Ibid., 6.
16. *A Quest for Excellence: Final Report to the President* (Washington, D.C.: Blue Ribbon Commission on Defense Management, June 1986), 4.
17. Ibid., xviii–xix.
18. Ibid., xxiv.
19. Ibid., 37–38.
20. Ibid.

21. Wickham, 18.

22. Congress, *United States Statutes at Large,* 99th Cong., vol. 100, pt. 1, Public Laws 99-241 through 99-452 (Washington, D.C.: GPO, 1986), 1007–8.

23. Ibid., 1005–9.

24. Don M. Snider, "DOD Reorganization: Part II, New Opportunities," *Parameters*, December 1987, 55.

25. Ibid., 57.

26. Congress, *United States Statutes at Large,* 1026.

27. Don M. Snider, "DOD Reorganization: Part I, New Imperatives," *Parameters,* September 1987, 95.

28. Ibid., 98.

29. House, Committee on Appropriations, *Department of Defense Appropriations for 1989,* pt. 6, 100th Cong., 2d sess. (Washington, D.C.: GPO, 1988), 471.

30. David A. Boutacoff, "WWMCCS Evolves to Meet Expanding Requirements," *Defense Electronics*, November 1984, 84–85.

31. Ibid., 85.

32. General Accounting Office (GAO), *Command and Control: Upgrades Allow Deferral of $500 Million Computer Acquisition,* IMREC-88-10 (Washington, D.C.: General Accounting Office, February 1988), 9, 15, 17, 21.

33. Ibid., 14–15.

34. Ibid., 10, 15, 19–20.

35. Ibid., 25.

36. GAO, *Programming Language: Status, Costs, and Issues Associated with Defense's Implementation of Ada,* IMTEC-89-9 (Washington, D.C.: GAO, 1989), 22.

37. GAO, *Command and Control,* 2.

38. Carol Hamilton, "Worldwide C^2 System Networks Strategic Data for Joint Chiefs," *Defense Electronics*, June 1988, 57.

39. Len Famiglietti, "Ada 'May Never Bring Savings,'" *Jane's Defence Weekly,* 29 April 1989, 761.

40. GAO, *Programming Language,* 21, 24.

Chapter 18

Defense Communications and the End of the Cold War

Throughout the 1980s, the Defense Communications Agency's involvement in WWMCCS was heavy and ongoing. In addition to the WWMCCS-related common-user systems it managed, the agency's Command and Control Technical Center supported the National Military Command System and its major command facilities. Its command center improvement program was intended to coordinate the application of state-of-the-art technology for improving the CINC's command centers. It provided systems engineering support for the Minimum Essential Emergency Communications Network. It was involved in the development of standard software for the WWMCCS computers and had major responsibilities for the WWMCCS Information System. Implicit or explicit to many of the agency's efforts during the decade was the effort to craft the Defense Communications System along the lines of doctrine—more survivable, enduring, and secure, with far greater connectivity—thereby eliminating the perception that the DCS was strictly a peacetime system. There were a number of implications to this movement to a "wartime" DCS, perhaps the most important of which was the pursuit of all-digital operations.

Digitization and Evolution

The advantages of digital communications were numerous; indeed, to the thinking of many, even overwhelming. Digital systems offered far greater flexibility for the communicator, since any signal that could be sampled and quantified could be accommodated.[1] In the area of communications security, digital encryption was both easier—bulk encryption of complete radio links would be feasible—and far more effective. Going digital meant that a high-quality secure voice capability, long an urgent defense requirement, could be achieved at last. Regeneration of signals was another digital plus, where at each terminal or relay point distortion in the digital signal could be removed and the signal retimed. The effect of this would be extremely high-quality transmissions almost independent of distance. With digital operations, communications channels could also be given substantially greater transfer capacity, a characteristic of major importance in computer-to-computer transmissions. Then there was the issue of cost. While the actual conversion from analog to digital would certainly be quite expensive, DCA officials pointed out that the efficiencies to be gained were such that the investment would be more than repaid.[2] Indeed, DCA was predicting a 50-fold increase in digital communications during the upcoming years, a veritable nuclear explosion of automated computer-based information and management systems.[3] The expectation was that economies of scale would bring down costs in direct proportion.

The inherent advantages of digital communications had long been known to DCA engineers.[4] Indeed, one of the agency's goals virtually from the moment it was established had been to integrate its general purpose switched networks into a single digital network carrying both voice and data and employing some sort of universal digital switch. But given the vast scale of the analog plant already in place and the huge investment it represented, officials also recognized from the outset that their desired all-digital system-of-the-future could not be achieved in a single step. DCA engineers began considering strategies for transitioning the system over a period of years, and they subsequently selected an approach to digitization that involved a so-called hybrid system during the period of transition. As a first step in this strategy, DCA was to begin a phased replacement of existing analog voice channels with digital channels. This necessarily implied that digital and analog capabilities would have to exist side by side during the transitional period, and the key to a felicitous hybrid marriage was pulse code modulation.

The key to pulse code modulation was the wideband modem. The AUTODIN network already had a limited capability of this sort, but it was based on the use of a single-voice-channel modem, which gave it limited capacity and speed. In contrast, the wideband modem operated over a far larger number of voice channels and as a result could accept digital data at rates up to one thousand times greater.[5] DCA officials explained how, using this approach, a large number of analog and digital channels could be linked together without any discernible degradation, using inexpensive conversion equipment readily available from commercial sources. Digital-analog transparency could thus be achieved throughout the DCS even while channels were being digitized. During this interim phase of operations, which DCA officials estimated would last some 10 to 20 years, the system would appear essentially unchanged; that is, users would continue to experience it as an analog system. And once all of the system's analog channels had been converted, it would be possible to rapidly complete the move to an all-digital DCS simply by replacing the system's analog switches with digital ones.[6] It was an ambitious effort, which DCA officials compared to the conversion of aircraft from prop-driven to jet engines. For present-day communicators as for pilots during that earlier era, what was in store was a costly period of transition filled with considerable pain and trauma, but in the end it would be worth it.[7] To the thinking of many defense officials, an all-digital future was as bright as it was inevitable.

All of this had the effect of even further blurring the distinction between tactical and strategic communications systems: "No one really can tell where tactical communications and strategic communications end or begin," DCA director Winston D. Powers noted. In bureaucratic recognition of technological reality, the Defense Communications Agency was merged with the Joint Tactical C^3 Agency (JTC^3A) at Fort Monmouth, New Jersey. This union, in Powers' words, should have taken place "eons ago." After all, DCA and JTC^3A had for years cooperated in efforts to link tactical and strategic planning on command and control, particularly through the provision of interfaces between tactical systems and the various DCS components. "It will be a working, breathing organization," Powers said. And it was quick to adopt both the spirit and the logic of Goldwater-Nichols, which called for the use of rapid prototyping and staged acquisition of assets.[8]

Although DCA officials were talking about how the DCS would evolve in the years to come, they had hardly become converts to the evolutionary approach to command and control system development. Throughout the 1980s, they continued to point out the evolutionary approach's many serious limitations: the fact that it did not specify the basis for subsequent evolution, that it lacked any notion of how to balance users' requirements with budgetary constraints, and that it ignored how the process actually worked and where subsystems were developed separately that then had to be interconnected to play together as a coherent whole.[9] Implicit in the evolutionary approach was a sort of flexible baseline, or programmatic moving target, such that programs would never be completed. As DCA officials were quick to point out, with programs kept in the development phase essentially forever, the result would likely not be savings but an open-ended, uncontrolled escalation of costs.

In recognition of this, the Packard Commission report and the Goldwater-Nichols defense reorganization had sketched out an alternative way, a sort of middle ground between the ad hockery of the evolutionary approach and the rigid closure of the weapons system approach. Known as a modular building block (MBB) architecture, it was developed by a DCA team and strongly supported by the assistant secretary of defense for C^3I. The MBB had as its underlying philosophy that systems acquisition would take place in a series of stages, or blocks. It was evolution but with a twist; now, a system would have to achieve specific goals in an identified developmental stage before approval would be granted to proceed to the next stage, something that had not been the case with the evolutionary approach.[10] But neither was it the weapons system approach, since the staged nature of development permitted, within limits, the modification of requirements to keep pace with advances in technology. With the program now conceived as a series of stages, unnecessary requirements could be dropped and new requirements added as the system developed.[11] MBB represented a compromise between centralization and decentralization and would leave its mark on a number of key WWMCCS-related systems developed by DCA during the 1980s. And so the cold war saga of WWMCCS ends where it began, with the infrastructure known as the Defense Communications System.

Toward the Defense Information System Network

A major interest of the DCA during this time was to realize its long-standing goal of transitioning to a single, integrated

Defense Communications System based on digital technologies. As DCA planned things, new technologies would be acquired incrementally, at first blurring the distinction between and then merging completely, the still-separate DCS common-user networks, all the while serving the voice and data requirements of the systems' users. An integrated DCS implied the use of a single integrated high-speed switch, which DCA had pursued from its inception. But pressure to rapidly expand the common-user networks during the early years meant that planning was necessarily oriented toward the use of commercially available automatic switches. For DCA, this meant a two-track approach: voice communications would use commercially available analog switches, while record transmissions would use commercial digital switches. On repeated occasions, agency officials had voiced the view that this was an interim strategy only, useful until integration of the common-user networks could take place.[12] But integration of AUTOVON, AUTODIN, and AUTOSEVOCOM required identification of the switching technique to be used: circuit switched techniques, which had always been used for voice communications, or point-to-point circuits, generally used for the transmission of data.[13] Once that choice was made, the two-track approach could be abandoned and the separate common-user networks it had engendered would be history at last.

DCA's plans, dating back to the end of the 1970s, first called for the primarily analog Automatic Voice Network to be phased out, replaced by a next-generation Defense Switched Network (DSN). The proposed DSN would provide common-user telephone service throughout the Department of Defense, accommodate the transmission of data, offer a range of special command and control features, and do all of this in a cost-effective manner.[14] Whereas the majority of AUTOVON users at the time were connected to a single backbone switch through a single set of access lines, DCA engineers judged that the best approach for the future network was an entirely different technological concept—a distributed network of numerous small, powerful switches to replace the existing population of fewer, larger switches. These new switches, collocated with their users at bases, posts, and other military facilities around the world, would be dual purpose, providing both local and long-haul communications services. They would be digital, since only digital switches would permit the message control functions to be distributed throughout the network in this fashion.[15] Such a network would be significantly more survivable and enduring than AUTOVON, it was believed, particularly in terms of its resistance to nuclear effects.[16] It would also be more secure, and provide a secure teleconferencing capability, the requirement for which had been identified by the WWMCCS Architecture. Because this would eliminate the need for an independent network for secure voice communications, DCA developed a secure voice transition plan to provide the necessary guidance for integrating AUTOSEVOCOM into the Defense Switched Network.[17] As the cold war drew to its close, considerable progress had been made toward implementing the DSN, with antiquated AUTOVON switches in the continental United States and overseas replaced by state-of-the-art digital-switching equipment.[18]

DCA's plans also called for the replacement of AUTODIN, the venerable Automatic Digital Network, with a new network intended as the primary vehicle for creating the total integration of the Defense Communications System's common-user systems that DCA had so long desired. This AUTODIN follow-on would also be a distributed network, based on the packet-switching technique pioneered in the ARPANET. Specifically, what DCA had in mind was taking a series of existing packet-switched networks—the ARPANET itself, the WWMCCS Intercomputer Network, the Intelligence Data Handling System, the Strategic Air Command Digital Information Network, the Community On-line Intelligence Network, and others—expanding and upgrading them, and then integrating them all into a single Defense Data Network (DDN).[19]

The schedule for integrating existing defense packet-switched networks into a single DDN consisted of several separate stages. First, the WWMCCS Intercomputer Network would be upgraded to the WWMCCS Information System. WIS would then be combined with the Intelligence Data Handling System, and the result would be called the Command, Control, and Intelligence (C^2I) network. Next, the ARPANET would be partitioned into two parts. One of these would be a classified research and development network that would not formally become part of the DDN. The other part of the ARPANET would be designated an unclassified military-user network called MILNET, a number of unclassified networks would be absorbed into it, and all of these would then be fully integrated into the DDN. Finally SACDIN (SAC Digital Network), which would initially be served by a dedicated top secret network using DDN components, would be fully integrated into the DDN.[20]

With the end of the cold war, the planning and development phases of DDN were essentially complete, and the network consisted of several hundred geographically distributed packet-switched nodes. The DCA's principal remaining task, it seemed, was to connect additional users to the network.[21] So, even as the international tensions that had spawned DCA receded, the communications infrastructure that it had worked on for almost 30 years, and upon which WWMCCS depended, was coming to fruition at last.

Postscript: The End of the Cold War

At the beginning of the 1980s, John Steinbruner observed that if a "constructive stabilization" of the political relationship between the United States and the Soviet Union could be achieved, then efforts to improve the efficiency and effectiveness of the United States's communications, command, control, and intelligence system would contribute substantially to overall security. Lacking such a political understanding between the superpowers, he argued, identical efforts might yield opposite results as the Soviets simply allocated additional weapons to new targets. In light of the acknowledged vulnerability of the US command structure at the time, political stabilization appeared to constitute a necessary precondition for any effective program of command and control.[22]

The Reagan administration's strategic modernization program was in large measure a consequence of, and response to, the Soviet arms buildup that took place during the 1970s under Leonid Brezhnev. Yet, that buildup, we now know, sapped critical energies from the Soviet economy and polity, cycled back to undermine the very military apparatus it was ostensibly created to advance, and contributed to the dramatic events of the late 1980s and early 1990s. As the curtain fell on the cold war, Aleksandr Solzhenitsyn pointed out in *Komsomolskaya Pravda* how the Brezhnev years "simply wasted our last resources on unlimited and unnecessary armaments . . . at a time when our own knees were trembling and we were about to fall down exhausted."[23] Since the giant has now fallen, since a constructive stabilization has been achieved, it seems obvious that a number of US defense systems conceived before the fall—considered integral to the national defense in an international climate in which the Soviet Union stood surrealistically tall in its ominous, artificial, and ultimately fatal final grasping for world influence and empire—would come in for review and reassessment. Such a reassessment obviously included those systems that are part of the World Wide Military Command and Control System. As a result, the 1990s have witnessed the demise (at least in a formal, bureaucratic sense) of several of the systems and agencies that had helped define the cold war era.

The first to go was the Defense Communications Agency. For just as the cold war had created the need for DCA, the end of that era brought with it the agency's end. On 25 June 1991, acting under the authority of Title III of the Goldwater-Nichols Defense Reorganization Act of 1986, the secretary of defense redesignated the DCA as the Defense Information Systems Agency (DISA). Although the new agency would operate under the authority of the assistant secretary of defense for C^{3}I, as it did before, the change in name was intended to recognize its broadened role in information systems management as well as communications.[24]

The advent of DISA heralded the demise of the Defense Communications System. DISA's primary function was specified to be management and operational control over the Defense Information System Network (DISN), the successor to DCS, and a formal definition for the new network was established by the ASD for C^{3}I in February 1994: "A subset of the Defense Information Infrastructure, the DISN is the DOD's consolidated worldwide enterprise-level telecomunications infrastructure that provides the end-to-end information transfer network for supporting military operations." In other words, DISN would include all of the assets that previously fell under the DCS umbrella, which would serve as the baseline capability from which future progress and programs would be measured.[25] As DISA explained things, the Defense Information Systems Network would do it all, constituting a "seamless web of communications networks, computers, software, databases, applications, and other capabilities that meets the information processing and transport needs of DOD users" under all conditions of peace, crisis, and war.[26]

So that this could take place, the new network had to meet a variety of effectiveness criteria. First, it had to be rapidly reconfigurable, capable of supporting joint task force requirements anywhere in the world. It had to provide fully interoperable communications between deployed forces and home bases, and between the communications assets of all relevant constituencies—the services, defense agencies, and America's allies. It had to provide the capacity to meet military needs, including adequate surge capacity for times of crises. It also had to provide a real-time management capability so that resources could be made available to users under all conditions of peace and war. As DISA phrased it, the network would permit war fighters "to 'plug in' and 'push or pull' information in a seamless, interoperable, and global battlespace," the point being to "assure dominant battlespace awareness from the warfighter's viewpoint."[27]

The DISN Joint Mission Need Statement, approved in early 1995, called for a smooth and incremental evolution away from the current system's reliance on defense-owned networks and toward maximum possible reliance on commercial services and technologies. It called for the DISN to be structured for modular and incremental evolution, allowing new technologies to be incorporated as they became available.[28] It called for DISN to provide the majority of communications requirements for WWMCCS's post-cold-war successor.

In 1992 personnel from DISA and the Joint Staff reviewed the WWMCCS automatic data-processing modernization program then in progress and found it wanting. That September they presented a plan for its termination, which was subsequently approved by the undersecretary of defense for acquisition, and funds were made available to effect the transition to a follow-on global command and control system (GCCS).[29] Like WWMCCS before it, and like the Defense Information

Systems Network with which it overlapped and shared resources, GCCS represented as much a concept as it did a set of assets; it was a sort of "umbrella strategy" intended to guide the evolution of defense communications for the decades to come. As DISA engineers described things, the GCCS would "enable the principles and concepts of the Joint Staff's C^4I [command, control, communications, computers, and intelligence] for the Warrior strategy" through of a set of evolutionary initiatives. Like the DISN, GCCS would emphasize maximum use of off-the-shelf technologies, its development structured so that new technologies could be incorporated incrementally, as they became available. In addition, like WWMCCS before it, the new system was intended to be a White House-to-foxhole system.[30] Unlike the case of WWMCCS, however, the GCCS might just be able to accomplish that goal, owing to the combination of the centralization effected by Goldwater-Nichols and the breathtaking pace of technological advance, especially in the area of automatic data processing. By mid-decade, the services and defense agencies had established GCCS program management offices to implement the new system, and the World Wide Military Command and Control System, a product of the cold war, had vanished along with the tensions of that bygone age.

Notes

1. James H. Babcock, "Defense Communications System Policy and Fiscal Guidelines and Areas of Future Emphasis," *Signal,* August 1977, 80.

2. Lee M. Paschall, "The Second-Generation DCS," *Air Force Magazine* 61 (July 1978): 73.

3. Herbert A. Schulke Jr., "Digital Systems in the Strategic Context and DCA Planning Activities," *Signal* 26 (July/August 1972): 55.

4. John P. Walsh, "R and D Objectives for Future Defense Communications System," *Signal* 23 (September 1968): 12.

5. Lee Blachowicz, "Wideband Modems—Toward a Digital DCS," *Signal* 26 (December 1971): 7.

6. Samuel L. Gravely Jr., "DCS at the Crossroads," *Signal* 33 (May/June 1979): 54–56.

7. Paschall, 73.

8. David T. Signori Jr., "The Command Center Improvement Program: An Innovative Approach," *Signal* 41 (March 1987): 65.

9. Samuel L. Gravely Jr., "The DCA—A Rock and a Hard Place," *Signal* 34 (April 1980): 8.

10. Donald C. Latham and David R. Israel, "A Modular Building Block Architecture," in *Principles of Command and Control*, eds. Jon L. Boyes and Stephen J. Andriole (Washington, D.C.: AFCEA International Press, 1987), 117.

11. Signori, 69.

12. Robert A. Bourcy, "AUTODIN—World Wide Automatic Digital Network," *Signal* 20 (March 1966): 16.

13. Robert R. Fossum and Vinton G. Cerf, "Communications Challenges for the 80s," *Signal*, October 1979, 20.

14. Martin A. Thompson and Tom M. Shimabukuro, "The Next Generation CONUS AUTOVON," *Signal* 34 (February 1980): 75.

15. Samuel L. Gravely Jr., "DCA's Route to Readiness," *Air Force Magazine* 62 (July 1979): 86.

16. "Encryption, Survivability Improve Defense Communications System," *Aviation Week & Space Technology*, 9 December 1985, 80.

17. Eugene F. Tighe Jr., "The Greatest Challenge to DOD Communications is to Maintain a Worldwide System," *Defense Systems Review* 2 (June 1984): 12–14.

18. House, Committee on Appropriations, *Department of Defense Appropriations for 1989*, pt. 6, 100th Cong., 2d sess. (Washington, D.C.: Government Printing Office, 1988), 472.

19. Stephen T. Walker, "Department of Defense Data Network," *Signal* 37 (October 1982): 45.

20. Ibid.

21. House, *DOD Appropriations for 1989*, 472.

22. John D. Steinbruner, "Nuclear Decapitation," *Foreign Policy*, Winter 1981–1982, 16.

23. "Excerpts From Solzhenitsyn Article," *New York Times*, 19 September 1990, A8.

24. General Accounting Office, *Defense Reorganization: DOD Establishment and Management of Defense Agencies*, NSIAD-92-210BR (Washington, D.C.: General Accounting Office, May 1992), 3.

25. DISA, *Strategy for the Defense Information Systems Network (DISN) – Executive Summary*; and www.disa.mil:80/info/pao041.htm.

26. Emmett Paige Jr., "Management and Life-Cycle Support for the Global Command and Control System," memorandum, 26 June 1995.

27. DISA, *Strategy for the Defense*; and www.disa.mil:80/info/pao041.htm.

28. Ibid.

29. Duane P. Andrews, memorandum to director, Defense Information Systems Agency, 6 October 1992.

30. DISA Internet Home Page, www.disa.mil/gcss/execsum.html.

Chapter 19

Organization, Technology, and Ideology in Command and Control

This work began by noting how three themes—(1) organization, (2) technology, and (3) ideology—dominated the development of the World Wide Military Command and Control System during the cold war era. This final chapter examines the influence of each of these themes in greater detail.

Organization: Subunit Domination

Of the three themes, organization appears to have been the most influential force in the development of WWMCCS, since from the very beginning WWMCCS has been an organization dominated by subunit concerns, emphasizing services' needs and requirements and not infrequently working to the detriment of the larger national interest. The result has been a multiplicity of problems and occasional major system failures when the system was called upon to function in a joint-service context. Not surprisingly, this has led to the widespread perception that WWMCCS is ineffective.

WWMCCS's subunits are of two general types. On the one hand are those WWMCCS entities, preeminently the military services, for whom overriding importance was attached to the fulfillment of their own missions. While themselves central to WWMCCS, the perception of the system held by the services has been that it is less than central to their own concerns and at times even antipathetic to them. The performance of these subunits and their pursuit and acceptance of command and control innovations have thus tended toward an emphasis on subunit autonomy and goals over the interests of the larger organization of which they are nominally a part.

On the other hand are WWMCCS subunits, such as the Defense Communications Agency and elements of the Office of the Secretary of Defense, whose concerns (often whose very existence)—were linked to the idea that WWMCCS is a centralized organizational entity—or at any rate should be. These subunits also pursue courses of action and accept or reject innovations, and their behavior in this is similarly selective. For them, however, the emphasis tends to be on those things that permit a higher degree of centralized control. Whether for the decentralizers or the centralizers, then, success almost invariably came at the expense of the other subunits' authority and autonomy. The result is that WWMCCS has historically been a locus of contentiousness, of goal dissensus, and of competing definitions of what constitutes adequate system performance.

In one sense this represents nothing more than a fundamental fact of organizational life: organizations have a division of labor. According to the specific nature of its task requirements, any organization will be horizontally differentiated into specialized subunits, and the requirements placed upon each will differ. Some subunits will have an internal mission, directing their attention toward intraorganizational matters, while the mission of others will be focused more externally. How well the organization functions will depend on the degree of interdependence that exists between the subunits and the extent to which they are linked into a cohesive whole.[1]

The central argument advanced in this book has been that WWMCCS's historical lack of an organizational center of gravity has resulted in a serious lack of coordination between its constituent elements. Specifically, the criteria for system effectiveness have been promulgated by two competing organizational factions, one whose interests and concerns lie with greater centralization of the command and control function and the other's in resisting that centralization. This structural ambiguity has meant that system elements have frequently worked at cross purposes, leading WWMCCS, an ostensibly rational system, to irrational outcomes—periodic breakdowns of control and even major system failures.[2] With such problematical performance, little wonder the system has frequently been viewed as ineffective.

The notion that apparently rational behaviors can produce irrational outcomes has a lengthy history and a distinguished pedigree in social scientific thought, and those who have wrestled with the issue suggest that different types of rationality exist, even in a single organization such as WWMCCS. The social theorist Max Weber posited two different types of rational action. First, there was the type he described as "formal," a means-oriented type focusing on procedure, emphasizing pragmatic, short-term calculations, and showing a concern for the coordination of means. The concern here is with efficiency, an entirely internal performance standard that involves what the organization is producing and at what cost. Efficiency is a standard that is relatively value-free, begging the larger question of whether the organization should actually be engaged in doing what it is doing—like a soldier who can rationally carry out an entire series of actions with no idea as to their ultimate end or of the place of his actions within the larger organizational framework.[3] Such actions are in fact quite rational in the sense that they involve definite goals that are appropriate to that level of action and organization. But where formal rationality dominates, the appropriateness of actions in some larger context is simply not considered. Weber contrasted this means-oriented formal rationality with what he called "substantive" rationality, a goal-directed type concerned with values that appealed to "ultimate concerns." Substantive rationality deals with effectiveness, the appropriateness of what is produced in light of some larger end. Weber was quite clear that however formally rational actions might be, they need not correspond to substantive goals.[4] Quite simply, efficiency and effectiveness represent independent evaluative standards.[5]

The suggestion is that an organization can be efficient but not effective, a place where a means-directed formal rationality can exist and predominate, while failing to serve substantively rational ends. For as Weber demonstrated, there is a tendency for formal rationality to supplant substantive rationality, with the means replacing the ends they were ostensibly designed to serve.[6] Indeed, the mantle of formal rationality devoid of any higher substantive purpose became for Weber an inescapable "iron cage."[7] It is a view that finds circumstantiation in a number of organizational studies, including Herbert Simon's discussion of welfare agencies,[8] Charles Perrow's examination of high-risk systems,[9] and Robert Jackall's account of managerial decision making.[10] In these studies we find a way to conceptualize WWMCCS's historical problems, its apparently endemic inability to perform effectively: much of the system was developed and deployed within organizational parameters where a practical "formal" rationality predominated. Specifically, the system was in large measure developed by organizational subunits whose missions and priorities never were perfectly aligned with the more substantive concerns of the broader WWMCCS macrosystem.

It is easy enough to understand why this was done. Subunits are interest groups that act to enhance their power and prestige relative to other organizational constituencies. They collect information to enhance their own value and to render other elements of the organization dependent upon their expertise. The information extracted from the environment and subsequently made available within the organization will be far from neutral, serving political as well as utilitarian functions, while other information gets ignored entirely.[11] Certain external and internal constituencies will figure more prominently in each subunit's evaluative calculus than others, and these will be attended to disproportionately. That is, a tendency develops for subunits and their members to evaluate all actions, by themselves or with others, exclusively in terms of their utility for the realization of subgoals, resulting in a contentious us/them sort of mind-set wherein the courses of action pursued are not at all rational or functional from the perspective of the larger organization. This is the phenomenon of suboptimization, a recipe for fractiousness and ultimately failure if the subunits' centrifugal impetus is not somehow held in check.

Most organizations employ a variety of devices, both carrots and sticks, to promote coordination between subunits, to secure adequate contributions from them on reasonable terms, and to see to it that whatever discrepancies exist between the subunits' goals and the larger organizational goals do not become overly large.[12] But what happens when these don't work? What happens when the organization lacks the ability to establish viable limits to subunits' natural self-interest? What happens when the influence of subunits becomes excessive, and when centralized decision makers, the organization's ostensible "dominant coalition," cease to be the ultimate arbiters of organizational performance? As the case of WWMCCS suggests, what occurs in a world of relentless, uncontrolled suboptimization is that the efficiency concerns of subunits prevail. What occurs also is that organizational performance can be judged decidedly ineffective when assessed by "substantive" criteria relevant to the organization as a whole, not just the criteria of concern to powerful subunits.[13] An organization that exhibits these characteristics might best be described as a "subunit-dominated organization."

A subunit-dominated organization is a place where the goal orientation of central decision makers does not determine subunits' orientation or actions with any precision or certainty. It is a place where the center is unable to impose effective oversight and control over lower-level parts of the organization. It is also a place where the idea of a centralized decision-making apparatus is simply no longer descriptive of the organizational reality that exists. Even more than loosely coupled systems, which assume a modicum of coordination and common purpose flowing from the top, subunit-dominated organizations are not fully cooperative systems. This does not describe a situation of the rational pursuit of optimal outcomes. It runs counter to rationalist models of bureaucratic functioning, which for military organizations involves centralization of policy making and strategic planning, as well as decentralization of actual operations.[14] Beset by an internal Balkanization, subunit-dominated organizations are political arenas in which subunits compete to advance their interests and where resources are distributed according to coalition bargaining power.

This work suggests that WWMCCS is a subunit-dominated organization, a place where not only has operational authority been decentralized but ultimate or true authority as well. Throughout the cold war era, the military services in many important instances remained independent entities with considerable bureaucratic power. Planning and force structure were predicated on unilateral service views of priorities and on how a future war might be fought. Views on the training and equipping and the support of forces logically followed, frequently at the expense of joint missions and overall combat readiness. Each service retained separate responsibility for its own budget and competed vigorously to increase its share of total defense dollars.[15] In other words, key WWMCCS system elements have operated in substantial measure beyond the influence of centralized guidelines or oversight. Their autonomy was such that they could often ignore or effectively resist central initiatives. Central or hierarchical officials lost the capacity to exercise effective oversight over these subunits, or they never had it to begin with.[16] What this meant was that the effort to create a worldwide military command and control system truly responsive to centralized control was resisted by the services, and when the WWMCCs could not be resisted, it was subordinated to the services' unique, mission-specific needs.[17] A lack of centralized control guaranteed that from the start WWMCCS would be more a locus of competition and conflict than a coherent single organizational entity—thus, the repeated characterizations over the years of WWMCCS as a "confederation," "loosely knit federation," and various other similar characterizations.[18] It was a condition guaranteeing that the system would have trouble, even experience major failures, when called on to operate in a joint service context.

Technology: Technological Push and User Pull

The second major theme in WWMCCS's development has been technology—specifically, the dramatic technological advances that began early in the cold war era and that thereafter never ceased to exert influence. WWMCCS was born within a context of ongoing research and development, an almost vertiginous pace of technological advance both in the military and civilian sectors that was continually altering the nature of warfare. What was the importance of technology in WWMCCS's evolution? Was its development driven primarily by a conscious, rational process that has been described as "user pull"? Or was it the result of the imperative of technological advance, of "technology push"?[19]

These questions are rightly seen as part of the larger debate about the role of technology in society, a debate that continues today and that has been characterized by considerable controversy. On one side are technological determinists who view technology as a major social force, arguing that as science marches onward, society necessarily follows.[20] In this view, research scientists fuel the fires of technological advance, and by extension military requirements, through their desire to pursue and bring to fruition interesting new technological concepts.[21] Major technological advances provide the impetus for their application to military systems, and military requirements issue from the flow of change, rather than from a priori assessments of requirements.[22] Concluding that social structures and their associated structures of belief are in part or in whole derivative of a technological imperative, these determinists represent one side in a wide-ranging debate about the relationship of technology to society—specifically, that the evolution of defense systems and policy are dominated by technological advances, rather than conscious human design.[23]

Critics of this point of view argue that the technological determinists have it backwards. Technology is not a social force, they say, but rather a social product. From a wide range of possible technical solutions to problems, those individuals and groups with greater influence will choose the ones that best serve their interests.[24] Not only that, this perspective holds that the range of possible solutions is circumscribed by these elites. It is a social determinist view, where the development of specific technologies is driven and shaped by power relations rather than by an essentially neutral process of scientific advance. Radical critiques from this perspective usually focus on the profit and power goals of the military-industrial complex.[25] More conventional analyses view the selection of specific technologies as the result of national strategic choice, where rational decision makers select specific technologies on the basis of precise calculations about national objectives, perceived threats, and strategic doctrine, all the while cognizant of budgetary constraints and other limitations.[26]

Which view is correct? The answer in the case of WWMCCS is a qualified "both." After all, revolutionary technological concepts and techniques do appear on the scene from time to time, affecting military relations. Witness the advent of atomic weapons, the development of the intercontinental ballistic missile, ballistic missile submarines, computers, and satellites during the years following World War II. Arguably, each of these new technologies revolutionized warfare, contributing thereby to the drive for a more responsive and centralized defense decision-making structure. Moreover, much of the thinking in the area of command and control has been highly technical, a hard engineering orientation, which resulted in the overwhelming proportion of research attention and financial investment going toward system designs that were technologically advanced—often magnificently so—yet that ultimately failed to meet the requirements of actual users because they failed to adequately consider the human and organizational context into which they will be introduced.[27]

In one way it all makes perfect sense. Given that the great watersheds in the history of warfare have always involved the application of new technologies, it is hardly surprising that a basic military mind-set has evolved that newer, more technologically sophisticated systems are by definition better ones. This logic has been validated by a Congress that frequently rewards the pursuit of more sophisticated technologies with higher funding and underscored by an American cultural emphasis on progress and a pervasive belief in technology as its guarantor.[28] WWMCCS's growth, then, as with the trajectory of the arms race more generally, arose from an institutionalized belief that a more capable technology is by definition a better one. Technology push has thus been a key process in the development of defense systems, producing most of the important, revolutionary new technologies.[29]

If both perspectives contain a measure of truth, it can be argued that the concept of revolution probably captures less well the dynamic of change in the domain of command and control, for few of the key developments in this area—the transistor, for example, or the laser or automatic data processing—truly revolutionized communications. Certainly each of these was an advance, producing improvements in communications capacity, reliability, and economy, but their impact was hardly as dramatic as the term *revolution* connotes. As these and other technologies arrived on the scene, they were assimilated into existing media, resulting in gradual, incremental change. Thus, while many new technologies were initially hailed as revolutionary, their integration into the workaday world of command and control generally proved to be less so, conditioned at every step by preexisting technologies and patterns of organizational relations and goals.[30] If there were a technological push, it was not necessarily an unqualified one.

Throughout the history of WWMCCS during the cold war, then, much of the impetus for technological development appears to have been closer to the "user pull" of the social determinists, "the institutional process by which users (notably the services) assess the adequacy of existing [systems] to meet military needs, and state the characteristics of the next generation of equipment desired to overcome identified inadequacies," as the Packard Commission described it.[31] All new technologies naturally carry within them the seeds of organizational change. But some types will be embraced as advantageous, whereas others will be rejected, and it is of interest to determine which is which. On the one hand, it has been suggested that innovations posing no threat to organizational routines, strategies, or "essence" are less likely to generate resistance and hence are more likely to be adopted.[32] Even innovations that challenge existing routines and strategies might be embraced when they are seen to promote desirable changes: for example, when an organization wishes to expand and when the innovation will permit that expansion to take place.[33] Obversely, technologies that presage unwanted changes—such as a reduction in organizational autonomy, or undesirable transformations in routines, customs, or allegiances—are likely to encounter resistance, including active hostility, even if in some objective sense they are inherently useful or appropriate.[34] This selectivity has led to the adoption of a wide range of new technologies, many of which have tended to focus on the needs and priorities of system subunits.[35] It meant that technologies were resisted if they were not perceived as being in line with the requirements of the services' military missions.

The Office of the Secretary of Defense has had only limited success in its efforts to have the services pay for systems designed in support of joint-service systems for strategic command and control. Funding for such high-priority programs as the Defense Satellite Communications System and the National Emergency Airborne Command Post were resisted by the services, for example, and received funding only under heavy pressure from the secretary. Lower-priority initiatives met with even less success. During the late 1970s the head of WWMCCS Engineering recommended the acquisition of 10 different types of communications equipment to improve national officials' ability to respond to urgent contingencies. Although the items included such relatively inexpensive equipment as transportable satellite earth terminals, the services declined requests to fund them.[36] In a subunit-dominated organization such as WWMCCS, it appears that social determinism, user pull, will be the dominant influence in terms of the acquisition of new technologies, although it will not determine outcomes with absolute certainty.

Ideology: The Evolutionary Approach

A subunit-dominated organization such as WWMCCS sounds emphatically ineffective. Its organizational structure, rife with fractiousness, appears like nothing so much as a centrifugal whirl straining to tear itself to pieces, subject to major system failures. A key question is thus how such a discordant assemblage of elements manages to hold itself together. It has been argued in this work that a major part of the answer in WWMCCS's case lies in a shared set of assumptions about the physical and social worlds—an ideology, in other words, that has permitted innovations to be pursued, technologies acquired, and the system's apparently implacably antagonistic subunits to function together more or less amicably over time. The ideology that has permitted these remarkable feats to be accomplished, one that has gained increased rhetorical and bureaucratic support with the passage of the years, is the "evolutionary approach" to command and control system development.

We have seen how the approach first gained a foothold in defense thinking in the early 1960s as an alternative to the then-dominant approach to system design and acquisition, the weapons system approach, whose governing idea was that all efforts should be directed toward the development of an identifiable target system that could be turned over to its users on a specific target date.[37] The problem was that by the early 1960s, a number of influential defense constituencies had concluded that the weapons system approach simply did not work when applied to command and control systems because they possessed characteristics not found in other complex systems. Perhaps most importantly, they were said to be dynamic, "information rich" systems highly dependent upon the information they contain and the demands placed upon them.[38] In contrast to more static systems whose very nature dictated that certain functions be performed repeatedly in a fixed order, command and control systems were viewed as characterized by an ever-changing configuration of individuals, functions, information requirements, and equipment that was inherently resistant to any simple formulaic ordering. Relevant actors could not always be identified, nor could operational requirements, and even when they could, they often did not remain fixed long enough to permit the development and deployment of appropriate system capabilities.[39] In developing WWMCCS, then, flexibility appeared essential.

A second general reason advanced for the uniqueness of command and control systems was that they were considered more "threat driven" than other military systems, meaning that they had to be uniquely sensitive to both qualitative and quantitative changes in enemy military capabilities.[40] The problem with the old weapons system approach was that by the time a command and control system had finally achieved full operational status, the military situation had frequently changed so much that the system was no longer appropriate to the threat. Making matters more difficult still was the rapid and accelerating pace of technological advance. With new and frequently competing technologies constantly being developed, locking in a specific system design all too often meant locking its user into a system that was obsolete by the time it was fielded.[41] Of course, it was possible to try to design systems to meet future requirements, but such exercises in prophesy proved to be expensive and were not infrequently frustrated by unpredictable international events and technological changes.[42] In the minds of many in the defense establishment, the unique fact that it appeared impossible to fully specify a command and control system's requirements at the time its development commenced called for an equally unique management approach.[43]

As we have seen, the alternative to the conventional model of management and planning that appeared most attractive allowed commanders to define, develop, and improve their own systems "naturally" over time, as circumstances warranted, as they were considered necessary to meet the changing requirements of their military missions.[44] "Changes in the command and control systems will be, of necessity, evolutionary," Robert

S. McNamara had declared, "and the systems must be flexible enough to adapt to changes in the world situation and U.S. strategy." Moreover, it was simply too expensive, indeed impossible, to develop complete stand-alone systems only to have to rip them out and replace them from scratch every time a new technology came down the pike or each time modifications were made to US strategic doctrine.[45] By advancing the idea that command and control modernization was an incremental, user-oriented process conducted under the broad cognizance of a central authority, the evolutionary approach held forth the promise of greater flexibility in a turbulent geo-political and technological environment.[46]

Adding new requirements and technologies as an ongoing series of modest improvements obviously meant that command and control systems would take longer to develop and thus cost more—but never mind. It was believed that unlike systems whose designs had been frozen earlier, those that emerged from the evolutionary process would possess greater capabilities, better reflect users' needs, and be more closely aligned with the requirements of national military policy. The result would be a "harmonious conglomerate of elements of different size, loosely but effectively federated."[47] Yet this phrase surely raises more questions than it answers. Precisely which elements should comprise this conglomerate? By what criteria should their effectiveness be assessed? Who should determine the answers to such questions as these? Far from rhetorical questions, they proved to be eminently real-world ones with real implications for system design, suggesting why they have repeatedly surfaced in the lengthy defense debate over the best way to develop, operate, and evaluate WWMCCS.

So the evolutionary approach had its down side, perhaps the most serious aspect of which was that it encouraged a lack of clear-cut responsibility. It permitted considerable laxity and carelessness in system specification. It did not require a full accounting of who the system's users would be, since these could always be identified later. For users who were identified, the approach failed to require a comprehensive specification of their requirements, in the belief that whatever deficits might exist could simply be addressed in subsequent phases of system evolution. It also failed to specify a point from which evolution would proceed, and it failed to demand a clear statement of the goals the system was supposed to accomplish, thus countenancing a lack of clarity as to how the system was supposed to function or be evaluated.[48] It was also financially wasteful, keeping programs in the development phase essentially forever, since all systems were to be subject to constant review in terms of their relationship to changes in the threat, to development of new weapons systems and other technologies, and to modifications in command structures or other organizational changes.[49]

Predicated as it was upon independence rather than interdependence of subunits, the approach ensured that subunits' actions would frequently bear little relationship to one another, resulting in unnecessary, costly duplication.[50] It also created strong pressures for contractor buy-ins, wherein artificially low prices would be given up front for a project, the contractor understanding full well that prices could be raised and raised again as the project "evolved."[51] To many critics this was not evolution but rather a profligate and essentially visionless process of ad hoc incrementalism, one whose end products would almost certainly require extensive work to make them play together as a coherent whole.

But if not evolution in fact, a key question is why did the evolutionary approach gain such wide currency in the first place? This work has suggested that the answer probably lies more in the bureaucratic utility of the approach than in its ability to create an optimal system for command and control. To those individuals and groups interested in advancing the cause of greater centralization in command and control, the evolutionary approach appears to have been initially attractive because it was a means to mollify the opposition to greater centralization in defense decision making. The evolutionary approach was also attractive to those groups, such as the services, whose interests were naturally antipathetic to greater centralization since it maintains that the decision-making process is situationally contingent and unknowable in advance. Centralized decision makers thus cannot adequately specify the sorts of information they require, with whom they might need to communicate, or precisely what type of system best suits their needs. In light of this ignorance at the center, the logical course of action is to devolve authority toward the periphery, providing greater flexibility for system development to lower-level system subunits.[52] With the evolutionary approach, subunits were able in substantial measure to co-opt the development of WWMCCS in ways favorable to their agendas and their interests.

Therefore, the evolutionary approach was able to prosper in the final analysis because of its bureaucratic utility and because it represented a way to meet the needs of both the proponents of centralization and decentralization simultaneously. It allowed the centralizers to commence building the defensewide command and control infrastructure they desired, something that might otherwise have been met with far more vigorous opposition. For those whose interests lay in decentralization, the approach offered a great deal of autonomy and considerable authority for WWMCCS's development and management, sweetening considerably the bitter pill of centralization. But as this work has shown, the compromise was hardly symmetrical, for the price of diminished service resistance ultimately proved to be the soul of the centralized WWMCCS concept.

In theoretical terms, the evolutionary approach allowed WWMCCS to be redefined from what we might call a "solution-driven" organization to one that was more "process driven." Solution-driven organizations are oriented toward the resolution of specific problems, problems that in turn tend to be closely aligned with an organization's official goals. Such organizations generally operate in a context of substantive rationality and are oriented toward some specific end or "ultimate value" as enunciated in public statements by their officials and as set forth in their public documents.[53] In the case of WWMCCS, the "ultimate value" is to provide the National Command Authorities with an ability to electronically orchestrate military responses to crises anywhere around the globe. The weapons system approach of command and control systems acquisition is consistent with this, involving as it does a rational process of identifying a population of system users, specifying their responsibilities, determining the various types of information they required, and developing specific technologies to provide it.

Obversely, the key for understanding process-driven organizations is that they are means-oriented, focusing on producing an ongoing stream of products or services. They are concerned with operative goals, with actual operating policies, and they reflect the actual needs of organizational constituencies. Process-driven organizations engage in activities with no clearly defined "ultimate" end state or specific solution toward which the organization is tending. Their goals, such as they are, turn on efficiency criteria with regard to the process itself, and thus they function within a context of formal rationality. With the organization's purpose no longer tied to the attainment of an ultimate goal or specific end state, any advancement or achievement, however substantial, is seen as simply a milestone on a road the length of which is no longer known with precision; indeed, a road whose length has been intentionally rendered imprecise.

It was in this way that the evolutionary approach subtly redefined the criteria for WWMCCS effectiveness. For the approach's proponents, the purpose of WWMCCS was no longer to attain a specific goal state by developing and deploying some identifiable set of command and control capabilities to meet specific performance criteria. Rather, the purpose was now modernization, evolution itself, with any technological or organizational innovations viewed as simply a part of this ongoing process. Proponents of the evolutionary approach have long tolerated a level of WWMCCS performance that others have considered marginal or substandard, and it is perhaps only in this light that their tolerance is understandable. To them, after all, WWMCCS's job was not to exhibit perfect performance, which is impossible in any case, but rather to be evolving toward some higher, however imperfectly conceived and understood, end state.

System Failures

Thus, we arrive at our final point of discussion: system failures. We have seen how WWMCCS's history has been punctuated by a series of serious, often spectacular breakdowns and failures. WWMCCS has, in more sense than one, truly been a child of crisis. The argument advanced here is that a subunit-dominated organization such as WWMCCS, focusing on lower-level interests and driven by concerns of process rather than results, could not avoid failure when viewed from a more substantive, systemwide perspective. To those process-oriented types who embraced the evolutionary approach, this was viewed as a regrettable but unavoidable part of evolution. To them, things were perhaps a bit problematical now, but the system was evolving after all, undergoing a process of continual improvement and enhancement. To them, the glass was emphatically (and at least) half full. But to other influential individuals and groups with occasion to assess WWMCCS's effectiveness, the Congress being perhaps the most conspicuous example, that glass appeared different. Wielding as they did a different, solution-driven evaluative yardstick, effective performance was viewed as perfect performance, a criterion that WWMCCS would fail to meet. From their perspective, the World Wide Military Command and Control System was not merely a child of crisis. In structural terms it was born to fail.

Notes

1. Johannes M. Pennings and Paul S. Goodman, "Toward a Workable Framework," in *New Perspectives on Organizational Effectiveness,* eds. Paul S. Goodman and Johannes M. Pennings (San Francisco: Jossey-Bass, 1977), 146–84.

2. Philip Selznick, *TVA and the Grass Roots: A Study in the Sociology of Formal Organization* (Berkeley: University of California Press, 1953), 258.

3. Karl Mannheim, *Man and Society: In an Age of Reconstruction* (London: Routledge & Kegan Paul, 1966), 54.

4. Stephen Kalberg, "Max Weber's Types of Rationality: Cornerstones for the Analysis of Rationalization Processes in History," *American Journal of Sociology* 85 (1980): 1145–79.

5. Jeffrey Pfeffer and Gerald R. Salancik, *The External Control of Organizations: A Resource Dependence Perspective* (New York: Harper & Row, 1978), 34–35.

6. Max Weber, *Economy and Society: An Outline of Interpretive Sociology,* vol. 1 (New York: Bedminster Press, 1968).

7. ———, *The Protestant Ethic and the Spirit of Capitalism* (New York: Charles Scribner's Sons, 1958), 181.

8. Herbert A. Simon, *Administrative Behavior,* 3d ed. (New York: Free Press, 1976), 201–2.

9. Charles Perrow, *Normal Accidents: Living With High-Risk Technologies* (New York: Basic Books, 1984).

10. Robert Jackall, *Moral Mazes: The World of Corporate Managers* (New York: Oxford University Press, 1988).

11. James G. March and Herbert A. Simon, *Organizations* (New York: John Wiley & Sons, 1958), 152–53.

12. Pennings and Goodman, 152.

13. Kim S. Cameron, "Effectiveness as Paradox: Consensus and Conflict in Conceptions of Organizational Effectiveness," *Management Science* 32, no. 5 (May 1986): 549.

14. Philip S. Kronenberg, "Command and Control as a Theory of Interorganizational Design," *Defense Analysis* 4, no. 3 (1988): 232–33.

15. Alain C. Enthoven and K. Wayne Smith, *How Much Is Enough? Shaping the Defense Program, 1961–1969* (New York: Harper & Row, 1971), 10–11.

16. Lawrence B. Mohr, *Explaining Organizational Behavior: The Limits and Possibilities of Theory and Research* (San Francisco: Jossey-Bass, 1982), 106-7.

17. Lawrence E. Adams, "The Evolving Role of C^3 in Crisis Management," *Signal*, August 1976, 60.

18. Comptroller General, *The World Wide Military Command and Control System—Major Changes Needed in its Automated Data Processing Management and Direction: Report to the Congress*, LCD-80-22, 14 December 1979, 2; Jon L. Boyes, "WWMCCS in Transition: A Navy View," *Signal*, August 1976, 72; and Hubert S. Cunningham and William E. Kenealy, "The Joint Chiefs of Staff and Command and Control," *Signal*, March 1975, 16.

19. *A Quest for Excellence: Final Report to the President* (Washington, D.C.: Blue Ribbon Commission on Defense Management, June 1986), 45.

20. For example, see Alvin W. Gouldner, *The Dialectic of Ideology and Technology: The Origins, Grammar, and Future of Ideology* (New York: Seabury Press, 1976); and Jacques Ellul, *The Technological Society* (New York: Alfred A. Knopf, 1964).

21. Solly Zuckerman, "Science Advisors and Scientific Advisors," *Proceedings of the American Philosophical Society* 124 (1980): 250–51.

22. John Steinbruner and Barry Carter, "Organizational and Political Dimensions of the Strategic Posture: The Problem of Reform," *Daedalus* 104 (Summer 1975): 143.

23. Ralph E. Lapp, *Arms Beyond Doubt: The Tyranny of Weapons Technology* (New York: Cowles, 1970).

24. See, for example, Karl Marx, *Capital: A Critique of Political Economy*, 3 vols. (New York: International Publishers, 1967); Harry Braverman, *Labor and Monopoly Capital: The Degradation of Work in the Twentieth Century* (New York: Monthly Review Press, 1974); and David F. Noble, *Forces of Production: A Social History of Industrial Automation* (New York: Knopf, 1984).

25. See, for example, C. Wright Mills, *The Power Elite* (New York: Oxford University Press, 1956); Seymour Melman, *The Permanent War Economy:*

American Capitalism in Decline (New York: Simon and Schuster, 1956); and James Fallows, *National Defense* (New York: Random House, 1981).

26. Graham T. Allison and Frederic A. Morris, "Exploring the Determinants of Military Weapons," *Daedalus* 104 (1975): 103.

27. I. B. Holley Jr., "Command, Control and Technology," *Defense Analysis* 4, no. 3 (1988): 270.

28. Warren G. Bennis et al., *The Planning of Change*, 4th ed. (New York: Holt, Rinehart and Winston, 1985), 17; and Gerhard Lenski and J. Lenski, *Human Societies* (New York: McGraw-Hill, 1974).

29. Blue Ribbon Commission, *A Quest for Excellence,* 45.

30. Gordon T. Gould Jr., "Trends in Communications—Revolution or Evolution?" *Signal,* May 1967, 60.

31. Blue Ribbon Commission, *A Quest for Excellence,* 45.

32. Matthew Evangelista, *Innovation and the Arms Race: How the United States and the Soviet Union Develop New Military Technologies* (Ithaca, N.Y.: Cornell University Press, 1988).

33. Barry R. Posen, *The Sources of Military Doctrine: France, Britain, and Germany between the World Wars* (Ithaca, N.Y.: Cornell University Press, 1984), 47.

34. Elting E. Morison, *Men, Machines, and Modern Times* (Cambridge, Mass.: MIT Press, 1966), 35–36.

35. Comptroller General, *The World Wide Military Command and Control System,* 12.

36. Bruce G. Blair, *Strategic Command and Control: Redefining the Nuclear Threat* (Washington, D.C.: Brookings Institution, 1985), 59.

37. J. J. Cahill, "Resource Management: A New Slant on C&C," *Armed Forces Management,* July 1963, 70.

38. Comptroller General, *The World Wide Military Command and Control System,* 14–18.

39. *TIG [The Inspector General] Brief,* "Directorate of Command Control and Communications (AFOCC)," 14 March 1969, 2.

40. William Morrison and David M. Russell, "C^3I Programs Demand Cost-effective Data Processing Solutions," *Defense Science & Electronics,* January 1987, 28.

41. Peter Grier, "Pentagon Arms Suffer From High-Tech Gap," *Christian Science Monitor,* 8 June 1989, 7.

42. J. P. McConnell, "Command and Control," *Sperryscope,* no. 2 (1965): 4.

43. Albert E. Babbitt, "New Communications and Technology in WWMCCS," *Signal,* August 1977, 84.

44. Comptroller General, *The World Wide Military Command and Control System,* 14–18.

45. John B. Bestic, "No More Confused Situations," *Signal,* March 1967, 53–56.

46. *TIG Brief,* 2.

47. "Dr. Fubini Stresses DDR&E's Desire for System Publication Compatibility," *DATA*, February 1965, 9.

48. Samuel L. Gravely Jr., "The DCA—A Rock and a Hard Place," *Signal* 34 (April 1980): 8.

49. McConnell, 2.

50. Pfeffer and Salancik, 70.

51. Bernard L. Weiss, "Keys to Success in C^3I," *Signal*, August 1982, 53.

52. J. S. Butz Jr., "USAF and the Computer Revolution," *Air Force Magazine*, March 1964, 35.

53. Charles Perrow, "The Analysis of Goals in Complex Organizations," *American Sociological Review* 26, no. 4 (December 1961): 855–56.

Epilogue

Editor's Note: The World Wide Military Command and Control System (WWMCS) was closed down in August 1996 and was replaced by the Global Command Control System (GCCS). To find out how the new system is currently working, the author interviewed Dr. Frank Perry, technical director of the Space and Naval Warfare Systems Command (SPAWAR). Dr. Perry is the chief technical authority within that command.

Interview with Dr. Frank Perry
20 March 2000

Pearson: Dr. Perry, please discuss your background and career, especially with regard to WWMCCS and its successor, the Global Command Control System.

Perry: I have been at SPAWAR for about a year and a half. For the previous three and a half years, I was the technical director—something like the chief scientist or chief engineer—at the Defense Information Systems Agency. That was the interval when Global Command Control System was developed. I had more than just a little bit to do with that whole development process, beginning in May 1995, when I went to DISA, on through August of 1998, when I left. So there was that association. There was also an earlier association with WWMCCS as well. In 1975 I spent a year as a programmer on the WWMCCS system, as a young lieutenant junior grade in the Navy, at what was then CINCLANTFLT, at an organization called the Atlantic Command Operational Support Facility. It basically ran the WWMCCS site there. So I had two encounters with WWMCCS, first in the mid-1970s, and then during the 1995–98 time frame at DISA.

Pearson: Let's go back a decade or so. In your view, what were the major themes, the major issues, which set the stage for the transition from WWMCCS to GCCS?

Perry: As I'm sure you know, WWMCCS was more than just a technology. It was not just ADP, but it was an entire process associated with deployment, planning, and execution—deployment meaning large-scale movement of forces in response to contingency operations. There certainly was a WWMCCS system, the ADP system, interconnected by a network called WIN (WWMCCS Intercomputer Network) that evolved back in the 1970s and went through the cold war era into the post-cold war era, based upon some very early networking technologies starting back in the mid-1970s. It started off with something called PWIN, the Prototype WWMCCS Intercomputer Network, and eventually evolved into WIN, predating a lot of the Internet kinds of technology that we have today. So it was ADP, plus the whole set of processes and people associated with how one plans for and executes large-scale movement of forces. So it's a whole bunch of processes, people, computers, and networks going all the way back to my earliest involvement with it in the mid-1970s.

The world began to change, especially with respect to technology. You talked a little bit about the world changing with respect to the end of the cold war. Well the world changed in perhaps as radical a fashion with respect to technology. As intercomputer networking became much more recognized, and Internet technology emerged on the scene, there was a very fundamental shift for a lot of people dealing with military technology. You went from the mainframe era, lasting through the early 1980s, through the minicomputer era, all the way down to the client server and desktop machines kind of architectures that we have today, leading off ultimately toward very thin client web-based architectures. Throughout the late 1980s–early 1990s, as those technological changes were starting to gain a foothold, the technology of WWMCCS with its by that time very proprietary and relatively low-performance wide-area intercomputer network and the mainframe technology base was increasingly nonresponsive. It could not adequately evolve to meet the changing characteristics and needs of command and control.

In the post-Desert Storm time frame as well, there was recognition within the senior leadership of the Joint Staff that one had to evolve this thing called WWMCCS into more than just a deliberate planning tool. There was the realization that one had to embrace more real-time execution aspects of command and control at the Joint Staff level, things like theater-level situational awareness—who's where, when—with respect to friendly forces, US and allied, as well as hostile forces. Those were challenges that the mainframe WWMCCS system never really had on its plate. And as you looked at the proprietary low-performing network and the mainframe-based hardware architecture, you really could not evolve that into some of the expanded missions dealing not just with deliberate planning for deployment but also with crisis planning and some execution-related functional capabilities.

Pearson: Does that mean that there is less of a "strategic" emphasis to the Global Command Control System? In the latter part of its history WWMCCS seemed to be moving in the direction of greater emphasis on the command and control of the strategic nuclear forces. Does GCCS continue that emphasis, or does it expand that emphasis to include theater and tactical operations?

Perry: I wouldn't say it is less of an emphasis on the strategic, I'd say it is more an emphasis on other things in addition to that. Certainly that major concern is something still there with global command and control. But the scope of global command and control has broadened to encompass support to additional capabilities in additional mission areas in the execution arena. If you look at warfare as defined in the joint service parlance from the national level down to the theater and down to the tactical level, you certainly start getting into the operational and tactical levels of warfare.

Pearson: We've been discussing some of the technological changes that gave rise to the Global Command Control System. How does GCCS differ organizationally from its predecessor?

Perry: In terms of the elements of the military departments that employ it?

Pearson: That, or have new organizational entities been created to administer the system? How does it look differently, organizationally speaking, from how things were under WWMCCS?

Perry: With respect to the mission of supporting command and control in crisis planning, it really has the same sorts of functional people in the unified and specified CINC organizations, who have the responsibility of putting those plans together. For example, supporting CINC organizations, like the US Transportation Command, play a major role in assessing transportation feasibility. For example, I want to move this brigade from point A to point B in 72 hours, but do I have the lift capacity to do that? You still have an awful lot of folks within the military departments acting as force providers who have to figure out, given a requirement for a light infantry brigade with a specific set of characteristics, where across the Army ground forces can I find such a brigade that meets the appropriate readiness criteria? All these topics were issues under WWMCCS; they all had constituencies that dealt with them across the unified and specified CINC organizations and the military departments, and none of that was substantively changed with the evolution to GCCS. The replacement of the technology was separated from any substantial re-engineering of the business processes. That re-engineering is something that is certainly of interest to the CINCs and Joint Staff, and there are initiatives underway to deal with it. But there was an overt decision not to tie fundamental re-engineering of those business processes together with the evolution of the technology in global command and control. It was difficult enough doing one without trying to do both at the same time.

With respect to the institutions to run them, you used to have, depending upon what point in time you looked at it, anywhere from 21 to 26 mainframe computer sites internetworked around the world with WWMCCS. Now we have shrunk down to a smaller number of GCCS major server locations internetworked around the world, and the size of the staffing pool to administer them dropped fairly dramatically in the process of moving forward to the more modern technology.

We also have slightly fewer locations, substantially fewer people required to keep GCCS running, but not a fundamentally new paradigm for how to administer GCCS. So organizationally, given that one did not fundamentally change the business processes for deployment planning, organizationally there was not substantial change.

Given that GCCS did start picking up missions that it did not previously support at the operational and even tactical level of warfare, (for example, maintaining tactical situation awareness), then you find evolution of doctrine occurring in terms of establishing something like a common operational picture. That doctrine addresses assembling a common operational picture across an area of responsibility such as the US Central Command AOR in and around the Arabian Sea for their forward-deployed forces, and how that whole picture would get relayed back to Central Command headquarters at McDill AFB in Tampa and then out to the stateside components in their service supporting organizations, like Ninth Air Force at Shaw AFB. In the event that something happened, everybody could be looking at the same picture and having a discussion in context about what's going on, rather than having a discussion with multiple independent mental images of the situation.

Pearson: How about DISA? Have there been increased organization responsibilities for DISA since the time of the creation of GCCS?

Perry: Yes. DISA produced GCCS and continues to evolve it to this day, so there is more of an initial development effort and an evolutionary development effort there than had been the case with WWMCCS, where WWMCCS was largely in catastrophic maintenance from a software point of view.

Pearson: Throughout its history, some people considered WWMCCS to be a "bureaucratic fiction," an organizing principle, more than it was a real system with a hard commitment of personnel, equipment, and funds. Give us an idea of why, in your view, this is not true for GCCS?

Perry: I didn't use the term *bureaucratic*, and I guess I wouldn't: it's not in my own mental picture of it all, neither in the earlier days of WWMCCS or in the later days of global command and control. My mental picture of it is more a set of business processes, a set of people to execute them, and ADP support. I think that the major change that has occurred with the introduction of GCCS is that the ADP support for the previous mission of deliberate planning has improved fairly substantially. And the evolved missions of crisis planning: since things don't happen on the same timelines any longer, the technology support for those new missions has been enabled.

Pearson: Going back over WWMCCS's history, one of the major dynamics shaping the system during the cold war era was a tension between the forces of centralization, represented primarily by OSD and such agencies as DCA, and the forces of decentralization, represented in the main by the military departments. Is that sort of tension still apparent in GCCS?

Perry: To some degree—probably to a slightly lesser degree than previously. I think that what you're referring to is in no small part a fundamental element of how we're organized within the Department of Defense. But as you look at the deployment planning mission, there is really not a lot of tension there because that really is a war-fighting commander in chief-focused job, and it really can't be done effectively in any other fashion. I think that if you go back to some of the very early days of WWMCCS, I would agree with you that a lot of that tension was there, and that a lot of service-specific extensions to WWMCCS were built. There are still some service-specific extensions built on top of Global Command Control System to meet some of the service-unique requirements associated with the deliberate or crisis planning process. But in my association with it recently, I have not really perceived that to be a tension. I perceive that to be more of a fact of life, that one has to add some service-unique capabilities on top of the joint capability in order to have it make sense to the independent services as force providers. Given the very different characteristics of deployment across the services, with the Army and the Air Force being garrison-based forces, and the Navy and the Marine Corps being, if you will, perpetually deployed forces, there are just some fundamental differences that must be acknowledged.

When you begin to look at some of the new missions that have been added to GCCS, and again I focus on things like situation awareness, there has been a broad embracing of those capabilities across the military departments. You will find today a GCCS Maritime within Navy, that takes the joint product one hundred percent and extends it where necessary, but it's not just a word game. The intent within Navy is to embrace the joint system, extend it where necessary to support maritime needs, and to deploy it as our mainstream GCCS Maritime command and control capability. We will not deploy something different and just anoint it, in bureaucratic faction, and call it GCCS Maritime. We actually deploy the same identical software. To greater or lesser degrees I think many folks within the services and service programs have begun to adopt this philosophy.

Pearson: Let's return to ADP for a moment. How have improvements in ADP technology over the past decade or two impacted upon the need for standardization of both hardware and software?

Perry: Well, the software for GCCS *is* standard, and with the exception of the service-specific extensions to it, it is all produced by DISA. Standardization of the applications is very important today to achieve the level of interoperability and interworking that is necessary to support the mission. The individual services don't go off and build their own alternative GCCS systems from the ground up. The philosophy is for them to take the base system from DISA and extend it where necessary. And some other technological constructs come into play, such as the Defense Information Infrastructure Common Operating Environment—DIICOE for short. In the early 1990s, when the whole push to try and move from WWMCCS ADP to GCCS emerged on the Joint Staff, the thinking was that there would be a common software baseline for this thing called

GCCS and it would be adopted by everybody. One of the things that changed in the 1995-96 time frame as we began to implement GCCS was a recognition that that common software framework, and a common integration approach, wasn't unique just to command and control but was applicable in other domains as well—for instance, in combat support and in intelligence processing. So DIICOE got constructed to abstract that away from the specific mission area of command and control into a more generalized integration philosophy, looking at where it's important to have identical common software and where it's not. But now today, GCCS is built on top of this environment called the DII Common Operating Environment, and many of the other service command and control systems are built on top of that same environment, to begin to achieve some of that same level of commonality.

Now with respect to hardware, you need to begin to differentiate between the commodity environment on the desktop and the large-capacity back room data servers or application server environment that an awful lot of the application software has to reside on. Today, DISA and the services basically provide the back end server architecture, but even that is fairly common at this point, and it has been constructed so that in many cases when you get to individual client desktops, commodity PCs are adequate to interact and accomplish the mission.

Pearson: Given those kinds of changes, what is the current status of the WWMCCS Intercomputer Network? What happened to it and where is it now?

Perry: When WWMCCS was shut down in August of 1996, the WIN was shut down as well.

Pearson: In what other ways has the computer revolution affected the ability of GCCS to serve its users?

Perry: How well it does off into the future in serving its users is going to depend upon how well it evolves. The new capability was put into place when WWMCCS was shut down in August of 1996. If you go back and look at the development process that got us there, it really started in earnest in November of 1994. And going pretty much from a standing start through worldwide deployment, to the point of being able to shut down WWMCCS and the WIN in August of 1996—a period of 21 months—was no small feat. There were some efforts earlier, but they didn't really start coming together as a cohesive and integrated development effort until about November of 1994.

As you look at what has transpired since the initial operational capability in August of 1996, there have been several incremental, and I believe one major, upgrades of the software capability. GCCS, like most any command and control system, is going to need to continue to evolve over time to meet evolving and in some cases expanding needs as articulated by the user community. The Joint Staff is the major user community representative, and that's really centered with the Operations Deputy on the Joint Staff, J-3. That's a very good thing, because the operators are the ones served by the system. The fact that the Joint Staff J-3 is engaged in managing and leading the requirements process, and in working with DISA and with the services to prioritize those requirements for implementation on some evolutionary development scale, is a very positive construct for GCCS to be viable. It will have to continue, and GCCS will have to continue to deliver.

Pearson: You have referred several times to the "evolutionary approach" to command and control system development. This approach holds that command and control systems are unique, requiring an equally unique management approach in which systems evolve "naturally" over time. What is the status of evolutionary thinking in GCCS today?

Perry: Well, inside of GCCS, and inside of DOD overall, there is a very recent revision to DOD 5000, which is essentially the acquisition policy and regulation for the Department of Defense. I think that over the years there has been progressively more and more realization that command and control is an information technology-based capability, and building that is fundamentally different than building a thousand Block 3 Tomahawk missiles, for example. In a lot of weapons programs the acquisition mind-set has been that you define your requirements, you lay out a program, you reach initial operational capability with the program, you finish it out, you reach final operational capability with it, and you're done. With command and control and information technology as it's evolving in today's world, you will never be done. In my view, the day you think you're done is the day you're on your way to a going-out-of-business plan.

Because the information technology you and I see on our desktops at home or in our office environment is changing so dramatically, we must provide capability commensurate with that in order to keep the users engaged and to keep it relevant. So the whole philosophy of "being done," from an acquisitions perspective, is one that does not fit in the information technology-based capability environment of command and control. I think there is progressively more and more recognition of that even in the formal acquisitions structures of the department. Change is always difficult, and this certainly represents change. So every once in a while there are assaults on attempts at change. But GCCS is reflective of the way that a lot of command and control and other IT-based acquisitions are moving and evolving within the department. As I said, if you look at some of the evolutionary and experimental features discussed in the most recent revision to DOD 5000, I think you will see progressively more and more of that becoming formally articulated as the policy of the Department of Defense.

Pearson: Effectiveness has always been considered the ultimate dependent variable in any organizational analysis. Throughout its history, WWMCCS was frequently characterized as "ineffective." I wonder if you could comment on how effectiveness is assessed inside GCCS, and, given the assessments, how effective is the system perceived to be.

Perry: I think as you look back over history, as WWMCCS first began to emerge, the more conventional contingency planning for deployments was fairly stationary. The geography of the Fulda Gap in Europe is what it is. The array of forces that were there in the cold war environment was pretty well understood. These days, a lot of the contingency operations that the department finds itself responding to, like a joint task force for a noncombatant emergency evacuation or something like that, are "come as you are," very short timeline events. The other aspect of the overall structure of WWMCCS that was always problematic is that you not only needed to plan for a large-scale deployment of forces, you also needed the ability to monitor the execution of that deployment. And where the WWMCCS process and technology had difficulty in the past was that there were not closed-loop data feeds to permit monitoring of the execution of the deployment. As you went into the first several days or weeks of the deployment of 500,000 men and women and all their associated materiel in support of something like a Desert Storm, you didn't have the closed-loop feedback mechanisms in place as a matter of course, either from the processes or the technology, to support that monitoring. So that issue, plus the evolving nature of deployment planning—going from less and less very deliberate planning to more and more crisis emergent sorts of deployment planning—those are certainly issues that affected WWMCCS and at least in part led to some of the issues associated with effectiveness that you have obviously heard about and are referring to.

As you begin to look at GCCS, the business processes haven't changed, and as a consequence those deployment execution monitoring capabilities have not been added to it. So some of those issues are in fact still issues, and they are in fact still being worked by the Joint Staff in cooperation with the military departments. Folks are now beginning to use the capability of the technology since the technology has changed in GCCS, and folks are starting to use that capability to evolve progressively more and more capability for transportation visibility and in-transit visibility of the various pieces of the deployment. So I guess as the bottom line, I wouldn't attempt to convey to anybody that we have achieved Nirvana in the deployment of GCCS, but I think that the evolutionary process and the modern technology implementations enable people to tackle some of the more fundamental process issues that were at the root of many of the effectiveness problems that you referred to in WWMCCS. Folks are starting to tackle those issues now, whereas the nature of the mainframe technology in WWMCCS made it next to impossible to tackle them in the past.

Pearson: Since the implementation of GCCS, what have been its major areas of success, in your view, and what are the major challenges that the system faces in the upcoming decades?

Perry: I think the major success was in the initial deployment and the subsequent evolutionary upgrades, basically taking something that had been in place since the early 1970s through 1996 and actually replacing it. That was a monumental task given the fact that over those 25-plus years the old system had extended tentacles throughout the military departments into so many different locations, in so many different organizational elements. I think it was a major feat to be able to move off of that old, obsolete implementation base into a more modern technology that permits addressing the issues that we've been discussing here.

Challenges for the future? As with any transition, change is always difficult, and gaining user acceptance has been a bit of a challenge because it's not the same old WWMCCS everybody was used to. I think most of those challenges are behind the department at this point. We've been operating on it for three and a half years now, since August of 1996 through today, and progressively more and more people are coming to accept the change, understand it, and, in many cases, embrace it because of the potential empowerment. I think if you went out and surveyed a whole bunch of former WWMCCS users you could still find a little bit of discord given that it is change, but I don't think you would find a lot of that today. And I think the real issues as we move off into the future are going to be the evolution of business processes for the deliberate crisis planning, deployment, and execution issues that we've talked about, as well as extending its combat execution monitoring and awareness kinds of capabilities. I think that the more significant issues are going to be those kinds of changes that are still ahead of GCCS and its employment across DOD.

Pearson: Any final thoughts with regard to the changes that have taken place, or any final concerns we should keep in mind?

Perry: No, none that we haven't already discussed, though as an interesting footnote I'll observe that you can now find pieces of WWMCCS in the Smithsonian Institution. When it was shut down in 1996, the department donated pieces of WWMCCS to the Smithsonian as a fairly significant element of at least the military side of national history, and they accepted it. Things like some of the computer equipment and terminals, or elements of it. Things like shopping carts that were used to wheel large stacks of WWMCCS printouts around the Pentagon were also a part of what was donated, and I'm not quite sure what they're doing with it right now. But as you indicated in your questions, WWMCCS was a lot more than that.

Pearson: Dr. Perry, thank you very much.

Index